Pesticide Formulation
Theory and Practice

Pesticide Formulation
Theory and Practice

Balraj S. Parmar
M.Sc., Ph.D., Dip. P.M.I.R., F.N.A.A.S., F.S.P.S.
Head of the Division

Sarman S. Tomar
M.Sc., Ph.D.
Principal Scientist

Division of Agricultural Chemicals,
Indian Agricultural Research Institute,
New Delhi - 110 012 (India)

CBS Publishers & Distributors Pvt. Ltd.
NEW DELHI • BANGALORE • PUNE • COCHIN • CHENNAI (INDIA)

Dedicated

to

The Fond Memory of

Late Prof. Sant Lal Chopra
Former Professor and Head,
Chemistry Department,
Punjab Agricultural University,
Ludhiana

ISBN : 81-239-1124-6

First Edition : 2004
Reprint : 2010

Published by Satish Kumar Jain and produced by V.K. Jain for
CBS Publishers & Distributors Pvt. Ltd.,
CBS Plaza, 4819/XI Prahlad Street, 24 Ansari Road, Daryaganj,
New Delhi - 110002, India.   • Website: www.cbspd.com
e-mail: delhi@cbspd.com, cbspubs@vsnl.com, cbspubs@airtelmail.in
Ph.: 23289259, 23266861, 23266867   • Fax: 011-23243014

Branches:
• *Bangalore:* Seema House, 2975, 17th Cross, K.R. Road,
  Bansankari 2nd Stage, Bangalore - 560070 Ph.: 26771678/79
  Fax: 080-26771680   • e-mail: bangalore@cbspd.com
• *Pune:* Shaan Brahmha Complex, 631/632, Basement, Appa Balwant
  Chowk, Budhwar Peth, Next to Ratan Talkies, Pune - 411002
  Ph.: 020-24464057/58   • Fax: 020-24464059
  e-mail: pune@cbspd.com
• *Cochin:* 36/14, Kalluvilakam, Lissie Hospital Road,
  Cochin - 682018, Kerala • e-mail: cochin@cbspd.com
  Ph.: 0484-4059061-65   • Fax: 0484-4059065
• *Chennai:* 20, West Park Road, Shenoy Nagar, Chennai - 600030
  e-mail: chennai@cbspd.com   Ph.: 044-26260666-26202620
  Fax: 044-45530020

Printed at : J.S. Offset Printers, Delhi

Foreword

Pesticides have played a key role in mitigation of plant and mammalian suffering but have a negative perception in public mind because they are intrinsically toxic. Their inappropriate use, not emphasized adequately, has also contributed to this perception.

Development and identification of new molecules has reached the stage of near saturation. It is also becoming uneconomical to develop new molecules because of high costs and increasing stringency of safety testing. This has brought the importance of pesticide formulation to the fore. The development of new formulations of the conventional and other pesticides has been recognized as a major advance in pesticide science.

Developments in the last three to four decades are witness to the dynamic nature of the subject of pesticide formulation. Several innovative products amongst solids (ex: water dispersible granule, floating granule, slow/controlled release products) and liquids (ex. suspension concentrate, emulsion concentrate, microemulsion, suspo-emulsion, soluble concentrate) have been reported. Besides, modern application technologies like smoke and vapour generators, aerosols, charged particle sprays etc. have also come into extensive use. Serious efforts are being made to explain variable deposits resulting from pesticide application in terms of pesticide formulation auxiliary composition *vis a vis* the physico-chemical nature of the target surface. A precise dose is aimed for a given situation. Advances have been witnessed in pesticide analyses too; which have become highly sophisticated. Besides estimating the principal active materials, these methods can simultaneously determine the impurities and the transformation products. The science of packaging has, likewise, become more systematic and has advanced tremendously. In addition to attractiveness and durability, packages are now designed to reduce interaction with active materials and promote handling and use safety. The development of water-soluble packs is a good example of innovation in packaging.

Much of the progress in pesticide formulation has taken place in the developed world. The developing countries are still struggling to bridge the gap. It is heartening to note that Dr. Balraj S. Parmar, Head of the Division and a pesticide formulation specialist and Dr. Sarman S. Tomar, Principal Scientist of the Division of Agricultural Chemicals, Indian Agricultural Research Institute, New Delhi have brought out a book on the subject. I understand that no textbook is available on this subject. The present publication will thus fill this void and serve as a standard reference for students, teachers, quality control units, industry and others concerned with pesticides.

I compliment the authors for an excellent and timely effort.

(V. L. Chopra)
Formerly Director General
Indian Council of Agricultural Research and
Secretary, Department of Agricultural
Research and Education, Govt. of India
President, National Academy of Agricultural Sciences

Preface

Pesticide formulation is both a science and an art. Whereas the science of it has progressed phenomenally during the past 3-4 decades, the art has remained a commercial secret. The development of dry and wet flowables and slow/controlled release products has been the most innovative landmark achievement. The subject has been discussed at several major international meets and constitutes a regular session at the various congresses on pesticide chemistry organized by the International Union of Pure and Applied Chemistry (IUPAC).

Several publications, based mostly on the deliberations at various meets, have appeared covering this subject. A text cum reference book has, however, eluded the researchers and students. There is no doubt that it is a Herculean task to attempt a text book on the subject, particularly in view of the diverse specialty areas such as physical, analytical, organic and inorganic chemistry, pest and human toxicology, chemical, mechanical and instrumental engineering and others, that are involved. Understandably, it is difficult to possess expertise in so many specialized areas. The authors are aware of the constraints and limitations but have still ventured to put together the information at their command for the benefit of one and all, particularly in view of the potential of this subject in future.

The subject matter has been presented under ten chapters. Starting with the basic definitions, classification, possibilities etc. as covered in chapter one on general aspects, basic information on solid and liquid formulations, that influences their preparation and performance, is reported in chapter two. It also provides information on the carriers and diluents and their interactions with the active ingredients.

Adjuvants constitute the backbone of pesticide formulation. Several of them, most notably the surfactants, are a part of most of the formulations to perform an array of functions, expected of a product. Chapter three discusses briefly the topics like surfactants, attractants, repellents, pheromones, synergists, photosensitizers, stailizers and others that find use as adjuvants in pesticide formulation.

Pesticide mixtures are in much talk these days. There are advantages as well as fears associated with their use. These multi-ingredient products may be present in a single formulation or two or more formulations be combined to obtain the mixtures. The subject has been treated briefly but comprehensively in chapter four.

Setting up of a formulation laboratory or a factory requires basic information and knowledge of the laboratory/factory wares and machinery and equipment. It is also necessary to understand the basic principles of formulation analysis before venturing into quality or quality control. Chapters five, six and seven are devoted to laboratory and principles of analysis, machinery and equipment and quality control and assurance and are expected to provide a broad idea on these topics.

Once a product of desired quality is produced, its packaging and labelling assume importance. Specifications are prescribed for the packaging and labelling of formulated products. These are briefly discussed in chapter eight.

Finally, the products are applied at use sites with an ultimate eye on efficacy. Different formulations need to be applied after a careful choice of the application equipment. Chapter nine gives an overview of the available options for application of different products. The tenth chapter attempts to explain bioefficacy in physico-chemical terms. It highlights the major thermodynamic parameters that influence efficacy and also provides an overview of the practical aspects related to bioefficacy.

The subject of pesticides in general and pesticide formulation in particular has been little understood by both the plant protection experts as well as the public at large. Often one comes across vague and misleading references being made to the terminology related to pesticides. Lately, this aspect has received attention of the international bodies and standard terms, codes, definitions have been proposed. These have been reported in glossary.

The present book may be considered as an humble effort by the authors to share some of the information available with them. It is expected to serve as a reference material for class room teaching on one side and for production and use of the product, on the other. It attempts to place before the readers the multitude of parameters that govern pesticide formulation. There will be several shortcomings and inconsistencies. It will be our endeavour to rectify the faults and mistakes over time. Criticism and suggestions for improvement would be most welcome.

The authors pay a respectful homage to the memory of Late **Professor Sant Lal Chopra,** Former Professor and Head, Department of Chemistry, Punjab Agricultural University, Ludhiana, who had inspired this publication. It was to be written jointly with him. Unfortunately, the cruel hands of destiny took him away untimely. **This book is dedicated to the memory of his revered soul.**

We would also like to place on record our indebtedness and gratitude to all the friends and colleages for their encouragement and suggestions, our wives and children for their patience, perseverance and support, the Indian Agricultural Research Institute for permission to take up the project, Mrs. Jasbir Kaur for secretarial support and all others who have helped in some way.

BALRAJ S. PARMAR
SARMAN S. TOMAR

Contents

CHAPTER 1

General Aspects

PESTICIDE FORMULATION

Pesticide product offered for sale. It generally comprises active ingredient(s), adjuvant(s) and other formulants combined to render the product useful and effective for the purpose claimed
(after FAO, 1986)

Most of the technical pesticides are formulated before use, by mixing the active ingredient(s) with inert(s) with inert(s) and/or other auxiliaries/adjuvants, to obtain a product which is effective, easy to handle and apply, possesses satisfactory shelf life and is devoid of undesirable side effects. The performance of the formulation, its interaction with the target and non-target organisms and with the various components of the ecosystem such as air, water, soil, plant, microorganisms, etc. including its fate in the environment, are the key aspects of pesticide formulation. For practical use, several solid (dust, water dispersible powder, granule, dry flowable, etc.) and liquid (solution, emulsifiable concentrate, wet flowable, etc.) products are available.

In order to make a good formulation of a toxicant, it is desirable to know its major physical properties, such as the melting and (or) boiling point, vapour pressure, specific gravity, solubility in various solvents, rate of hydrolysis, the degradation by ultraviolet radiation, etc., and the biological activity against the target. Likewise, it is important to know the formulation characters such as viscosity of the spray solution, the particle size of the spray drop, the foaming properties, etc. The various properties need to conform to the specificat s prescribed by the official Standards Institutions of the land. This knowledge will facilitate a proper choice of formulation ingredients and the operational parameters.

Before a formulation is put up in the market for sale, some data on its retention on the plant and pest, penetration and translocation, crystal size of the deposit, residues and half life, mode of action, fate in the environment etc. should be generated.

Key Objectives

There are several objectives of formulating a technical pesticide, the key ones being dosage, cover, mass and momentum. Formulation enables application of proper dosage to provide the requisite cover to the target crop and the mass and momentum to the product, which are vital factors in pest control. For example, application at an assumed rate of 1 kg ha^{-1} (syn. 10^9 µg of an active ingredient per 10^8 cm^2 on a flat surface), will imply a uniform dose of 10 µg per cm^2. At this rate, a material of density of 1 g ml^{-1} would result in a thickness of 0.10 µm (roughly the thickness of a thread of a spider's web).

A well grown crop has a leaf area that is 10-20 times the area of the land on which it stands. It is owing to the wrinkles on the leaves, with hills and valleys several microns in size. Several of the modern high potency chemicals are applied at doses far lower than 1 kg ha^{-1}. Thus, most chemicals may have to be spread at less than one hundredth of a micron thickness (0.01 µm) to provide a proper cover to the crop.

A spherical particle of 100 µm diameter has a volume ($4/3\pi r^3$) of about 5.2×10^{-7} cm^3. With a density of 1 g ml^{-1}, quite often, even a single particle per 20-25 cm^2 may not be available at dose levels of 0.1 kg ha^{-1} or less. This particle being the key player in pest control, a particle size of lower than 100 µm may have to be resorted to for providing a reasonable cover for pest control.

Reduction of particle size may provide a better coverage to the crop through an efficient distribution of the material, but it has repercussions on the mass and momentum of the particles. For example, a 100 µm diameter particle is reported to fall in air viscosity at a rate of 30 cm sec^{-1} (Green et al., 1977). It takes about 1/10th of a second and a fall through about 3.5 cm, to reach this speed. A 50 µm diameter particle falls at only one quarter of this speed, leaving it at the mercy of the slightest wind.

Need Based Products: Several pest control situations make it obligatory to formulate a product in a particular form. Seed dressings, rodent baits, pest traps, smoke generators etc. are some of the examples. Many toxic and hazardous chemicals like carbofuran are formulated as encapsulated granules so as to enable their safe handling. For use in water scarcity areas, dust formulations are preferred. Thus, pesticide formulation is undertaken to develop a product that enables its need-based application.

Economic, Efficient and Safe Use: Pesticide formulation enables economic, efficient and safe use of pesticides. The spill over loss and hazard of an active ingredient will be far less from its formulation than those from the use of concentrated technical material. Likewise, handling a formulation (diluted product) is safer for the workers than handling a technical (concentrated) product. Safety is also improved through visibility of the deposit on the treated mass. Several adjuvants e.g. attractants, synergists and crop safeners are added to the formulations to make them safer by reducing the amount of the active ingredient required for bioeffectiveness and/or by improving selectivity. These are also used to economize on the formulation cost. Besides, the synergised products broaden the spectrum of activity of the toxicant, delay or overcome the problem of pest resistance and enable an understanding of several basic processes involving the biochemical mechanisms i.e. resistance, mode of action, etc. in pests.

The Process

The formulation process comprises of certain distinct steps, to be performed preferably in a described sequence, as follows:

1. *Physical Treatment:* It comprises of simple physical treatment of the active ingredient and/or the carrier/diluent to bring them in a form that can be conveniently processed. For example, grinding of the active ingredient or the solid carrier to make powder, liquefaction of gases by pressure, etc.

2. *Combination with other Toxicants or Adjuvants:* With the aim to broaden the spectrum of bioactivity or improve bio-effectiveness through synergism, it is intended to employ two or more active ingredients or an active ingredient along with an adjuvant *viz.* synergist(s). All of these are combined at this stage.

3. *Combination with Insecticidally Inert Materials:* It implies addition to or mixing with the solid or liquid carriers/diluents, in order to make a preparation of the desired strength.

4. *Addition of Adjuvant for Special Effects:* At the final stage of formulation, adjuvants are added for special effects (wetting, suspending, emulsifying, dispersing, sticking, stabilizing, etc.).
   Due mixing or blending is carried out at various stages of the formulation process, as per requirement.

Types and Product Spectrum

In general, a pesticide formulation exists only as a solid or liquid type. Gaseous products, as a rule, are not employed in fields. The science and art of pesticide formulation address primarily to putting various active ingredients in the above two forms. A large number of formulation possibilities have been identified under each formulation type. These are listed in Table 1.1 along with examples of formulations (in parentheses). The major formulation type codes are given in Table 1.2.

TABLE 1.1. PESTICIDE FORMULATION TYPES AND POSSIBILITIES

Solids
1. **Single solid toxicant**
   Without auxiliaries/adjuvants
   a. Volatile solid to be vaporised as a fumigant spontaneously or by heat e.g. camphor, naphthalene balls, other vapour yielding solids (FG, FT, FU, GE, MG, PT, PW, SM, TB, TP, VP)

b. Powdered solid for direct application as dust (usually too concentrated for direct application) e.g. carbaryl, 2,4-D, etc. (DP, DS, PW, SM, TC).

c. Solid, liquid, or gas combined chemically to form a solid e.g. chemically reacted controlled release products (FG, FP, FT, FW, GB, GF, GG, GP, GR, MG, NB, PB, PT, PW, RB, SB, SM, SP, SS, TB, TC, TP, VP).

With or without auxiliaries/adjuvant

d. Powdered solid with or without wetting agent for dissolution in water e.g. water soluble powder concentrate (PW, SG, SM, SP, SS).

e. Powdered solid with or without wetting and suspending agents for suspending in water e.g. dispersible toxicant powder (SM, ST, WP).

2. **Solid or liquid toxicant combined with an inert solid**

a. Solids mixed mechanically e.g. powder (BB, CB, DP, DS, GE, GP, IM. IW, IS, MG, PB, PR, PS, PT, PW, RB, SB, SP, SS, ST, TB, TP, VP, WG, WP, WT).

b. Solid toxicant combined mechanically with a carrier e.g. straight dust (BB, CB, DP, DS, GE, GP, IM. IW, IS, MG, PB, PR, PS, PT, PW, RB, SB, SM, SP, SS, ST, TB, TP, VP, WG, WP, WT)

c. Solid or liquid toxicant impregnated on particles of inert solid e.g. impregnated dust, granule, polymeric products (BB, CB, DP, DS, GE, GP, IM. IW, IS, MG, PB, PR, PS, PT, PW, RB, SB, SM, SP, SS, ST, TB, TP, VP, WG, WP, WT).

d. Types *a*, *b* or *c* as concentrate diluted mechanically with further inert solid for use as dusts e.g. dust or powder concentrate.

e. Types *a*, *b* or *c* with wetters and other adjuvant(s) for use as wettable powder e.g. water dispersible products, wettable powder, water dispersible powder.

f. Volatile solid or liquid mixed with combustible mixture for use as vapour or aerosol when ignited e.g. smoke generators, coils, etc. (AE, FD, FP, FT, FU, FW, GE, GF, HN, KN, NB, TP).

g. Volatile solid or liquid combined physically with solid carrier for use as tablet e.g. celphos or aluminium phosphide tablets (BB, CB, FU, GE, IM, IW, IS, PR, PS, PT, PW, RB, SB, SM, TB, TP, UP).

Liquids

1. **Single liquid toxicants**
Without auxiliaries/adjuvants

a. Fumigant gas liquefied by pressure e.g. liquid ammonia (FU, GA, GE, KN, LF, NB, OL, PO, VP).

b. Volatile liquid to be vaporised as a fumigant spontaneously or by heat e.g. carbon tetrachloride, chloroform (FU, GE, KN, LF, NB, OI, PO, VP).

c. Liquid for direct application (usually too concentrated for good dispersion) e.g. liquid active ingredients like malathion, methyl parathion, neem oil, etc. (AE, HN, LF, LI, LS, MS, NB, OI, PD, PO, PY, TC, UL, VP).

d. Liquid for dilution with water or other solvent just before application e.g. oil or water soluble liquid toxicants, oil or aqueous concentrates, etc. (AE, AS, KN, LF, LI, LS, MS, OI, OL, PD, PO).

With auxiliaries/adjuvants

e. Liquid with emulsifier for suspending in water as emulsion e.g. emulsifiable oil concentrate (EC, LI, MS, SN, UL).

2. **Solution of solid or liquid toxicant in liquid solvent**

a. Mixture of fumigants, to reduce fire hazard, give joint action, or facilitate manufacture e.g. ethylene dichloride-carbon tetrachloride, EDCT mixture (AE, FU, GE, KN, LF, PO, PY, SN).

b. Solution in liquefied gas to produce an aerosol e.g. toxicant solution in chlorofluoroalkanes (AE, FU, GE, KN, LF, PO, PY, SN).

c. Solution in toxicant liquid for joint action e.g. synergised products (AE, EC, LF, LI, LS, MS, OI, OL, PD, PO, PY, SN).

d. Solution in inert liquid for easier mechanical application e.g. solution, microemulsion (AE, AS, LF, LI, LS, MS, NB, OI, OL, PD, PO, PY, SN, UL, VP).

e. Concentrate for further dilution with liquid before application e.g. solution concentrate, microemulsion concentrate (AE, AS, EC, EM, EO, EW, LF, LI, LS, MS, NB, OF, OI, OL, PD, PO, PY, SN, UL, VP).

f. Solution with emulsifiers for suspending in water as emulsion e.g. emulsifiable concentrate (AE, EC, GE, LF, LI, LS, MS, OI, PO, PY, SL, SN, UL)

g. Solution in water-soluble liquid (e.g., acetone) to give toxicant suspension on dilution e.g. solution of solid toxicant such as carbaryl in water miscible solvent (LJ, LS, PY, SL, SN).

3. Suspension of solid toxicant in a liquid

a. Liquid emulsified in another liquid, usually as concentrate for further dilution e.g. emulsion concentrate (EM, EO, EW, LI).

b. Solid suspended in liquid e.g. paste, cream, suspension concentrate (CM, DS, LJ, LP, OF, PA, PD, PO, FS, SC, SV, WS).

Refer Table 1.2 for formulation type codes

TABLE 1.2. INTERNATIONAL FORMULATION CODES

Code	Type of formulation	Code	Type of formulation
AB	Grain bait	LI	Liquid
AE	Aerosol	LP	Liquid paste
AS	Aqueous solution	LS	Liquid seed treatment
BB	Block bait	MC	Microcapsule suspension
BR	Briquette	MG	Microgranule
CA	Coating agent	MS	Mist spray
CB	Bait concentrate	NB	Fogging concentrate
CG	Encapsulated granule	OF	Oil-miscible flowable concentrate
CM	Cream	OI	Oil
CR	Crystal	OL	Oil-miscible liquid
CS	Capsule suspension	PA	Paste
DP	Dustable powder	PB	Plate bait
DS	Dry seed treatment	PD	Poison drink
EC	Emulsifiable concentrate	PO	Pour-on
EM	Emulsion	PR	Plant rodlet
EO	Water-in-oil emulsion	PS	Seed coated with a pesticide
EW	Oil-in-water emulsion	PT	Pellet
FC	Liquid cream	PW	Powder
FD	Smoke tin	PY	Pump spray
FG	Fine granule	RB	Bait (ready for use)
FP	Smoke cartridge	RS	Ready-to-use suspension
FS	Flowable concentrate for seed treatment	SB	Scrap bait
FT	Smoke tablet	SC	Suspension concentrate
FU	Fumigant	SE	Suspo-emulsion

FW	Smoke pellet	SG	Water-soluble granule
GA	Gas	SL	Soluble concentrate
GB	Granular bait	SM	Solid material
GE	Gas-generating product	SN	Solution
GF	Smoke granule	SP	Water-soluble powder
GG	Macrogranule	SS	Water-soluble powder for seed treatment
GL	Gel	ST	Seed treatment
GP	Flo-dust	SU	Ultra low volume suspension
GR	Granule	TB	Tablet
GS	Grease	TC	Technical material
HN	Hot fogging concentrate	TP	Tracking powder
IC	Impregnated collar	TW	Twin pack
IM	Impregnated material	UL	Ultra low volume liquid
IW	Impregnated strip	VP	Vapour-releasing product
IS	Impregnated wiping cloth	WG	Water-dispersible granule
KN	Cold fogging concentrate	WP	Wettable powder
LA	Lacquer	WS	Slurry for seed treatment
LF	Liquid fumigant	WT	Water-soluble tablet

UNIDO (1992) Integrated International Safety Guidelines for Pesticide Formulation in Developing Countries, United Nations Industrial Development Organization, Vienna.

Classification Based on Aggregate Condition, Physical System and Mode of Dilution

Each of the aggregate condition i.e. solid or liquid offers several physical systems, whose mode of dilution before application exemplifies a variety of products that have been arrived at on the commercial scene. Key examples are listed in Table 1.3. It is seen that most of the common pesticide formulations can be suitably grouped in this manner. However, there can be some isolated examples, which have not been covered in this scheme. For example, dust concentrate, a powder that is diluted with a solid diluent to bring it to field strength before application. The ready to use emulsion products may be developed for use as such without dilution, and so on. Such products will need to be appropriately placed in the Table on invention.

TABLE 1.3. CLASSIFICATION OF FORMULATIONS BASED ON AGGREGATE CONDITION, PHYSICAL SYSTEM AND DILUTION MODE

Aggregate condition	Physical system	Diluent for application		
		None	Organic solvent	Water
Solid	Powder	*Dust, dry seed dressing, vapour/ smoke forming powder	Oil dispersible powder, oil soluble powder	Water dispersible powder, wettable powder, water soluble powder, slurry dressing powder
	Granule	Granular products		Dry flowable (water dispersible granule), floating granule

	Polymer	Polymeric products (slow/controlled release)		Sprayable polymeric products
Liquid	Solution	ULV product, liquid seed dressing, oil solution, aerosol, fog, foam forming product, vapour generator (VP)	Oil concentrate or oil miscible solution concentrate	Water soluble concentrate, emulsifiable or invert emulsifiable concentrate
	Emulsion	Microemulsion (ready to use)		Emulsion concentrate
	Suspension			Wet flowable (suspension concentrate), sprayable polymeric suspension concentrate
	Suspo-emulsion			Suspo-emulsion (Suspo)

Note: *Dust concentrates are diluted with solid diluents before use.

Basic Parameters Affecting Preparation and Performance

Irrespective of whether an active ingredient exists as a solid or a liquid, it can be formulated in either of the solid or the liquid forms. Powdered solid active ingredients were probably the earliest used form of the toxicants till the use of various solid or liquid carriers or diluents found application in pesticide formulation. The choice of the form of the pesticide product will depend on the purpose of pesticide use i.e. to control insects, fungi, weeds, nematodes, mites, rodents, algae, etc. and the manner in which the control is to be affected i.e. as killer, attractant, repellent, antifeedant, metamorphosis disrupter, growth regulator, etc. The application mode i.e. soil incorporation, spray, dusting, bait, etc. will also govern the choice of the form. The physico-chemical properties of the active technical pesticide, the weather conditions of application, economic considerations, ready availability of formulation auxiliaries, etc., will all govern the choice of the formulation and the use form.

REFERENCES

Green, M. B., Hartley, J. S. and West, T. F. (1977). *Chemicals for Crop Protection and Pest Control.* Pergamon Press, pp. 180-181.

CHAPTER 2

Preparation and Performance

SOLID FORMULATIONS

FACTORS AFFECTING PREPARATION AND PERFORMANCE

THE preparation of most solid powders depends upon the particle shape, size, relative hardness of components, etc. since these govern the homogeneous mixing of particles mechanically. Chemical compatibility of the components and non aggregation (non-clumping) on keeping are the favourable attributes for their shelf life. The shape and size along with the surface of the particles, and sticky materials govern their adhesion on the applied surfaces. During application, the powder must be free flowing and readily dispersible in air or water, and provide the desired deposit taking care of drift and other factors. Wetting, dispersion and settling have to be taken care of during application of solids suspended in water.

The factors affecting the preparation and performance of solid formulations are briefly discussed below.

Particle Shape

The particle shape governs the flow, settling, adhesion and exposure of the surface of the powders. Uniform shape may be seldom attained and its variation influences the particle size determination. A single dimension is necessary to represent the size as it simplifies averaging or expression of size distribution of a heterogeneous dispersion.

Particle Shape of Diluents Used in Solid Formulations

Powder particles may be spherical, polyhedral, tubular, spicular, irregular or of other shapes. Usually, diluents for the production of pesticide dusts are hydrophobic minerals of the talc and pyrophyllite type and rarely chalk, gypsum, kaolin, kieselguhr, tripolic earth, silica gel or various other clays. The best diluents for most preparations used for dusting plants are pyrophyllite and talc which have a lamellar structure and adhere well to the plant foliage (Gunther and Gunther, 1971).

Bertlett (1951) reported particle shapes of different carriers *viz*: sulphur-ragged fractured crystals; calcite-ragged edged crystals; silica (synthetic)—smooth edged amorphous grains; diatomite—sharply fractured diatoms; talc (varieties)—smooth and amorphous grains, sharp crystals and grains, sharply fractured grains; pyrophyllite (varieties)—sharply fractured plates and grains, ragged edged fractured grains; montmorillonite (bentonite)—sharp edged laminated crystals; attapulgite—spicular; pumice (frianite)—sharply fractured grains; wood—fibrous; bark—fibrous; shell—amorphous grains. Polon (1973) reported particle shapes of attapulgite—needle shaped; talc—fibrous and platy; pyrophyllite—thick platelets; kaolins—thin platelets; diatomaceous earth—irregular.

Parameters Affected by Particle Shape

(a) *Flowability:* Polon (1973) related flowability to the shapes. Talcs lose flowability on impregnation and attapulgite becomes fibrous in crystal habit. Irregular materials tend to flow poorly, thin plates tend to adhere well, and needle shaped particles abode the surface.

(b) *Tenacity:* Tenacity, 'the resistance to weathering of a protectant', may be inherent or imparted by means of stickers etc. Inherent tenacity is influenced by particle size, shape, nature of surface and solubility. A round particle which has only a small point of contact with the leaf surface, would adhere less well than a flat, rectangular or oblong shaped particle providing a large contact surface. Manufacturers of diluents for dusts that have particles shaped like flat plates or saucers claim that this shape makes the diluents adherent because of a close interlocking of gross surface irregularities.

(c) *Bioactivity:* From the biological point of view, a sharp irregular particle would be most likely to abrase the integument while every particle making a large area of intimate contact should be efficient in adhering and in passing on toxic material to the insect.

Dusts and the particles resting on a plane surface appear to cover surfaces without having a large area of close contact. If the spherical particles that are themselves inert are coated with a thin layer of toxic material, very little poison may be transferred to the surface on which they rest unless some additional process such as surface diffusion or spread of a mobile solvent from the substratum over the particle surface, takes place. No clear evidence that particle shape is an important factor in insecticidal action is available.

The insecticidal effect of a dust is enhanced if the surface of the particles is crystalline or angular in character, while it is diminished if the surface is amorphous or rounded. The flaky form of powdered glass kills insects more quickly than the smooth spherical form and is also more abrasive. Particle shape may at times result in abrasion of wax layer causing the killing of insects. It can also influence absorption on the waxy layer.

Measurement

Sphericity: Particle shape is generally measured in terms of sphericity i.e. resemblance of the particle to a perfect sphere of same size or volume.

"Sphericity" = S'/S,

where, S = measured surface of the particles

S' = surface of a sphere of the same volume

The sphericity is less than one for any shape other than a sphere.

Particle Shape in Colloids: In colloids, Brownian agglomeration is an important factor affecting the behaviour of liquid or gaseous suspension of sub-micrometer particles. Based on the assumption that all particles are spherical, various parameters have been evaluated by solving the population balance equation. No complete explanation could be given for coagulation behaviour of solid particles as they necessarily form non-spherical aggregates. Usually, aggregates from Brownian agglomeration are made up of (*i*) uncharged particles which have random shapes (cluster aggregates) or (*ii*) charged aerosols from nearly linear aggregates (chain agglomerates)

Okuyama *et al.* (1981) stated that motion of fine particle aggregate is a function of Stokes radius, which is a function of dynamic shape factor, Stokes radius being used to characterise the drag force. Probability of an agglomeration given a collision is nearly unity, only the hydrodynamic and perhaps inertial forces are important and these can be characterised adequately with the Stokes law. Accordingly, it is assumed that the nonspherical shape of the agglomerates is sufficiently accounted for by the dynamic shape factor. It was concluded from the studies on behaviour of fine solid particles undergoing Brownian coagulation that evolution of particle size distribution with time depended strongly on $K(N)$ where, $K(N)$ is the dynamic shape factor of an aggregate composed of *n* primary particles. The various shape factors have been defined for identification of properties of non-spherical particles correlating with particle size by Kousaka *et al.* (1981) *viz.* surface shape factor (ϕS), volume shape factor (ϕV) and resistance shape factor (K_R), also known as dynamic shape factor.

Dynamic Shape Factor: It is defined as "the ratio of the drag force exerted on an aggregate particle to that exerted on a sphere having an equivalent volume". Dynamic shape factor plays an important role in the motion of an aggregate particle in a fluid. It is given by the formula:

$$K = \left(\frac{6}{\pi}\phi V_{\text{agst}}\right)^{\frac{2}{3}}[1-\Sigma]^{-\frac{1}{3}}$$

Volume shape factor is given by

$$\phi V_{agst} = \frac{(d_{pl})^3}{(d_{agst})} \frac{\pi}{6} \frac{n}{(1 - \Sigma)}$$

where, Σ = porosity of the aggregate, V_{agst} = volume shape factor of aggregate, (st = denotes Stokes), d_{pl} = diameter of primary particle, d_{agst} = diameter of aggregate including the effect of porosity, n = number of primary particles.

Volume shape factor depends on the shape of the aggregate which is $\pi/6$ if the aggregate is spherical. Porosity depends on orientation of primary particles.

Cluster model aggregates have a porosity ranging from 0.476-0.259 for cubic and rhombohedral packing. In straight chain model aggregates, it is a non-spherical particle with no porosity. So, dynamic shape factor is equal to volume shape factor.

The value of Σ will increase with n up to 20 and then it will assume a certain value at $N \geq 20$. The porosity at $N \geq 20$ will depend on the manner in which primary particles are oriented and their shape. For aggregates composed of large primary particles (≥ 1 μm), the porosity is around 0.5. Porosity is more for aggregates composed of fine particles.

Specific Surface: Indirect measure of shape may be obtained by measuring 'specific surface' for a given quantity of powder. Specific surface depends upon the shape of the particles. Specific surface *vis à vis* standard shapes are known and an idea about shape is made. If the property chosen is one that is important in practice, this method may be very useful.

Electron Microscopy: A beam of electrons like a beam of light, gets deflected in electromagnetic field. The electromagnetic lense orients the electron rays as parallel beam. Electron scattering gives an image magnification up to 50,000-100,000 times. The particles measuring 1Å or 10^{-8} cm are observed as large as 10^{-3} cm and this helps distinguish the shapes and sizes of different particles.

Effects

Particle shape influences the physico-chemical and biological properties and the shelf life of the formulation. Some of the effects are as follows:

(i) Materials of irregular particle shape tend to be harder to mix and grind.

(ii) The larger exposed surface of the irregular shaped powders enables them to sorb (adsorb and absorb) more liquid or sticky material without caking. They tend to be more bulky.

(iii) The irregularity in shape also increases adhesion because of the greater surface of the particles available for contact with the second surface.

(iv) Dusts containing irregularly shaped particles tend to flow less easily and break agglomerates less easily during application.

(v) Irregular particles settle less rapidly in either air or liquid than the spherical particles of the same diameter.

(*vi*) If different components of the formulation have different shapes, they may tend to segregate in the air stream and also on settling.

Particle Size

Like spherical droplets, the particle size of solids is, for convenience, expressed in terms of the 'diameter' of a sphere which is equivalent to the particle in question, in some chosen respect. This basis of equivalence is as important as the method of averaging. The choice is set in some cases by the method of measurement, while in others more leeway is possible.

The American Mosquito Control Association has given a classification of sprays and dusts based on size (Table 2.1). Some of the important size and shape related parameters of particles are summarized in Table 2.2.

TABLE 2.1. PARTICLE SIZE CLASSIFICATION OF SPRAYS AND DUSTS

Particulars	Particle or droplet size (microns)
SPRAY	
Coarse	400-above
Fine	100-400
Mist	50-100
Aerosol and fog	0.1-50
Fume and smoke	0.001-0.1
Vapour	< 0.001
DUST	
Coarse	175-above
Medium	45-175
Fine	44-below

TABLE 2. 2. KEY PARAMETERS RELATED TO SIZE AND SHAPE

Circumference of a circle	$2\pi r^2$
Area of a circle	πr^2
Area of ellipse with semi axis a and b	πab
Surface of a sphere	$4\pi r^2$ or πD^2
Volume of a cylinder	$\pi r^2 \times$ height
Volume of a sphere	$4/3\pi r^3$
Volume of a cone	$1/3\pi r^2 \times$ height
Volume of a pyramid	1/3 area of base $\times$ height
Volume of a prism	Area of base $\times$ height

r = radius, D = diameter

Like particle shape, the particle size also has far reaching effects on the physico-chemical properties, shelf life and biological performance of the formulations. The key effects are described below:

Volume Per Particle (∝ D^3, where, D = **diameter**): The volume of a sphere is given by $4/3\pi r^3$ (r = radius) or $4/3\pi(D/2)^3$. Volume shape factor depends on the shape of the aggregate and it is $\pi/6$, if the aggregate is spherical. Porosity depends on orientation of primary particles. An increase in the size of the particle results in an increase in the surface area of the particle. Consequently, it reduces the number of particles per unit mass and is, therefore, a step in the wrong direction in case of powders.

The dosage available per particle, even though more, reduces the effectiveness of the total deposition. The larger particles also cause abrasion within equipment, which is reduced with size reduction.

The increased volume per particle may become advantageous, as in granular products, where the overall release per unit mass will be reduced due to a lesser number of particles resulting in lower specific surface per unit mass. This also reduces the pesticide—carrier interaction by reducing the total exposed surface per unit mass.

Number of Particles per Unit Weight ($\propto 1/D^3$ or 1/Volume of the Particle): When particles are of smaller size, the number of particles that becomes available for effectiveness is large. This will ensure a thorough and uniform deposit on a surface using powders, leaving far less uncovered area. The probability of such particles hitting the target is increased. The even coverage of the treated surface obtained with a large number of small sized particles is of paramount importance in an effective pest control.

Surface Per Particle ($\propto D^2$): The surface of a sphere ($4\pi r^2$) is directly proportional to its size. However, as stated above, increase in the diameter of a powder particle is a step in the wrong direction because the volume or mass of the particles increases still further. Also, as per Stokes law, other parameters remaining constant, the velocity with which the particle falls down increases with an increase in particle size.

Specific Surface: (Surface per unit weight of material, $\propto D^2/D^3$ or $1/D$)

$$\text{Specific surface of a sphere}\,(S) = \frac{\text{Surface area of sphere}}{\text{Volume of sphere}}$$

$$= \frac{4\pi r^2 \text{ or } \pi D^2}{\dfrac{\pi D^3}{6}} = 6 / D_{sv}$$

where, D_{sv} = average diameter on volume surface basis

Increase in specific surface owing to a reduction in particle size improves adsorption and absorption, the latter due to the fine capillary spaces between particles. A uniform thinner coating of the toxicant is obtained due to a greater number of the fine powder particles. These are the important considerations in providing good flow, storage and dispensing properties to dusts and wettable powders.

A reduction in size generally causes poorer flow because of the greater number of surfaces in contact. However, a film of adsorbed air will make very fine powders more bulky and aid in their flowability.

The coverage of a surface by a fine powder is more even because of the larger number of particles. It is also more complete because of the greater specific surface. The greater surface per unit weight of the material also improves adhesion of fine particles on various surfaces.

Weathering due to mechanical disturbances tends to be less in case of fine particles because of their greater adhesive property. However, losses due to volatility of solution or chemical degradation of the product are greater with an increase in the surface area. The effective toxicity of formulations with finer particles is higher than otherwise, particularly in absence of other factors like lack of deposition on target.

Particle size is often expressed as mass average or better mass median diameter which is the diameter of a particle whose volume is measured from the number of particles. It should be greater than 75 microns so that it falls on foliage and is retained by it for activity. If mass median diameter is 10 microns or less, then almost all particles slide down. Therefore, size of particles is to be appropriately manipulated.

It is apparent, therefore, that by altering size of particles, widely varying characteristics of the formulations can be obtained. Development of dust (driftless, fine and regular), water dispersible powder, granule, fine granule (ordinary and *F*) etc. may be cited as the cases amongst solid formulations and emulsion, microemulsion, wet flowable etc. as examples among the liquid ones where particle size has been exploited in the development of products. Rate of release of active ingredient has been advantageously controlled by varying particle size.

Measurement

Determination of actual particle size and its distribution offers considerable difficulties. The available methods only permit the determination of degree of disaggregation of particles and rarely of particle size. They all are based on determination of an apparent diameter of spherical particles with the exception of direct visual observation using an electron microscope. Electron microscopy has demonstrated only the plate like structure of the clay particles. The apparent spherical diameter so obtained is only an approximation on account of lack of a more accurate determination.

There are many laboratory methods for determination of particle size of solid particles but there are only limited methods for liquid droplets. The choice of a method depends upon the important properties and also on the equipment available.

Methods Based on Measurement of Linear Dimension

Sieve Analysis: This is one of the earliest, simplest and most common methods of separating or classifying coarse particles. It gives a direct separation into weighable fractions of limited size ranges. Sieving is a convenient procedure for segregating particles coarser than 0.05 mm (50 μm). The size of the finest workable sieve is 325 mesh (with an opening of about 44 microns). This is larger than most insecticide powders and due to this measures only a small content of very coarse particles and aggregates. Sieves of finer openings, if available, are difficult to operate.

The probability of a particle passing through the aperture of a given sieve during a fixed shaking period depends upon nature of the particles and the mesh size of the sieve. For example, shape of the particle controls its passage only in a certain orientation and this leaves a limited chance of passing through, except on prolonged shaking. The sieve openings are often unequal in size too, requiring extensive shaking before all particles have had the opportunity of approaching the largest opening to pass through. The requirement that sieving be continued to 'completion' can rarely be achieved during practical periods of shaking. Good reproducibility requires careful standardization of procedure.

Most sieves have square openings but a few may have circular or rectangular too. There are several standard series of sieves with standardized openings. Several countries of the world e.g. USA, Britain, France, Germany, India and others have standard sieves. Quite often, the sieve opening size is expressed in 'mesh' which provides the number of openings to a linear inch. Table 2.3 gives width of apertures for some common British Standard Sieve (BSS) mesh numbers.

TABLE 2.3. RELATION BETWEEN WIDTH OF APERTURE AND COMMON BSS MESH SIZE

BSS Mesh No.	Width of aperture	BSS Mesh No.	Width of aperture
5	2.85 mm	44	355 microns
6	2.80 mm	52	300 microns
7	2.36 mm	60	250 microns
8	2.00 mm	72	212 microns
10	1.70 mm	85	180 microns
12	1.40 mm	100	150 microns
14	1.18 mm	120	125 microns
16	1.00 mm	150	106 microns
18	850 microns	170	90 microns
22	710 microns	200	75 microns
25	600 microns	240	63 microns
30	500 microns	300	53 microns
36	425 microns	325	44 microns

Note: The opening in mm can be fairly well approximated on the assumption that the opening is 0.63 of the mesh interval, whence the size of opening in mm = 16/Meshes per inch e.g. 100 mesh = 0.16 mm.

Motion imparted to sieves greatly affects the efficiency of sieving. Several mechanical shaking devices have been developed to provide rotatory or vibratory motion. Frequently, screening is done for five minutes for 100 g of the sample.

Materials may be sieved dry or wet. Such materials that form aggregates during shaking, and in case of those which are already suspended in water, wet sieving is practiced. For wet sieving, sieves are fitted in frames to prevent loss of material. It greatly increases total amount of material passing through the sieve.

Microscopic Measurement: It finds universal application in measurement of liquid droplets and is frequently used in case of solids. The method permits considerable latitude, and gives both the average, as well as the distribution of size. Choice of the most significant dimension of particles of irregular shape is difficult because the thinnest dimension is almost always aligned vertically, so that a two dimensional measurement may give a wrong impression. The method tends to be tedious and may not give a good measure of desired property unless great care is taken in manipulation.

Designation of diameter of particles of irregular shape is also problematic. Both the 'maximum' and the 'minimum' values may not be representative of properties of such particles.

It is not easy to determine the mean size of particles because of many concepts upon which a mean value may be based, and also because every method has a limit on the lower side, which leaves a fraction, sometimes of considerable proportion, which is not sufficiently well characterised to accurately calculate the mean value.

Size distribution is determined on the assumption that sizes are normally distributed on a logarithmic basis. The mechanical grinding does help us produce finer particles than this distribution can account for, but this situation is usually ignored.

Mass-median diameter used for droplets is less useful in case of powders because bulk of a solid particle is seldom available to a surface or an insect; only the exposed surface is available. Consequently, the most generally useful average is one which gives diameter of a sphere that would expose the same surface per unit weight (or volume) as the particles measured (surface, $S = 6/D_{SV}$). D_{SV}, when expressed in microns, is numerically equal to cubic centimetre of material needed to cover one square metre surface (Gooden, 1944).

The D_{SV} can be calculated from microscopic data ($D_{SV} = \Sigma nD^3 / \Sigma nD^2$. If volume or weight fraction (F) instead of numbers are available, the harmonic mean based on volume gives the same average ($1/D_{SV} = \Sigma F/D$, since $F = nD^3/\Sigma nD^3$; $F/D = nD^2/\Sigma nD^3$; $\Sigma F/D = \Sigma nD^2/\Sigma nD^3$. So, $D_{SV} = \Sigma nD^3/\Sigma nD^2$).

Microscopic data averaged in this way and corrected properly for shape should be comparable with air permeation and gas adsorption methods which are used to measure the same property more directly.

Methods Based on Measurement of Specific Surface of the Particles

Gas Adsorption: Monomolecular layers of adsorbed gas permit calculation of total area of a known weight of a powder from the known area of single gas molecules. The specific surface, S, is determined directly ($S = 6/D_{SV}$). The technique is somewhat difficult. Being direct and involving less assumptions and variables than most others, this method finds wider application. This is an accurate method for determination of average size based on surface area.

Air Permeation: This is another method based on measurement of specific surface. It helps to determine mean surface diameter of the particles and measures, in principle, permeability of a packed column to air, from which total surface area could be deduced; the calibration being such, however, that the value obtained is diameter of uniform spherical particles corresponding to total surface area. Density and degree of packing being constant, the resistance of a packed column of a powder to a stream of air or liquid is a function of specific surface. Its chief disadvantage, as in case of gas adsorption, lies in lack of any measure of range of sizes as this also measures the average size. However, the method is rapid and can be handled by special equipment.

Methods Based on Stokes Law

Sedimentation: Sedimentation methods give both average value and particle size distribution. These are based on principle of Stokes law which states that the resistance offered by a liquid to the fall of a particle varies with radius of the sphere (particle). It is expressed as follows:

$$V = \frac{2}{9}\frac{(d_p - d)gr^2}{\eta}$$

where, V = fall velocity (cm sec^{-1})

g = acceleration due to gravity (cm sec^{-2})

d_p = density of particles

d = density of liquid

r = radius of particles in cm

η = absolute viscosity of the liquid

This law makes certain assumptions such as the particles are spherical, the dispersion is complete, without agglomerates. Turbulence, convection, or wall effects are absent. Although the assumption regarding the sphericity of particles is admittedly not correct, yet the results are reported in equivalent spheres. The rate of fall is not correlated exactly with any easily calculated property, but is related roughly to surface. The results are, therefore, approximately comparable to other methods, if averaged on D_{sv} basis. The other assumptions can be approximated quite well to obtain comparable results.

Even though the measurements through sedimentation methods are not very accurate, they demonstrate the behaviour of powders in suspension better than other methods because of similarity of conditions.

Numerous experimental methods of measuring sedimentation of particles are available. Some of these are mentioned below:

(*i*) *Pipette Method:* It is governed by the fact that sedimentation eliminates through the depth (h) in a time (t) all particles having velocities greater than h/t, while retaining at that depth the original concentration of particles having settling velocities less than h/t. The withdrawal of a small volume (say 50 ml) by a pipette at a depth (h) at time (t) furnishes a sample from which all particles coarser than a particular diameter (x) as per Stokes equation have been eliminated and in which all particles finer than that size are present in the same amount as initially. The volume element at depth (h) has, in effect been 'screened' by sedimentation, so that the ratio of the weight (W) of particles present in that volume at time (t), divided by the weight (W_o) of particles present in it initially, is equal to $P/100$, where, P is the percentage of particles, by mass, smaller than x. Now, the ratio W/W_o can also be written as the concentration ratio C/C_o, giving $C/C_o = P/100$. This equation connects the concentration (C) of the pipette sample in $g\,l^{-1}$ to the parameter P of the particle size distribution, C_o being the weight of solids in the entire sample divided by the volume of suspension.

(*ii*) **Hydrometer Method:** It involves measurement of concentration of the suspension at a given depth at different intervals of time by measuring its density with a sensitive hydrometer. The method is similar to the pipette method except for the manner of determining the concentration of solids in suspension.

Let ρ represent the density of suspension, P_L the density of liquid and P_s the particle density, all in $g\,l^{-1}$, then $\rho = P_L + (C/1000)(1 - P_L/P_s)$, where, C is the concentration of suspended solids in $g\,l^{-1}$. Although the buoyant force on a hydrometer is determined directly by the suspension density, hydrometer scales can be calibrated in terms of C for particular values of density of liquid P_L and particle density P_s. The large size of hydrometer bulb necessary to give adequate sensitivity reduces the depth discrimination of the instrument. This limitation can be overcome by applying a correction factor.

The hydrometer method, like the pipette method, depends fundamentally upon Stokes equation, which may be written as:

$$x = \frac{\theta}{t^{\frac{1}{2}}}, \text{ where, } \theta = 1000\left[\frac{30\eta h}{g(P_S - P_L)}\right]^{\frac{1}{2}}$$

θ being the sedimentation parameter

The value of θ is not constant during sedimentation because the depth of immersion of hydrometer bulb is variable. By equating the depth (h) to the distance in cm from the surface of the suspension to the centre of the hydrometer bulb, it is possible to relate measured values of (h) to the stem readings (R) making θ, a determinate function of R.

(*iii*) **Manometer Method:** It employs a manometer at a fixed depth in the suspension to measure the density and thereby concentration of the suspension down to that depth.

(*iv*) **Others**

(*a*) **The Oden Balance:** Not in use now, the method catches the settling material directly on a balance pan immersed at a measured depth and records the size periodically.

(b) **Settling in Towers/Tunnels:** Fractions can be caught in dust towers and tunnels after settling in air, as a rough measure of size distribution. Stray air currents tend to make such measurements rough. For more precise measurement, scanning photo-sedimentographs are available these days. These monitor the particle settling and give particle size values based on computerized models.

(c) **Sedimentation in the Centrifugal Field:** In case of particles of less than 5 μm, the settling rate is very slow. For example, a quartz crystal (density 2.64 g cm^{-3} measuring 1 μm sinks only 1 cm in water in about 30 hours. In such situations, centrifugal fields are used. The centrifugal acceleration merely has to be inserted in the Stokes formula and in practice the glass cylinder is replaced by a rotating measuring drum. The calculation of mass is carried out at a certain radius, dependent on time. Size calculations are suitably carried out mathematically.

(d) **Elutriation:** It involves the same principle as sedimentation, except that the medium instead of being static is moving upward to carry with it the falling particles. The blown out particles settle at different distances based on size. Air is the usual medium. This method has a great advantage of permitting the collection, direct weighing and later use of particles of narrow size ranges.

Miscellaneous Methods

Optical Methods: These are based on the relation of concentration to the size of the paricles and depend upon transmission of a beam of light through a suspension of particles. The transmission depends upon both the size and the shape of particles. The method is most useful for comparing different samples of the same material where the shape is nearly constant. It provides only the average size of the particles. Both turbidimetry and nephelometry are employed for the estimation. In the former, the intensity of the transmitted light is measured while in the latter, the light scattered by colloidal suspensions is measured employing suitable standards as reference.

Laser Diffraction Spectroscopy: Laser diffraction spectroscopy has become a powerful tool for measurement of particle size. It is based on measuring total diffraction spectrum at given angles of all particles in the sample volume. The method covering a superb dynamic range, is accurate, quick and reproducible. Since this is the latest concept in particle size determination, it is discussed in some detail.

When an electromagnetic beam meets a spherical body, it is scattered in all directions in space. The intensity of the scattered light is a function of the wavelength of light *vis à vis* the diameter of the sphere. Three different ranges as mentioned below are identified.

(i) **Rayleigh Scatter Range (Upper limit 0.02 μm):** Sphere diameter is very small compared to wavelength. In this range, relationship of intensities of incident wave to the scattered wave is proportional to sphere diameter to the power 6. The angle of distribution of scattered light is symmetrical and independent of sphere diameter. Rayleigh scattering is also often called dipole scattering due to dipole structure of the polar diagramme.

(*ii*) **Mie Scatter Range (0.2 μm):** The sphere diameter is smaller than wavelength. In such cases, the polar scatter diagrammes lose their symmetry as particle size increases. A maximum of the scattered light intensity concentrates increasingly and narrowly around the scatter angle $\theta°$, as a forward beam. The scattered light intensity that depends on scatter angle, loses the symmetry of the Rayleigh scatter; the intensity increasing in the forward range. The forward beam traces back due to diffraction, becomes increasingly thinner as particle diameter increases, and maxima and minima begin to form. The maxima and minima which already appear in the Mie scatter range explain why it is possible to carry out measurements of size distribution even in the boundary area (particle size only a little smaller than wavelength).

(*iii*) **Geometric Optics Range:** The sphere diameter is greater than wavelength, due to which the situation stabilizes. If a spherical particle is hit by parallel, monochromatic light, a diffraction pattern, called a 'Fraunhofer diffraction pattern' (Fig. 2.1), appears in the focal plane of a lens placed behind it in the light beam due to an interference by the light waves diffracted at the edge. The diffraction pattern of a round disc or an opening consists of alternate light and dark concentric rings. The radius R_0 of the first dark ring can be calculated using the equation

$$R_o = \frac{1.22 f \text{ wavelenth}}{\text{Disk diameter}}$$

As the diameter of the disc increases, R_o decreases. If monochromatic light and an optical arrangement of given focal length are used, the size of the disc can be determined from the position of R_o.

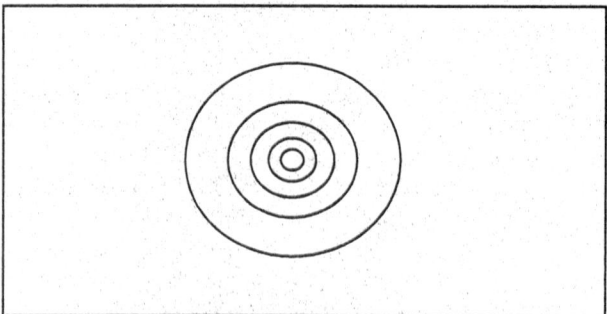

Fig. 2.1: Fraunhofer Diffraction Pattern of a Sphere

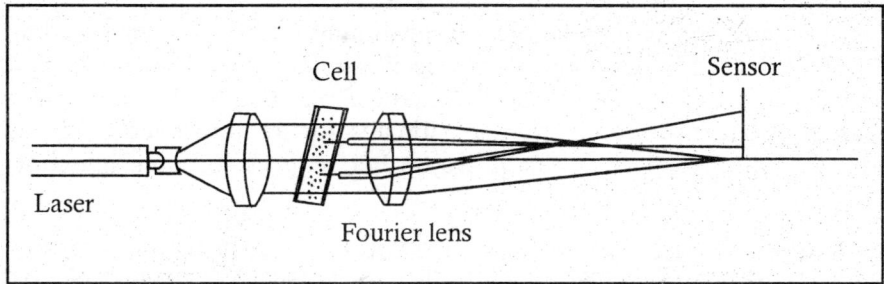

Fig. 2.2: Particle Sizer Using Parallel Beams Principle

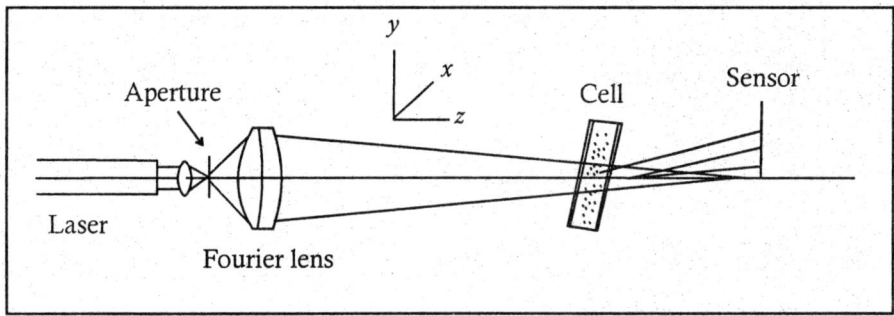

Fig. 2.3: Particle Sizer Using Convergent Beams Principle

In units known today, the forward scattering of monochromatic light is used so that the particle is captured within a parallel light beam and the laser beam behind it is focussed by a lens (Fig. 2.2). To analyse the intensity distribution of a Fraunhofer diffraction pattern (Fig. 2.1), a Fourier transformation lens is normally used, with a receiver unit located at its focal point (Fig. 2.3). The focal length governs the actual measuring range of the system. The intensity of the light scattered in differnt angles is measured by the radially placed individual sensor elements. The information is transformed into size distribution with the aid of mathematical algorithm.

Sorptivity

Sorptivity is the capacity of a powdered inert material to maintain its form on the addition of a liquid in a quantity up to but not exceeding the transition point between dryness and plasticity of the total mass. Sometimes, the term 'sorption index', expressing the weight of the technical material which can be absorbed by 100 g of the inert powdered mineral up to the point of plasticity, is used. The sorption index is like the oil absorption value, but the two values differ depending on the liquid chemical used (Table 2.4). If liquids vary in density, the sorptive capacity varies inversely with the density. If the linseed oil absorption value is known, the toxicant sorption index can be estimated through multiplication with the ratio of the higher to the lower density.

Both the absorptive and adsorptive properties are encompassed in the term sorptivity. It is affected by particle shape, size, structure as well as density of the carrier.

The strengths of most formulations are invariably far below their sorption index (Table 2.4), because of consideration of free flowability, easy workability and other allied aspects. Most of the formulations contain in addition to the active ingredient, deactivators, surfactants and other auxiliaries for an improved shelf life and performance. Addition of such auxiliaries further reduces sorption index for active ingredient.

TABLE 2.4: SORPTIVITY OF SOME INDIAN PESTICIDE FORMULATION CARRIERS BASED ON LINSEED OIL ABSORPTION (SPATULA RUB-OUT, ASTM 1966) AND SORPTION INDICES BASED ON TECHNICAL (96.96%) MALATHION

Carrier	Sorptivity (%w/w)	Sorption index (% w/w malathion)	Workable malathion dust (% w/w)
Attapulgite	100.4	20 - 30	20
Bentonite	65.1	25 - 30	25
Celite	223.2	20 - 25	20
Fuller's earth	119.0	25 - 30	25
Hydrated calcium silicate	372.0	>50	50
Kaolinite	88.4	20 - 30	20
Silica gel–H	186.0	40 - 50	40

Halder (1982), Halder and Parmar (1984)

The solid formulation materials are grouped into 'carriers' or 'diluents' based on their soprtion index. Some examples along with the sorption index values are listed in Table 2.5.

TABLE 2.5: SORPTIVITY OF COMMON PESTICIDE CARRIERS AND DILUENTS

Mineral	Typical sorption index (% w/w, chlordane, rub-out method)
Carrier	
Silica (synthetic)	400
Diatomite (salt water)	270
Vermiculite (expanded)	250
Attapulgite	230
Diatomite (fresh water)	200
Perlite	200
Montmorillonite (non-bentonoid)	190
Kaolinite	160
Diluent	
Pyrophyllite	90
Bentonite	80
Pumice	78
Talc	73
Silica (natural)	64
Limestone ($CaCO_3$)	50
Gypsum ($CaSO_4$)	50

Flanagan (1983)

Measurement

Linseed Oil Absorption Test: Amount of linseed oil that can be taken up by a powder before it becomes completely wet. There are two common methods, the ASTM rub-out method and the ASTM Gardner-Coleman sorptivity test. Principle of determination in both the cases is same except that in the former the end point is given by the curling off of the material with a spatula and in case of the latter, the material can be rolled into a ball with a stirring rod.

Irregular materials like the diatomaceous earths are highly absorptive but have poor properties in other respects. They may be used in mixtures to improve absorption but are seldom used alone. Bentonite is highly adsorptive. Often, it may hold toxicants so tightly as to make them less effective by inhibiting their release.

Density

Density is mass per unit volume ($d = M/V$). Specific gravity is density of the material as compared with a chosen standard. Actual density is that of the material which makes up the particles. Bulk density is of the whole powder including the pore spaces and void volume of the particles. Apparent or particle density is weight per unit volume of the solid part of the inert and its pores.

A material of high density has less surface per unit weight, reducing adsorption and adhesion capacity and covering power. For the same shape, the bulk is less and the feed through a duster is faster on a weight basis. Particles of high density settle more rapidly in either air or liquid. Pronounced difference in the densities of components may lead to segregation during storage or application.

The bulk density depends on the particle density, size and shape and upon the sample history. Fluffing and trapping of air lower the bulk density and vibration packing or simply storage tends to increase it. Even the shape and size of the container affect the bulk density. Arbitrary fixed conditions are used for comparative measurement.

Importance

Bulk density is important in formulation. The degree of coverage of pesticide over an area, wind drift, penetration of foliage, sinking in water, handling through production and application equipment, cost of packaging and transportation etc. are all the factors affected by it. The lighter materials are more absorptive and cake less. These, however, will tend to flow poorly and feed too slowly through dusters. Lighter materials are usually of irregular particle shape and tend to settle more slowly. This will be an advantage or a disadvantage, depending upon intended use. For example, these may be carried away from target in situations of drift and may provide a thorough coverage of treated plant in still situations.

For many commercial pesticides, the bulk densities provided by attapulgites and montmorillonites ($30 - 40$ lb ft^{-3}) are ideal. Inerts like vermiculite are not used in commercial pesticide market because of difficulties in wide scale application but are well received in household pesticide markets where light weight product bags have appeal. When inerts heavier than 40 lb ft^{-3} are used, the products do not provide good coverage and difficulties in passing through the application equipment are encountered (Polon, 1973).

According to Oulton (1967), true density, apparent or particle density and bulk density are often used indiscriminately. Bulk density is not directly related to particle size, though (within certain limits) finer dusts tend to be less dense. Most powders vary in particle sizes and since the smaller particles fit into the voids between the larger ones, the bulk density is greater than that of a uniform dust. The particle shape greatly influences the degree of packing. With powders composed of particles of very irregular shape, bulk density will be low. In case of diatomaceous earths where the particles themselves are porous (due to internal spaces), the bulk density will be at a maximum. David and Gardiner (1950) observed that 5% w/w preparation of DDT on various carriers revealed no difference in toxicity to certain stored grain pests; the variation in the bulk density being from 1.2 cc g^{-1} (calcium carbonate) to 5.1 cc g^{-1} (almicide). Similarly, Harlow (1957) could not establish any direct relationship between bulk density of ten dusts varying from 0.76 to 8.08 g cc^{-1} and their efficacy as carriers.

Density of Solid Carriers and Diluents

Some examples of low and high bulk density materials are given below:

Low Bulk Density: Silica gel, hydrated alumina, calcium silicate, fuller's earth, diatomaceous earths.
High Bulk Density: Pyrophyllite, talc, calcium carbonate, kaolinite.

Sarup (1967, 1970) and Sircar (1975) determined the bulk density of some Indian diluents before compaction (loose bulk value), compaction after different tappings and after total compaction. After total compaction, China clay (Indian clay) was the most bulky followed by fuller's earth, kaolinite, steatite, attapulgite, talc, tripolite, selenite, pyrophyllite, rock phosphate, bauxite, gypsum, feldspar, quartz, silica, dolomite, magnesite and calcite. Halder (1982) reported bulk density (before and after compaction, g 100 cc^{-1}) and particle density (g cc^{-1}) respectively of some Indian carriers as follows: Attapulgite 65, 87, 2.87; bentonite 92, 113, 2.79; celite 40, 52, 2.63; fuller's earth 66, 87, 2.91; hydrated calcium silicate 34, 41, 2.48,; kaolinite 75, 89, 2.71 and silica gel-H 50, 62, 2.42.

Measurement

The bulk density of carriers and diluents can be measured by two different techniques.

(i) **The Loose-Packed (or aerated) Bulk Density Technique:** It gives the bulk density in its fluffed form as determined with a volumeter. This is a measurement of randomly oriented particles that are allowed to fall a minimum distance and to settle without orientation. It is a useful index in estimating the maximum amount of carrier or diluent which might be added to a dry blender. When a blender is in operation, the powdered material which it contains is continually being circulated so that aeration and disorientation are at a maximum.

(ii) **The Packed Bulk Density Technique:** It gives the weight of a volume of powdered material after it has been vibrated to maximum orientation. The material is completely deaerated and the particles are permitted to settle in their most stable geometric alignment. Packed bulk density is an index of the maximum weight of the powdered material, which can be packed in a container of any, given size.

The ratio between density of loosely and that of the vibrated packed material varies according to specific gravity, particle shape and particle size distribution.

Actual Density

Actual density of a powder is usually determined by change in volume of a liquid caused by addition of a weighed amount of a solid. The liquid fills the space between particles and displaces adsorbed air, so that only the actual solid is measured. Pycnometer or a vessel of accurately known volume, is weighed when filled with pure liquid, then filled with suspension of known strength and weighed again. A hydrometer can be used for the measurement if settling of particles is very slow. Errors can be caused by incomplete wetting, incomplete removal of air, porosity of particles, solubility of particles or adsorption on the surface.

Reaction

Reaction is expressed as H^+ ion concentration or activity in dilute aqueous solutions. It is an index of the acidic or alkaline nature of the material.

Importance

Depending on sensitivity of a toxicant towards reaction and moisture content of environment, the pH of carriers and diluents employed in their formulation affects stability. The effect of pH is pronounced in presence of moisture.

pH of Solid Carriers and Diluents

Fowkes et al. (1960) reported pH of several carriers based on 1% diluent in water as follows: kaolinite 4.6, montmorillonite group 5.1, attapulgite (diluex) 9.5, attapulgite (atta clay) 7.2, diatomite 5.2, pyrophyllite 6.2, silica (Hi – Sil 101) 9.4. The pH of similar suspensions of some Indian carriers was: Fuller's earth, Indian clay and kaolinite 3.5-4.6, bentonite and pyrophyllite 7.6, attapulgite 9.1. The pH of quartz, rock phosphate, calcite, bauxite, selenite, gypsum, steatite, talc, tripolite, feldspar, magnesite, silica and dolomite ranged between 7.8-8.9 (Sarup, 1967). Sircar (1975) also reported similar values except that the kaolinite and China clay were found to be neutral and diatomite alkaline. Halder (1982) reported the pH of 1% carrier suspensions in water as : attapulgite 9.3, bentonite 8.4, celite 8.7, fuller's earth 9.2, hydrated calcium silicate 9.7, kaolinite 9.1, silica gel-H 6.6. Thus, besides the source, a lot to lot variation in the pH of samples is observed.

Measurment

Colorimetric Methods: These employ a range of indicators or a multiple range indicator or indicator test papers, to have an approximate idea about the pH of a solution. The standard charts or colour comparators are employed for a comparison and pH known.

Potentiometric Methods: The modern methods employ pH meter which is essentially an electronic millivoltmeter and operates directly from the *AC* mains. It has pH, millivolt and Δ pH scales, enabling a direct record of pH and measures quite small changes with accuracy.

Surface Acidity

Since the general adoption of Arrhenius theory, the usual quantitative measure of acidity has been the hydrogen-ion concentration or at times, the hydrogen ion activity. This measure is based entirely upon behaviour of dilute aqueous solutions and there are serious difficulties in non-aqueous or concentrated aqueous solutions. The measurement of acidity with a basic indicator, as suggested by Hammett and Deyrup (1932) is unambiguous, historically reasonable and measurable. They have discussed this problem in a series of publications and described a new acidity function H_0 as a measure of acidity. It is characterized by a reaction in which a proton is transferred to a neutral molecule (e.g. urea + water, $H_2NCONH_2 + H_2O \rightarrow H_2NCONH_3 + OH^-$; $H_2SO_4 + H_2O \rightarrow H_3O^+ + HSO_4^-$) salt formation or ionization of the base.

$$H_o = -\log a_{H^+} \frac{fB}{fBH^+}$$

where,
a_{H^+} = activity of the hydrogen ion
f = activity coefficient
B = base
BH^+ = conjugated acid of the base.

For measuring this acidity, Hammett and Coworkers developed a series of indicators which were capable of absorbing a proton at various degrees of acidity. These indicators are Bronsted bases, which are transformed into their conjugated acids by absorbing a proton. A change in colour as a consequence of this proton transfer is essential for using the substance as an indicator. Some examples of H_o values (Gould, 1962) are: (+7.0) water, (–1.5) 6 molal hydrochloric acid, (–2.63) molal sulphuric acid, (–5.54) 70% sulphuric acid, (–10.6) 100% sulphuric acid.

The acidity function had been used originally by Hammett and coworkers to study the dependence of acid catalysed reactions in liquid media upon acidity. It was extended to catalytic reactions on solid surfaces by Walling (1950) who considered these as Lewis acids. For this, the equation is slightly altered as:

$$H_o = -\log a_A \frac{fB}{fAB}$$

where, a_A represents the activity of the Lewis acid or of the electron acceptor.

Hammett Indicators

Hammett's original scale of indicators has been extended by various workers to the range of lower acidities to enable an expanded application. Therefore, indicators are now available for whole scale of acidity and their transition points are only short steps apart (Table 2.6).

TABLE 2.6. IMPORTANT INDICATORS FOR DETERMINING SURFACE ACIDITY

Indicator*	pKa of the indicator	Colour of neutral base	Colour of conjugated acid	Weight % sulphuric acid
Neutral red	+ 6.8	Yellow	Red	8×10^{-8}
4-Phenylazo-1-naphthylamine	+ 4.3	Yellow	Red	5×10^{-5}
4-Dimethylaminoazobenzene	+ 3.3	Yellow	Red	3×10^{-4}
p-Phenylazoaniline	+ 2.8	Yellow	Orange	—
4-o-Tolylazo-o-toluidine	+ 2.0	Yellow	Red	—
4-Phenylazodiphenylamine	+ 1.5	Yellow	Purple	0.002
N,N-Dimethyl-p-1-naphthyl-azoaniline	+ 1.2	Yellow	Blue	—
4,4,4-Methylidynetris (N,N-dimethylaniline)	+ 0.8	Yellow	Blue	—
4-Nitro-4'-nitrophenylazo-diphenylamine	+ 0.4	Orange	Violet	0.3
Dicinnamalacetone	− 3.0	Yellow	Red	48
1,9-Diphenyl-1,3,6,8-non-atetraen-5-one	− 2.2	Yellow	Red	50
Chalcone (Benzalacetophenone)	− 5.6	Colourless	Yellow	70
Anthraquinone	− 8.2	Colourless	Yellow	90

*For indicator preparation, dissolve 10 mg indicator in 100 ml of anhydrous benzene or 200 mg in 100 ml of 2, 2, 4-trimethylpentane (*iso* octane)

Surface Activity of Solid Carriers and Diluents

Disproportionate distribution of electric charges in the clay surface leads to development of positively charged centres (acid centres). These are electrophillic in nature and their strength varies depending on chemical composition of the surface and the degree of distortion in the lattice structures (isomorphous substitutions, broken edges, hydrolysis etc.).

It is expressed in terms of pKa with a numerical value ranging from +7 to −8. The optimum pKa of a carrier for a toxicant can be worked out by formulating the toxicant with different carriers and conducting the degradation rate studies.

Both acid strength at a site as well as number of acid sites in a given mass or weight are important. For example, the most active acid sites of kaolinite probably have a pKa equal to −8 but it has only 1/3rd as many acid sites as montmorillonite for a given weight. Ullrich (1964) reported that pH of attapulgite mineral "Diluex" varied from 8.1-9.5 but the Hammett indicators provided its surface acidity equivalent to 70% sulphuric acid.

Benesi (1956) and Matsumoto *et al.* (1957) reported that kaolinite had the highest acid strength (H_o = −3.0 to −5.6) followed by montmorillonite, attapulgite, diatomite, pyrophyllite, talc, silica and calcite (+3.3, transitional).

A poor correlation between the pKa or H_o and pH values of aqueous slurry of the carrier exists because the H_o values measure the strongest acid sites on the clay surface whereas the pH is result of hydrolysis of all portions of clay surface (Matsumoto *et al.*, 1957, Fowkes *et al.*, 1960). In a clay like Diluex (referred above, attapulgite with 7% magnesium oxide), alkaline portions of the surface produce more than enough hydroxyl ions to neutralize the hydrogen ions originating from the acidic portions of the surface, resulting in alkaline reaction in water even though it is strongly acidic to Hammett indicators. Similarly, bentonite having slightly alkaline reaction in water was found to be acidic to Hammett indicators. Dolomite having alkaline reaction was least acidic to Hammett indicators (Fowkes *et al.*, 1960).

The effect of different cations on H_o values and catalytic activity of kaolinite revealed that the natural clay and sodium kaolinite were identical. Hydrogen kaolinite and aluminium kaolinite were more strongly acidic and much more active catalysts, promoting ten fold faster rates of reaction (Fowkes *et al.*, 1960).

Surface acidities of some Indian diluents have also been reported. The strongest acid sites were on the surface of fuller's earth (H_o = –5.6) followed by kaolinite = Indian clay (–3.0 to –5.6), bentonite = attapulgite = steatite (+1.5 to –3.0), tripolite = talc = pyrophyllite = magnesite (+1.5 transitional), rock phosphate = silica (+3.3 to 1.5), gypsum = selenite = feldspar (+3.3 transitional) and bauxite = calcite = dolomite = quartz (+4.0 to +3.3) (Sarup, 1967). However, Halder (1982) reported that attapulgite, bentonite, fuller's earth, hydrated calcium silicate, kaolinite and silica gel–H revealed *pKa* of +3.3 only. Both these authors did not observe any correlation between pH and pKa of the carriers.

The generality of clay catalysis of decomposition reactions in pesticidal dusts has been amply demonstrated. Matsumoto *et al.*, (1957) reported that carriers with *pKa* less than 3.0 were unsuitable for formulating organophosphorus compounds without their deactivation. Using malathion, Matsumoto *et al.* (1957, 1958) found a very low rate of degradation at pKa $\geq$ 3.0 which varied with the carrier in a decreasing order: kieselguhr > acid clay > clay > talc. With 0.7% polyoxyethylene alkyl ether as stabilizer, the order was : acid clay > kieselguhr > talc > clay. No stabilizer worked with acid clay, suggesting its unsuitability with malathion. The same relationship was observed with methyl parathion (Matsumoto *et al.*, 1958). The decomposition appeared to be due to acid catalysis (hydrolysis by H^+ ions). Matsumoto (1958) investigated the decomposition rate of malathion in dust formulation in relation to surface acidity, pH, substitution acidity and Thomas acidity of the carriers. The substitution acidity of carriers followed the order: acid clay > clay > kieselguhr > atta clay > silica sand. Thomas acidity of carriers was irregular but some correlation between surface acidity and substitution acidity existed although the correlation between pH, substitution acidity and Thomas acidity was irregular.

It has been reported that limestone, bentonite and diatomaceous earths of pKa 1.5 can be used to formulate HCH without any appreciable loss of insecticidal activity. Endrin loses insecticidal activity on materials having pKa less than 3.3.

Deactivators

Acid sites of carrier can be deactivated with certain organic chemicals which preferentially share their electrons with the site to form a covalent bond stronger than the bond formed between pesticidal chemical and the acid centre. Compounds containing oxygen in an ether linkage (ethers, glycol ethers etc.) or amine derivatives are effective for this purpose. The deactivators need to be tested for compatibility with the pesticidal ingredients and surface active minerals. Up to 10% deactivator (based on weight of inert) has been reportedly used for deactivation. This, as expected, increases the formulation cost and reduces the carrying capacity of the carrier for an active ingredient. The other properties and characteristics of carriers and diluents being satisfactory, one with the least surface acidity should be used. Lot to lot, source to source and carrier to carrier variations need to be monitored.

For each carrier-toxicant-deactivator combination, the deactivator amount is worked out with the help of indicators. With exceptions, more sorptive carriers are generally more active and require larger amounts of deactivator.

Common Deactivators for Acid and Basic Sites

Acid Sites: Urea [$OC-(NH_2)_2$], sodium hydroxide (NaOH), lime [$Ca(OH)_2$], sodium carbonate (Na_2CO_3), monoethanolamine ($H_2NCH_2CH_2OH$), ethyleneglycol ($OHCH_2CH_2OH$), diethylene-glycol ($OHCH_2CH_2OCH_2CH_2OH$), triethyleneglycol ($OHCH_2CH_2OCH_2CH_2OCH_2CH_2OH$), acetylacetone ($CH_3COCH_2COCH_3$), isopropanol ($CH_3CHOHCH_3$), hexamethylene tetramine [HMT, $(CH_2)_6N_4$], sodium ligno-sulphonates etc.

Basic Sites: Basic sites on surface of clays can be deactivated with certain weak acids such as tall oil and rosin acids, certain alcohols, ethers, ketones, glycols, etc.

Measurement

Colour Change of Adsorbed Indicator: A quick and reasonably approximate idea of pKa can be had by use of Hammett indicators (Table 2.6). Colour of the neutral base changes when its conjugate with acid is formed. By matching the colour change with the anticipated shade, an idea about pKa of the carrier is made.

Volumetric Method: Amine titration (Benesi, 1957) has been employed to quantitatively estimate acid strength of catalytic surfaces.

DDT, aldrin, dieldrin, heptachlor, chlordane, endrin, aramite are all decomposed by acid sites. Malathion, thiophosphates are decomposed by basic sites. With heptachlor, diethylene glycol; with aldrin and dieldrin, urea, ammonia solution or ammonium bicarbonate; with endrin, HMT, polyethylene compounds in acetone; and with ethion, diethylene glycol and dipropylene glycol have been reported as effective deactivators. Dipropylene glycol is a good deactivator for carriers (ex. kaolin) used for making dry formulation of endosulfan. Talc and calcium carbonate do not require any deactivator. Maracarb N (a sodium lignosulphonate) and urea are good deactivators of carriers for formulating aramite.

Organophosphates like malathion, parathion, methyl parathion, chlorthion etc. are susceptible to alkaline hydrolysis. In such cases, proper method is to neutralize the alkaline active sites on carriers with the help of weak acids such as tall oil, rosin acids, glycols etc. Alcohols, ethers, ketones, glycols have been used as stabilizers for dry formulations of parathion and methyl parathion. Diethylene glycol and dipropylene glycol have been suggested deactivators for dry formulation of ethion. Diazinon has been found to decompose on all mineral carriers. For its formulations, walnut shell is a better carrier. Glycols have been suggested as deactivators for phorate. Phosdrin decomposes on alkaline carriers with high moisture content. Talc, pyrophyllite, calcined gypsum and volcanic dust are suitable for its formulation without any deactivator. Talc to which maleic acid has been added as an anti-decomposition agent, was the most compatible material with malathion.

Basic sites on mineral surface contribute to the decomposition of thiophanate type toxicants. Copper catalyses this hydrolytic degradation though nickel, iron, calcium, magnesium and zinc have no effect. The presence of copper needs to be suitably taken care of.

Carbamates are also susceptible to alkaline hydrolysis and must be formulated only with acidic carriers.

Flowability

Flowability of a formulation or a carrier is the rate at which a material can be poured, moved or displaced. An ideally formulated product should have the same flowability as the inert. Particle size, shape, density, humidity, presence of sticky constituents etc. affect flowability. Fundamentally, it is more related to the particle shape (Polon, 1973) though it is somewhat modified by particle size. As a rule, upto a limit, the coarser the particle size, better the flowability. Agricultural diluents with particles that tend to agglomerate, as a rule, have poor flowability even in the larger particle size (40 μm) range.

In the formulation process, the power requirement for blending or working the material decreases with increase in flowability. Flowability of the inert is important for ease of handling and storage as a raw material, flow of material through blenders, minimising clogging in the grinding mills and packaging of the final product. Product flowability is also important during application e.g. flow out of bags, field applicators, shaker cannisters etc.

Malina (1960) defined flowability index as the flowability of a dust or dust formulation relative to the flowability of unimpregnated attapulgite (100) and measured it as follows:

$$\text{Flowability index} = \frac{t_a d_a}{t_f d_f} \times 100$$

where,

t_a = time required for 5 g of attapulgite to pass through 60 mesh screen

t_f = time required for 5 g of test material/formulation to pass through 60 mesh screen

d_a = light pack density of attapulgite

d_f = light pack density of test material/formulation

Following this, flowability indices for five inerts were reported as follows: attapulgite 100, talc 112, pyrophyllite 89, kaolin 12 and diatomaceous earth 8. Following an adjustment for differences in density, the flow index of each inert was calculated with respect to attapulgite taken as 100. Talc rapidly lost its flowability when impregnated. Depending on the flowability index, it was concluded that attapulgite and unimpregnated talcs have good flowability, and that the flowability of clay decreased as an increased amount of oily substance was impregnated on it. Kaolinite and diatomaceous earths have poor flowability.

Measurement

The flowability has been measured by purely laboratory methods, such as the angle or slope, or by purely practical means such as rate of flow through an actual duster. The distance moved on dusting under constant conditions will provide an idea of flowability.

Electrical Charges

Electrokinetic Charges: These are caused by preferential adsorption of ions from solutions, and cannot be separated from opposite charges in close association. These charges are permanent as long as liquid and solid are in contact. These may be important in the deposition of particles from suspensions and in adhesion.

Electrostatic Charges: These arise from contact or rubbing between particles and applicator surface or between particles. An actual transfer of electrons takes place, and the opposite charges can be widely separated e.g. an insulated piece of application equipment can be left with a very high charge after an oppositely charged dust has been discharged out of it.

Similarly charged particles repel each other and may be strongly attracted to oppositely charged surfaces. Under favourable conditions, the charges may substantially increase deposition. The electrical charges are easily lost in damp air or by contact with surfaces, so adhesion is not affected as it is by the electrokinetic charge.

Plant materials tend to assume a negative charge, while silicates and arsenicals become positive. Most dust clouds may contain particles of opposite charges as well.

Electrostatic charges are so variable, so temporary and so extremely dependent upon weather, mechanical details and other variables, that no good evaluation of their importance under practical conditions seems possible. Some attempts have been made to introduce such charges intentionally, but little success has been achieved.

Hardness

Hardness is a measure of the strength and plastic characteristics of materials and is a convenient factor for comparison and quality control purposes. This function depends upon the composition of material. Generally, soft materials get coated over the harder one when two such materials are blended together. Therefore, the carriers or diluents need to be and generally are harder than the crystals of organic insecticides.

Effects of Hardness

Abrasive character of carriers and diluents is primarily governed by their hardness. It results in wear of processing equipment, metering devices and orifices of application equipment. Hard materials like quartz occurring as an impurity in pyrophyllite, are known to cause serious abrasion and wear in application equipment. It is also due to the hardness that the chemically inert carriers and diluents are able to cause abrasion of water impervious coating of insect and the consequent desiccation. Abrasive materials include pyrophyllite, pumice, silica and diatomite while non-abrasive materials include kaolinite and talc.

Measurement

Hardness is measured by the indentation produced by a hard body on surface of the materials. 'Brinell hardness' H_B is measured by the indentation produced by a hardened steel ball (dia. 10 mm) under a specific load P (kg): taken normally as 30 d^2 for steels and 10 d^2 for non-ferrous alloys (d = diameter of the ball in mm). This hardness is taken as the ratio of the force P to the area of spherical indentation produced.

'Rockwell hardness' is defined as the depth of the indentation produced by a spheroconical diamond Brale penetrator with 120° cone angle and 0.200 mm end radius or hard steel ball of 1.588, 3.175, 6.350 or 12.700 mm diameter, under the action of a specific force. There are three scales of hardness (A, B and C) corresponding to three values of the applied force 60, 100 and 150 kg respectively, the hardness being denoted as H_{RA}, H_{RB}, H_{RC}.

A method similar to Brinell test is used for the dynamic testing of materials. In this, a ball either falls freely or forms a striking surface of a pendulum. The analysis of the indentation formed enables estimates to be made of the effect or rate of strain on the plastic characteristics of a material.

Even simpler equipments can yield most reliable results. For example, for tensile tests, a calibrated wire with the top held in a vice and a weight hanger attached to the bottom can be used. In case of a wire of diameter 2 mm, made of steel (Cr-5, δ_s = 2800 kg cm^{-2}, δ_B = 5500 kg cm^{-2}), the force required to cause failure is of the order of 160 kg. This force can easily be applied by weights placed on the hanger. If the length of wire is 3 m, the overall extension when the yield point is reached will be 3 mm. The extension (the movement of the bottom of the wire or the weight hanger) can be measured with a high degree of accuracy by means of dial gauges reading to 0.01 mm. The advantage of the method is that the loads are applied directly to the specimen. It is particularly a useful method for testing highly elastic materials like rubber, nylon etc.

Water Content

Water exists in several forms in mineral carriers though the boundaries of each form are not clearly defined, there being considerable overlapping. Searle and Grimshaw (1959) have recognised the principal types of water as mentioned below:

(*i*) Combined water or water of constitution, an essential part of the crystal lattice with hydroxy groupings as the usual form.

(*ii*) Water of crystallization or hydrate water which differs from the combined water in as much as the original substance can be readily reformed by treating the dehydrated material with water.

(*iii*) Broken bond water which is linked to the unsatisfied valencies occurring at the edges of crystals.

(*iv*) Absorbed water which is associated with the exchangeable cations in certain clay minerals and is taken up within the crystal lattice on exposure to humid air (montmorillonite).

(*v*) Absorbed water, hygroscopic water or moisture, related only to the exposed surface area of mineral. If and also depends on humidity of atmosphere.

Effects of Water

A decrease in acid strength by presence of water and acetone has been attributed to the firmly bound water due to which hydrated silica-alumina and silica-magnesia may have the structure of 'polyacids' of considerable strength (Walling, 1950). Effect of water absorption is either to decrease colour intensities of the absorbed indicators or to cause a shift to lower acid strength (Benesi, 1956).

Rate of decomposition of toxicant has been reported to be slower on clays containing more absorbed water. The indicator dyes showed that clays with H_0 values of around -5.6 were neutralized to about $+1.5$ by water vapour alone though neutralization with water vapour was neither sufficiently complete nor permanent for safe storage (Fowkes *et al.*, 1960).

Water Content of Solid Carriers and Diluents

Sarup (1967) reported the per cent moisture content by gravimetric method in different carriers as follows: selenite (22.3), gypsum (11.4), bentonite (10.9), attapulgite (10.6), fuller's earth (7.1), tripolite (2.6), kaolinite (1.6), rock phosphate (1.4), China clay (1.0), steatite (0.61), magnesite (0.4), silica (0.3), bauxite (0.3), pyrophyllite (0.2), quartz (0.3), feldspar (0.2), calcite (0.1), talc (0.1) and dolomite (0.1). By the same method, Sirkar (1975) reported moisture content as: attapulgite (10.9), gypsum (9.8), fuller's earth (7.1), bentonite (6.2), diatomite (3.8), kaolinite (1.1), China clay (0.7), magnesite (0.2), pyrophyllite (0.2), talc (0.1) and quartz (0.0). Halder (1982) obtained the values : attapulgite (2.1), bentonite (8.8), celite (2.1), fuller's earth (8.2), hydrated calcium silicate (4.8), kaolinite (4.3), silica gel H (5.1).

The variation in moisture content of some of the samples may be attributed to variation in the sample lot and the environmental conditions at the time of analysis.

Measurement

Karl-Fischer Reagent: A known weight of the test material is treated with freshly prepared Karl Fischer reagent and the volume of the reagent consumed is determined by back titrating with standard water-methanol solution (Vogel, 1964). One ml of Karl-Fischer reagent = 6.66 mg of water.

Drying to Constant Weight: A weighed amount of the sample is placed in an oven at 110°C and dried to constant weight. The loss in weight gives the amount of water lost in the process and the per cent content is calculated.

CARRIERS AND DILUENTS

In preparation of solid pesticide formulations, the active ingredient is mixed with different kinds of inert materials, termed "carriers". These are usually used in preparation of powder concentrates like dust concentrate, water dispersible powder etc. Inerts possessing low sorptivity and high bulk density are called "diluents" and are usually used in preparation of direct application powders like dusts etc. Proper choice of these carriers and diluents is of paramount importance. A wrong selection can make useless product of a good pesticide.

Inert materials used in pesticide formulation may be grouped into three classes:

1. Inorganic materials
2. Natural organics or botanicals
3. Synthetic materials

1. Inorganic Material

Silicate clay minerals constitute the bulk of inorganics used in pesticide formulation. In pure form, these are endowed with a characteristic lattice structure. However, there are disorders due to isomorphous substitution of cations and anions and also by an irregular stacking of crystalline layers.

Clays, in general, include all organic and inorganic soil particles which are crystalline or amorphous. These are formed directly by crystallisation from solutions of silicates and aluminates (genesis) or by alteration (diagenesis) of rocks and minerals by direct changes both in solid and solution phases. The solid phase in clays is polymer formed due repeated molecular units (dia ~ 3 μm or more) and may be crystalline or amorphous depending on the regularity of structure.

Structurally, clay minerals consist of silicon tetrahedral or aluminium octahedral units existing as singlets or joined together to give chain like, sheet like or three dimensional network system. Silicon tetrahedron comprises of a silicon atom coordinated by four oxygen atoms and aluminium octahedron comprises of an aluminium atom coordinated by eight oxygen atoms. The coordinating atoms may be isomorphously substituted by atoms of similar atomic size. If the charge of the substitution atom is different, then the mineral is rendered electrically charged. The charges are neutralised by absorbing oppositely charged ions from the medium.

The electrical charges, besides arising from isomorphous substitution, can also arise at broken edge bonds of clay lattices. The primary bonds are broken and the valencies of the exposed lattice atoms are not completely compensated, leading to the origin of electrical charges at the edges, where cations are attracted to neutralize the charges.

Some properties of the major clay minerals are given in Table 2.7. It is seen that there is a wide variation in cation exchange capacity (CEC) and specific surface area of the clay minerals. These properties are important in determining their adsorption characteristics, which are the most important with regards to their use in pesticide formulation. A schematic presentation of the structure of important clay minerals is shown in Fig. 2.4.

TABLE 2.7. IMPORTANT PROPERTIES OF MAJOR CLAY MINERALS

Clay mineral	Lattice structure	Swelling properties	CEC (me 100 g^{-1})	Surface area ($m^2 g^{-1}$)
Montmorillonite	2 : 1	Expanding	80-120	750 - 800
Vermiculite	2 : 1	Limited expanding	100-150	500 - 700
Illite	2 : 1	Non-expanding	10-40	50-125
Kaolinite	1 : 1	Non-expanding	3-10	10-50

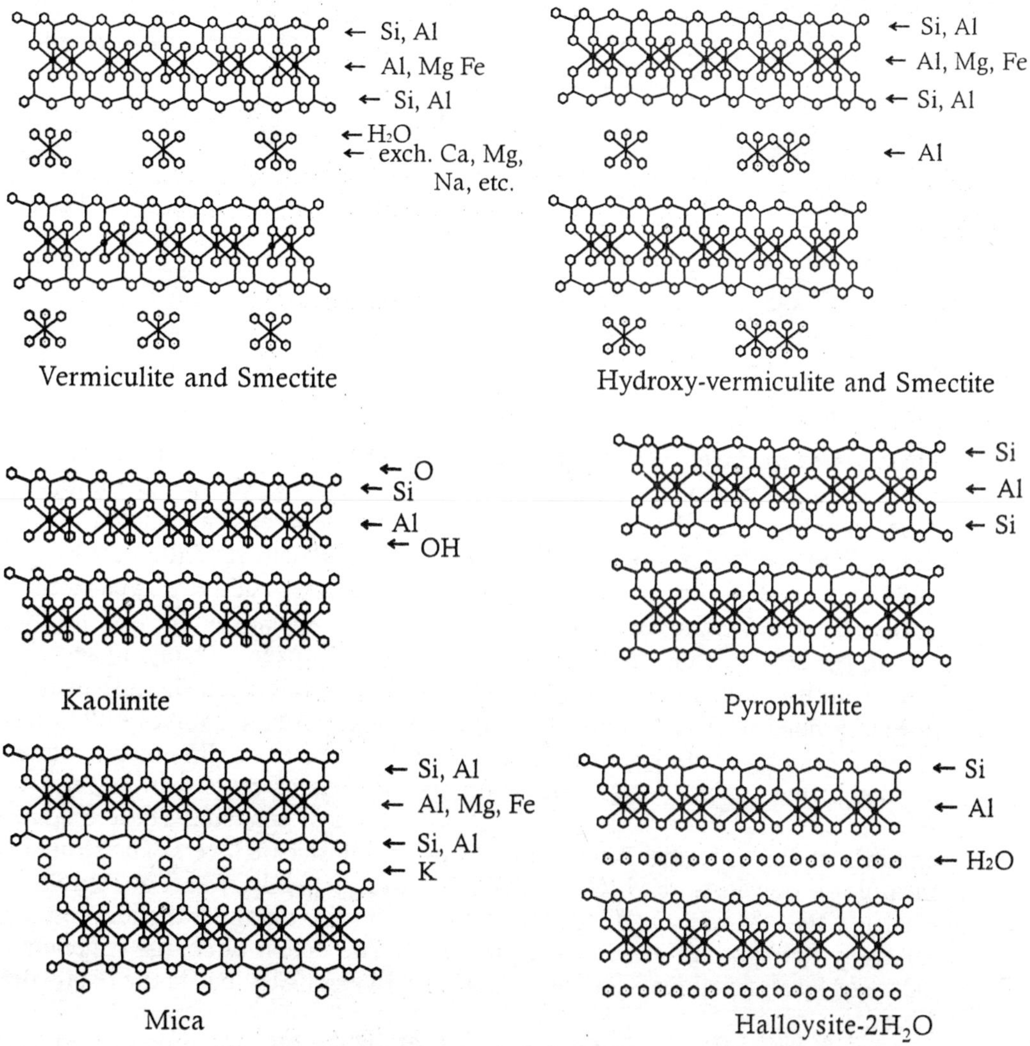

Vermiculite and Smectite

Hydroxy-vermiculite and Smectite

Kaolinite

Pyrophyllite

Mica

Halloysite-2H_2O

Fig. 2.4: Structures of Some Important Clay Minerals

Silicate Clay Minerals

Silicate clay minerals used in pesticide formulation may be classified into four groups as follows:

A. Polygorskite group
  (i) Attapulgite or polygorskite
  (ii) Sepiolite

B. Kaolinite group
  (i) Anauxite
  (ii) Dickite
  (iii) Kaolinite

C. Montmorillonite group
  (i) Beidellite
  (ii) Montmorillonite
  (iii) Nontronite
  (iv) Saponite

D. Mica group
  (i) Illite
  (ii) Vermiculite

A. Polygorskite Group

(i) **Attapulgite and Sepiolite:** Attapulgite is an alumino-magnesium silicate with about equal proportion of Al and Mg while sepiolite is a Mg-silicate containing lesser quantity of Al^{3+} ions. These minerals have a fibrous morphology which results from bands elongated parallel to the C-axis and consist of alternate ribbons with a 2 : 1 structure. The ribbons consist of five octahedral positions in polygorskite and eight positions in sepiolite. These appear as the justaposition of two pyroxenic chains in polygorskite and three pyroxenic chains in sepiolite. The structure has a continuous plane of atoms with (i) tetrahedral position primarily filled with Si^{4+} atoms and (ii) octahedral position primarily filled with $Mg^{2+}Al^{3+}$ ions alternating to form open channels of fixed dimension running parallel to the chains. Polygorskite contains channels of cross section 3.8 × 6.3 Å whereas these channels are 3.8 × 9.4 Å in csase of sepiolite. These channels contain cations and water molecules in between the silicate layers. The structural formula of polygorskite is $(OH_2)_4 (OH)_2 (Mg_5Si_8O_{20} \cdot 4H_2O)$ and of sepiolite is $(OH_2)_4 (OH)_6 (Mg_9Si_{12}O_{30} \cdot 6H_2O)$. Substitution in the tetrahedral sheet remains relatively unimportant in both these minerals. The deficit of charge is generally balanced by an equal replacement of Mg^{2+} by trivalent Al^{3+} ions in the octahedral sheet.

CEC of polygorskite is in the range of 5-30 me per 100 g and that of sepiolite 20-45 me per 100 g. The dissociation of K and Ca from polygorskite is reported to be somewhat greater than from many other clay minerals. Sepiolite is reported to decompose readily in an acid medium whereas polygorskite is more resistant.

Due to open packing, these minerals have a large specific surface area—polygorskite–140 m^2g^{-1} and sepiolite 392 m^2g^{-1}.

(*ii*) **Talc and Pyrophyllite:** Talc is trioctahedral, hydrated Mg–silicate composed of two tetrahedral sheets with a central octahedral sheet in which all octahedral positions are occupied by Mg^{2+}. Pyrophyllite is dioctahedral aluminosilicate with a crystal structure similar to talc, but only 2 of every 3 octahedra are occupied by Al^{3+}. Each layer is bound to another by relatively weak Van der Waals forces. The structural formula of talc is $Mg_3Si_4O_{10}(OH)_2$ while that of pyrophyllite is $Al_2Si_4O_{10}(OH)_2$. The lack of layer charges precludes chemical reactivity of talc or pyrophyllite.

Both minerals have prominent basal cleavages as a result of the weak bonds between adjacent layers. Consequently, talc and pyrophyllite have a softness and flexibility caused by ready sliding of successive layers over one another.

B. Kaolinite Group

The minerals of this group consist of unit layers formed by the linkages of one Si(Al, Fe)—O tetrahedral sheet with one Al (Fe, Mg)—OH octahedral sheet in regular succession i.e. the 1 : 1 type. Two third of the octahedral positions are occupied by Al^{3+} ions and the tetraderal positions by Si^{4+} atoms. Halloysite, dickite and nacrite contain the same basic layers as kaolinite but the stacking sequence of layers is different in each mineral.

Hydrogen bonding has formerly been considered as the major force binding kaolinite layers together. Recent investigations have shown that OH--O bonds are predominantly electrostatic where both oxygen atoms are bonded to other cations.

Kaolinite has characteristic platy morphology and fine particle size. The plates retain their morphology even on heating to 1000°C. The structure of kaolinite crystal is altered by dry grinding.

(*i*) **Kaolinite (Al_2 $(OH)_4$ Si_2O_5); $Al_4Si_4O_{10}(OH)_8$:** Kaolinite may be formed by the weathering of *K* and *Na* feldspars from rocks or by the hydrothermal action of carbonic and sulphuric acid solutions on feldspars and micas. It may also be formed by silicification of hydragillite by silicic acid solution. Surface areas of kaolinites range from 5.0 to 14.5 m^2g^{-1}. CEC values lie between 1.25 and 4.75 me per 100 g.

(*ii*) **Halloysite ($Al_2(OH)_4Si_2O_5 \cdot 2H_2O$):** The lattice structure of halloysite is more disordered compared to that of kaolinite. It also contains interlayered weakly bound H_2O molecules.

C. Montmorillonite Group (Smectites)

This group includes a few important 2:1 expanding type of minerals. These minerals consist of a series of successive layers, each formed by one Al (Mg, Fe, Cu, Cr, Zn,

Li, Mn)—OH octahedral sheet enclosed between two (Si, Al)—O tetrahedral sheets, due to which these are indicated as 2:1 type. Due to the weak interlayer charge, the large surface areas (400-800 m^2g^{-1}), the high CEC (10-200 me per 100 g) and poor ordering of several layers, smectites will swell in water or other liquids of high polarizability. Adsorbed cations also influence the swelling property. Higher the hydration level of the cation adsorbed (*Na, Li*), higher is the swelling.

(*i*) *Montmorillonite* [(Al, Mg)$_2$, (OH)$_2$ Si$_4$O$_{10}$ (Ca^{2+}, Mg^{2+}, Na$^+$) nH$_2$O]: Montmorillonite is an important raw material employed in various industries. It is used as a catalyst for cracking petroleum and oils, as a decolourant for oils and for clarifying oils and wines, as a carrier for catalysts in general, besides being a carrier in solid pesticide formulations. Its activity can be largely increased by treatment with NaOH, or strong acids; called activated bentonites, which are commercially available under several trade names.

(*ii*) *Nontronite* [(Fe, Mg)$_2$ (OH)$_2$ (Si, Al$_4$O$_{10}$)(Ca^{2+}, Mg^{2+}) nH$_2$O]: This member of the group is rich in iron (Fe$_2$O$_3$, 20-30%).

(*iii*) *Saponite* [(Mg, Al, Fe)$_3$ (OH)$_2$ (Si, Al)$_4$O$_{10}$ (Na$^+$, Ca^{2+}, Mg^{2+}) nH$_2$O]: It is the Mg rich smectite (MgO ~ 8.25%).

D. Illite Group (Mica Minerals)

Minerals of the mica group consist of layers, each with two (Si, Al)—O tetrahedral sheets enclosing one Al (Mg, Fe)—OH octahedral sheet. The layers are limited by K, Mg, Ca or Na cations which compensate the surplus of negative charge resulting from isomorphous replacement of trivalent Al by bivalent Mg or Fe in octahedra, or the tetravalent Si by trivalent Al or Fe in the tetrahedra. The succession of the layers is quite regular, and this is also indicated as 2 : 1 mineral.

Mica minerals may be divided into two groups; the dioctahedral and trioctahedral. In dioctahedral, one third or nearly one third of octahedra holes are unoccupied. In the trioctahedra, nearly all the octahedral holes are occupied by cations.

(*i*) *Illite* [(Al, Fe, Mg)$_2$ (OH)$_2$/(Si, Al$_4$O), (OH)$_{10}$ (K, K$^+$, H$^+$)]: The term has been used for mica like minerals occurring in sedimentary materials. Their exact place is under debate.

(*ii*) *Vermiculite* [(Mg, Al, Fe)$_3$ (OH)$_2$/(Si$_3$Al)O$_{10}$ Mg^{2+} (H$_2$O)$_4$]: Two different vermiculites are common and are based on ideal mica structures. The trioctahederal vermiculites are an alteration of biotite whereas dioctahedral vermiculites are alteration products of muscovite. Most soil vermiculite is probably dioctahedral with properties assigned to it by extrapolation from coarser trioctahedral vermiculite or from *K*-depleted muscovite.

The coarse Mg-rich mica mineral is expanded to 14 Å because of strong hydration of Mg^{2+} present in the interlayers. It contracts to 10 Å when heated at 550°C or when treated with a concentrated solution of highly polarisable cations (K^+, NH_4^+).

Other Mineral

Besides the silicate clay minerals, several other inorganic minerals have also been used as carriers and diluents in pesticide formulation. These include the elements (eg. sulphur), certain carbonates (e.g. calcite, dolomite etc.), sulphates (e.g. gypsum), oxides (e.g. calcium or magnesium oxide, lime, silicon oxides like diatomite, tripolite etc.), phosphites (e.g. apatite) and others like pumice. Some of these materials are briefly described below and some information on their properties/chemical composition is given in Tables 2.8, 2.9, 2.10.

Sulphur: Elemental sulphur has been used for a long time as a dust diluent for insecticides, particularly when a simultaneous fungicidal action is also desired. Sulphur acts as a strong repellent to certain pests.

Gypsum: $(CaSO_4 \cdot 2H_2O)$: Chemically hydrated calcium sulphate, it is inert and neutral. It is often not preferred due to its high bulk density. Being salt of a strong base and a strong acid, it is stable enough to be chemically inactive. It tends to absorb moisture to form cake during storage. Its hardness is about two.

Selenite: It is another crystalline variety of gypsum and in pure form has pearly lustre. The material is hygroscopic and has a tendency to cake.

Carbonates: Limestone $(CaCO_3)$ and slaked lime are less active than quicklime or hydrated lime $[Ca(OH)_2]$ but they react readily with salts of acids that are stronger than carbonic acid and also with nicotine sulphate. In such a case, mixture of dolomite $(MgCO_3, CaCO_3)$ and hydrated lime, is recommended. The carbonates provide dusts of slightly alkaline nature. Yet they are important as carriers of insecticides due to their low cost and availability. Magnesium carbonate is a light powder that tends to form aggregates particularly when wet. The particles are round and uniform in size ranging from 1-3 μm in diameter.

Oxides

(a) *Bauxite* is not a simple mineral but a term used to describe a massive formulation which is rich in alumina. Bauxite is a mixture of mono- and tri-hydrates of alumina with little clay.

(b) *Neosyl* is a form of silica (SiO_2) consisting of very fine particles, below 1 μm size. It readily forms aggregates. Neosyl has good abrasive properties, and is very hygroscopic.

TABLE 2.8. CHEMICAL COMPOSITION (%) OF MAGNESITE, DOLOMITE AND ROCK PHOSPHATE

Mineral composition	Diluent		
	Magnesite	Dolomite	Rock phosphate
SiO_2	0-5	3.92	2.60
SO_3	—	—	2.62
Al_2O_3	0-1	0.12	0.40
P_2O_5	—	—	30.30
F	—	—	2.83
Fe_2O_3	0-1	0.38	0.47
Cl	—	—	0.06
MgO	85-95	21.14	0.62
CaO	0-2	31.23	49.66
Na_2O	—	—	1.23
K_2O	—	—	0.05
CO_2	—	42.85	6.07

Searle and Grimshaw, (1959).

TABLE 2.9. PHYSICAL PROPERTIES OF SOME NON-CLAY INORGANIC DILUENTS

Diluent	Specific gravity	Bulk density (g ml^{-1})	Particle size range (µm)	Specific surface (cm^{2}g^{-1})	Hardness (on Moh's scale)
Almicide	3.22	5.1	<2	29460	~ 9
Gypsum	2.37	1.4	<40	5500	1.6-2
Neosyl	2.20	7.8	<1	52790	~ 7
Sil-o-cel	2.00	11.0	<10	42710	—
$CaCO_3$ (ppt.)	2.65	1.2	<10	600	—
Lamp black	1.74	—	—	—	Indeterminate
Carbon black	1.90	—	—	—	Indeterminate
Glass (Flaky)	—	—	—	2-20	~ 5
Glass round	—	—	—	2-20	~ 5

David and Gardiner, (1950).

TABLE 2.10. SOME PROPERTIES OF NON-SILICATE CARRIERS / DILUENTS OF INDIAN ORIGIN

Diluent	Surface acidity (pKa)	pHw(1%)	Bulk density (g 100 ml^{-1})	Specific surface (cm^{2}g^{-1})	Moisture content (%, dry wt. basis)	Hardness (Moh's scale)
Steatite	1.5- –3	9.3	96.7	3931	0.58	1.0
Magnesite	1.5	8.6	160.7	1489	0.39	3.5
Rock phosphate	3.3-1.5	7.9	122.7	1617	1.37	—
Gypsum	3.3	8.2	136.3	6627	11.43	2.0
Selenite	3.3	8.2	113.6	6651	22.28	2.0
Bauxite	4.0-3.3	8.0	127.0	658	0.26	1.3
Calcite	4.0-3.3	7.9	227.4	677	0.13	3.0
Dolomite	4.0-3.3	8.9	151.0	823	0.08	3.5

Sarup, (1967)

(c) **Sil-o-cel** is also largely silica, SiO_2, but unlike neosyl, it is a diatomaceous earth and consists largely of porous diatom shells of a variety of shapes. Usually, their practical size is up to 10 μm size but larger dimensions are also encountered.

(d) **Crystalline silica**, a very finely ground sand, has 98% SiO_2, 0.2% moisture and small amount of Al, Fe, Mg, Ca salts. It is a white powder weighing about 25 kg ft^{-3} and has a specific gravity of 2.68. Particle shape is irregular and sharp edged. Particle size ranges from 1-14.7 μm with an average of 2.9 μm in diameter.

(e) **Amorphous silica** is a form of siliceous earth obtained from diatoms. It contains 93-95% SiO_2 and a small amount of Al, Fe impurities. It is alkaline in nature. The material is light buff in colour and is very light and bulky, weighing 3.5 kg ft^{-3}. Particles are round and porous and have considerably inert surface with specific gravity of 2.3.

(f) **Phosphates:** Rock phosphate is a complex calcium phosphate $[Ca_5 (F, Cl) P_3O_{12}]$ with specific gravity 3.2. It is dirty white in colour.

Road dust: Fine quality road dust is being often employed as a carrier for dust in the pest control operations.

Carbon black: Carbon in the form of lamp and acetylene black is used as a pesticide carrier. The particles are small, about 1 μm in diameter and show a marked tendency to form aggregates.

Glass: Powdered glass is in the form of sharply angular thin transparent flakes. It is also used as spheres formed by heat treatment. Both materials contain particles from 2-50 μm diameter, with the majority in 2-20 μm range. It is not appreciably hygroscopic.

Fly Ash: Fly ash, a by-product of thermal power plants, has been reported as a useful carrier for pesticides (Rao and Parmar, 1986, Parmar and Srivastava, 1986).

Natural Organics or Botanicals

A number of natural materials, particularly of botanical origin, have been used as pesticide carriers or diluents. These include the citrus pulp, corn cobs, ground food grains or flours, rice husk, soybean, tobacco, walnut shell, wood powder or saw dust, pyrethrum marc and other similar materials. The use of these materials is based on their meeting the specific requirements like adsorption, natural lures etc. or their being available in plenty as cheap alternatives. However, considering the overall pesticide consumption, the use of such materials is highly specific and limited in quantity.

Synthetic Material

Inorganic compounds such as precipitated hydrated calcium silicate, precipitated calcium carbonate, precipitated hydrated silicon dioxide are also used in pesticide formulation. A material, almicide, is an artificially prepared alumina (Al_2O_3). It is highly abrasive and hygroscopic. Under the microscope, it reveals a mosaic of cracks. These flakes may be up to about 120 μm long but they are readily disintegrated into particles of about 0.25-2 μm diameter sizes.

Indian Clay and Clay Minerals Used in Pesticide Formulation

In India only a limited type of inerts namely, talc, steatite, soapstone, pyrophyllite, kaolin or China clay, bentonite, hydrated calcium silicate etc. are used. Talc, a hydrous silicate of magnesium (H_2O_3 MgO_4SiO_2), is commonly used. Steatite is a purer variety of compact and massive talc. Pot stone, a little impure and hard variety of soapstone, is used for carving out pots. Soapstone is soft greenish rock containing high percentage of talc. Most of the talcs are alkaline with pH 8-9 and bulk density 0.3-0.7 g ml^{-1}. The composition of talc, however, varies considerably from place to place and from deposit to deposit. The average per cent composition is: SiO_2, 58.98; MgO 30.92; Al_2O_3, 1.03; Iron oxide, 1.03; P_2O_5, 0.12; TiO_2, CaO, $MnSO_4$, Na_2O (alkalies), 0.27; ignition loss of water (%) at 100°C, 5.54. Rajasthan, Andhra Pradesh, Bihar and Madhya Pradesh are the chief states producing this mineral. Rajasthan alone accounts for about 88% of the total Indian production.

Pyrophyllite and talc are identical minerals so far as their physical properties and utilization aspects are concerned. Pyrophyllite is essentially a hydrous aluminium silicate ($Al_2O_3 \cdot 4SiO_2H_2O$) and is a dust diluent like talc. In general, pyrophyllites are mildly acidic to neutral in reaction with a pH of 6.0 to 7.0. Extensive deposits occur in Rajasthan (Udaipur District) and Uttar Pradesh (Jhansi). The production from Udaipur district is utilized in the pesticide industry alone and that from Jhansi is utilized by pesticide and other industries. General per cent composition of pyrophyllite is: SiO_2, 75.0; Al_2O_3, 19.3; Fe_2O_3, 1.3; CaO, 0.1; alkalies, 0.4; ignition loss, 3.9.

Kaolin, which is popularly known as China clay, is used as a carrier for water dispersible powder (W.P.) formulations in India. The particle size of kaolin is finer than that of mica, talc or pyrophyllite. The kaolin particles have larger amount of porosity than the above materials. In formulation with liquids or amorphous toxicants, kaolinites of finer particle size will take up higher percentage of toxicants, before their sorptive capacity is satisfied. Thus, formulations containing higher amount of toxicant can be made which will still retain the desired good quality of powdered material. Important deposits of high purity are mostly confined to Rajmahal and Patharghatta hills in Bhagalpur, Rajmahal hills in Santhal Parganas and near Hatgamaria in Singhbhum, Bihar; Vailpur in Trivandarum and Kundara in Quilon, and Sabarkantha in Gujarat. Different samples of China clay have been reported with bulk density: 0.7-0.8 g ml^{-1}; pH 7.2-7.5; iron content 0.15-0.2%, both plastic and non-plastic aqueous slurries, wettability in water 5 seconds, moisture 1-2% and material lost on ignition from 5-13 per cent. Washed China clay is also produced.

Vermiculite exists in a wide range of colours varying from black, passing through various shades of brown to yellow. Vermiculite's hardness ranges from 1.5 to 3, specific gravity of the crude material is about 2.5. It has low bulk density.

Fuller's earth is a natural bleaching clay of montmorillonite group possessing high adsorption property. The main difference between fuller's earth and calcium bentonite is in the mode of occurrence. Bihar, Rajasthan and Gujarat are the chief producing states of bentonite and fuller's earth.

Though the suitability of each carrier for a specific use has to be worked out, in India, generally finely divided porous clay, kaolin and diatomaceous earths are stated to be used as carriers for dust concentrates and talc, kaolin, pyrophyllite and soapstone for field strength dusts. China clay, either alone or in combination with other materials, is used for water dispersible powders. The granules are mostly prepared on China clay, bentonite, quartz sand, gypsum and natural $CaCO_3$.

MAJOR FORMULATIONS

POWDER

Dustable Powder (DP or Dust)

Free flowing powder suitable for dusting —(GIFAP, 1989)

Dust consists of an active ingredient combined with a carrier or a diluent with or without certain adjuvants for special effects like adhesion/retention etc. It may be prepared as 'dust concentrate (syn. dust base) of strength between 25-75% active ingredient, to be used after suitable dilution, or as ready to use dust, containing usually between 0.5-10% of the active ingredient. The particle size is usually below 200 mesh (~75 μm), preferably in the range of 30-50 microns.

Dusts may be 'straight' or 'impregnated', the former being a blend of the solid/liquid active ingredient with the carrier obtained by simple mixing and the latter involving impregnation of the active ingredient as such or from its solution, on the carrier or diluent. It is the easiest and most convenient formulation form for most active ingredients. It is also one of the earliest and has been the most widely used. Initially, the role of carrier(s)/diluent(s)/other formulant(s) on stability of active ingredients was not considered important since these auxiliaries were regarded as inert. However, with the realization of existence of active centres on surface of the carriers and diluents, its formulation has been attempted in a more scientific manner taking into consideration the active ingredient—carrier (or other formulant(s)) interaction and possible effect of the same on stability of the actives.

Drift of dusts at various stages of production and application has been another aspect of serious concern, hindering their promotion. With only 10-40% of the applied mass reaching the target, a uniform distribution of dusts with most of the modern dusters is not possible.

Due to various shortcomings, use of dusts has been discontinued in certain advanced countries. However, with due caution coupled with innovation, it can continue to serve as an important formulation, particularly in the developing world.

Design

The particle size distribution, flowability and bulk density are the three critical parameters of dusts. To build these traits in the product, the choice of an appropriate carrier/diluent is the most crucial. Most formulators opt for cheap auxiliaries. However, formulant-active ingredient interaction is of utmost importance in finalizing the composition. Silicate minerals have been the most preferred carrier/diluent employed in formulating dusts though other materials of natural or synthetic origin can also be employed.

A narrow particle size range helps to minimize segregation of particles during settling. It also checks formation of layers of different size ranges during handling, storage and transportation. Toxicant concentrations too vary on different sized particles and the variation may be large if the size range is too wide. Also, the smaller particles are conducive for more bioactivity, volatilization and drift, the aspects that need to be given due consideration before finalising the optimum size.

The flowability of dust should be as close as possible to that of the carrier used for its formulation. Its loss is indicative of the excessive amount of the active ingredient, over the limit of sorptivity. Excess of toxicant may also promote sticking or caking characters in a product.

The bulk density of the dust will be affected greatly by the nature of the carrier and/or toxicant employed. Products in the range of 30-50 lbs per cubic feet are satisfactory.

Specifications and Analysis

Specifications and test methods, as published by the national organizations of various countries as well as the international bodies, are applicable. Salient international organizations include the World Health Organization (WHO), Food and Agriculture Organization (FAO), Collaborative International Pesticides Analytical Council (CIPAC) etc., the last named describes only the analytical procedures. In India, such specifications, are developed by the Bureau of Indian Standards (BIS, formerly Indian Standards Institute, ISI).

Key requirements of dusts along with recommended test methods for their determination are summarized in Table 2.11 (ISI, 1982, WHO, 1985; FAO, 1999; CIPAC, 1970).

TABLE 2.11. SOURCES FOR THE KEY REQUIREMENTS OF DUSTS AND THEIR TEST METHODS

Requirement	Test method	
	FAO/CIPAC	ISI
Active ingredient	As per specification of individual pesticide	
Acidity/alkalinity	MT 33	11.3, 12.3
pH range	MT 75	—
Bulk density	—	—
(before compaction)	MT 3	12.2
(after compaction)	MT 33	12.2
Flowability	MT 44	—
Dustability after accelerated storage	MT 34	—
Sieve requirement	MT 59.1	12.1
Accelerated storage stability	MT 46.1.1	—

Dusts need to be free flowing products to enable uniform application. The particle size requirement prescribes that at least 97% of the material passes through a 200 mesh sieve. The bulk density of the powder after compaction (dropping 20 times from a height of 15 cm) must not exceed the value obtained before compaction by more than 60 per cent. The active ingredient must remain sufficiently stable to meet the shelf life requirement. The reaction of the product must also remain within the prescribed values. A carrier/diluent that decomposes an active material, is either to be replaced or suitably deactivated to obtain stable products. The deactivators need to be carefully chosen keeping the product economics and performance in view.

Preparation

Preparation of dusts can be divided into three separate steps:

  (*i*) Pre-blending of toxicant and carrier
  (*ii*) Pulverization of the mix
  (*iii*) Post-blending of the pulverized mix

If a solid or semi-solid toxicant can be readily ground, it may be used directly. Otherwise, its solution is to be prepared. The liquid toxicants of low viscosity can be sprayed directly on the carrier and impregnated. Otherwise, heating or dissolution in a solvent may be required.

Dusts of less than 300-mesh size are formulated by mixing the requisite quantity of the active ingredient with the carrier. Initially, a concentrate can be prepared which is subsequently diluted. When solid technical materials are used, an impact mill such as an atomizer or micron mill is used. With very hard technical materials or when milling to small (μm) size is required, a jet mill is recommended. In case of liquid active ingredients, mixing is first carried out by using the ribbon blender equipped with a spray, followed by milling in an impact mill e.g. pin mill, and re-blending using ribbon blender.

Innovative Dusts

Air Mixed Oil Coated Dust

The atomised dust and oil are mixed with a special device and the dust particles are coated with 20% or more of the oil. The dust is applied from a considerable distance by means of a blower so that these particles neither break up nor coalesce. The dust gives a better deposit in the case of heavy dust blown over with a low velocity. For example, sulphur particles and lead arsenate dust particles are large and heavier and form group of particles, giving better deposit. Light deposit and poor adherence of dust particles fail to control the insects.

Dust–Driftless and Fine

Driftless (DP–DL) and fine (DP–FD) dusts have been developed in Japan (Uejima, 1983) to combat pests under green house conditions. The former is aimed at overcoming the drift problem associated with the conventional dusts. In a simple innovation, the mean diameter of the particles has been increased to 20-30 μm as compared to that of 10-12 μm in case of conventional dusts. The larger particle size range is achieved by using larger than 10 μm size carrier dust and by adding flocculating agent. The dust with a mean diameter of less than 5 μm is designated as fine dust. Due to good floatability, it can cover even the underneath portions of leaves and other plant parts. A comparison of the physical properties of the DP–DL, DP–FD and regular dusts is given in Table 2.12.

TABLE 2.12. PHYSICAL PARAMETERS OF DRIFTLESS, FINE AND REGULAR DUSTS

Property	Driftless dust	Fine dust	Regular dust
Fineness (mesh)	> 300	>300	> 300
Mean diameter (μm)	20-30	5	10-12
Bulk density (g ml^{-1})	0.7-1.1	0.05-0.10	0.45-0.65
Floatability*	< 20	>85	40-60
Particles under 10 μm	< 20%	–	About 50%

*An index for indication of floating particles. A large index number indicates high floatability.

The DP–DL dusts are less floating during operation, reduce environmental pollution, provide good coverage of target and are conveniently applied with power dusters. The DP–FD can also be applied with dusters. These require less time for application and in addition do not lead to staining or contamination of treated surfaces, provide good control and offer no application limitation as in smoking or fogging.

The driftless dust formulations of organophosphorus, carbamate and cartap insecticides are registered in Japan since 1978 to control stem borer, hoppers and other insects, which attack rice crop. Those of organophosphorus, antibiotics and other fungicides are registered for control of rice blast, sheath blight and other diseases of paddy. DP–FD formulations of chlorothalonil, thiophanate methyl, chinomethionate, copper hydroxide, copper oxychloride, procymidon, methidathion, salithion and phenthoate were registered for the control of several pests in Japan.

Water Dispersible Powder/Wettable Powder (WP)

Pesticide in a dry form with surfactant, often mixed with, or coated on, a fine solid carrier, for dispersion in water to form a suspension   —(GIFAP, 1989)

The two terms WDP and WP have been often used synonymously. However, WP is now the internationally accepted code. Wettability and dispersion are the key characters of water dispersible powders.

WPs are cheap to produce, employ no organic solvent(s), are unaffected by low temperatures and involve lower packing cost. These, however, produce dust, are difficult to quantify, have poor efficacy and rainfastness and may, at times, block lines and nozzles.

The formulation essentially consists of an active ingredient, surfactant (dispersing agent) and a carrier. Often, wetting agent (surfactant) that helps in reducing the interfacial tension between particles and water; and adjuvants such as stickers or adhesives, antifoam agents etc. are incorporated to improve the performance or retention of deposits on the treated surfaces. Two typical compositions are shown in Table 2.13.

TABLE 2.13. TYPICAL COMPOSITIONS OF WATER DISPERSIBLE POWDERS

Ingredient	Solid (a.i.)	Liquid or waxy (a.i.)
Active ingredient(s) a.i.	Upto 90 %	Upto 50 %
Wetting agent(s)	1-5 %	1-5 %
Dispersing agent(s)	3-10 %	3-10 %
Stabilizer(s), antifoam(s), stickers(s) etc.	0-5 %	0-5 %
Precipitated silica	0-15 %	Upto 40 %
Filler	To make up final mass	To make up final mass
Total	100 %	100 %

In appearance, WPs resemble dust bases but differ from them in that these are applied on dilution with water. Two crucial traits of such formulations are: (*i*) rapid wetting and (*ii*) good suspensibility on dilution with water for field application.

The size of the particles, which is finer than 325 mesh, (~ 44 µm) helps in obtaining good suspensions. Particles of one to three microns are ideal for better suspensibility. The addition of dispersants helps in preventing the agglomeration of particles, thereby decreasing the rate of sedimentation.

WPs of 25-75% active ingredient strength are generally ideal and universally accepted. Use of low bulk density and high sorptivity carriers, is, therefore, essential. Diatomaceous earths and synthetic silica are often used to upgrade the sorptive power of the carrier(s).

Design

The physical state and physico-chemical properties of the active ingredient guide the choice of the formulation auxiliaries. Liquid or waxy materials or solids with low melting points would need absorbent carriers. For heat labile and melting prone actives, cooling during the grinding process may be necessary. Photolabile products will need to be processed in dark. Formulant-active compatibility and interaction must be known before actual formulation and this information should guide control of parameters like temperature, pH etc., as per need. Since most actives are hydrophobic in nature, proper choice of surfactants for wetting and dispersion is necessary.

The most commonly used carriers are kaolin (China clay) and synthetic silicates though attapulgite, montmorillonite, diatomaceous earths, talc, pyrophyllite are also used. The suspensibility, pack volume and other characters of a WP are greatly influenced by the carrier(s). Since the density and the diameter of the particles affect suspensibility, use of light variety of kaolin and/or other carriers is recommended. The carrier also needs to be inert to the active for a long shelf life of the formulation. Sometimes, gums, sodium salt of carboxymethyl cellulose, polyvinyl pyrrolidone and other water soluble polymers are added to WPs to increase the viscosity of the suspensions for improving suspensibility.

The wetting agents used are generally sodium salts of alkyl benzene sulphonic acids or sometimes the non-ionics such as polyethylene oxide derivatives of alkyl phenols, alcohol sulphates, sodium sulfosuccinates etc. Dispersing agents are lignosulphonic acids and sulphonated polyphenols. Generally, 1-2% of each of the wetting and dispersing agents are used. The exact recipe is generally worked by trial and error.

It is also customary to add stabilizers (sodium bicarbonate, magnesium carbonate etc.) to the formulations where free acids (ex. hydrochloric acid) are likely to be generated during storage, to neutralize the acid. Antifoams are also added, wherever required, to overcome excessive foaming.

The WPs need to comply with the requirements of a good shelf life during which period their active ingredient, acidity or alkalinity, suspensibility, wettability and size must meet the prescribed requirements. The suspensibility is observed in standard hard water of prescribed hardness, usually 342 ppm as $CaCO_3$. A fairly accurate prediction of the shelf life is obtained by subjecting the powders to the accelerated storage test ($54 \pm 1°C$ with 25 g cm^{-2} pressure for 14 days). After storage, the sample must comply with various prescribed requirements.

Specifications and Analysis

As in case of dusts, the specifications and test methods for WPs too are published by the national organizations of various countries and the international bodies. Table 2.14 lists the principal requirements and the test methods prescribed by the key organizations.

TABLE 2.14. KEY REQUIREMENTS OF WPs AND THEIR TEST METHODS

Requirement	Test method		
	WHO	FAO/CIPAC	ISI
Active ingredient	As per specification of individual pesticide		
Water	–	MT 30	–
Acidity/alkalinity	M/3	MT 31	11.3
pH range	–	MT 75	–
Sieve	M/4	MT 59.3	11.1
Suspensibility	As per specification	MT 15.1	11.2
Wettability	–	MT 53.3	11.4
Flowability of powder	–	MT 44	–
Foaming	–	MT 47	–
Accelerated storage stability	As per specification	MT 46.1	11.1.2.1

Note: For materials packaged in sealed water soluble bag, dissolution of the bag (MT 176), suspensibility (MT 15.1, MT 177) and persistent foam (MT 47.2) are prescribed (FAO 1999).

Preparation

Formulation of WP can be divided into five steps, namely pre-blending, pulverizing, re-blending, milling and post-blending. If the auxiliaries/adjuvants are in powder form, the steps 2 and 3 can be avoided irrespective of whether the active ingredient is a solid or a liquid.

During manufacture, the solid technical material is first milled with the carrier as pulverizing aid to yield very fine particles of the size of about 1-5 micron. This is done by employing a jet mill. The pulverized technical material is then mixed with additives and carriers to get the product of desired strength. With liquid technical materials, milling may be done with a micron mill, hammer mill or a micronizer. WPs can also be formulated by single mixing in a Henschel type mixer or plough mixer.

The most suitable mill to achieve proper particle size distribution is fluid energy mill or jet mill. The pre-mix of active ingredient, filler and surfactant is fed through a hopper and intercepted with angle jets. As the particles travel around the periphery of the grinding-classifying chamber, there is a high energy release and a high order of turbulence which causes the particles to grind upon themselves and rupture into finer fragments. The particles are classified and the oversized ones are returned to the chamber for further grinding.

On cost consideration, jet mills are not employed by all manufacturers of WPs. The alternatives include Raymonds mill, hammer mill, or micropulveriser. ACM-Pulveriser (Mikropul) is commonly employed by Indian formulators. It is a pin mill in which the feed is carried through the rotating pins/bar hammers and recycled through an attached wane classifier.

The material to be ground is conveyed to the grinding chamber through the feed hopper by the feed screw, the speed of which is infinitely variable. The bar type hammers mounted on the rotary disc provide high impact on the material at high velocity, thus causing disintegration of the material. The primary air stream flowing up the inner wall of the grinding chamber carries the material particles through the dispersion ring into the separator wheel. In this section, each ground particle is affected by two forces opposite to each other i.e. (*i*) the suction force caused by the flowing air, and (*ii*) the centrifugal force caused by rotation of the separator.

The suction force prevails if the particles are adequately fine and discharges them, with the primary air stream as finished product. If the particles are not fine enough, the centrifugal force prevails and these will be rejected by separator and thrown into the grinding chamber by means of secondary air stream. The fineness of the finished product can be adjusted by manipulating the speed of separator wheel, air volume, rotor speed etc.

Granule (GR)

Solid formulation comprising particles of defined size > 80 μm diameter, for application without further dilution, usually to soil —(IUPAC, 1998)

A free flowing solid formulation of a defined granule size range ready for use —(FAO, 1999)

Granules consist of an active ingredient extended on a carrier and then transformed into granules of 4-80 mesh size, with or without a binder. These can also be prepared by impregnating or coating the active ingredient on the pre-fabricated blank granules. The concentration of active ingredient in granules varies from 1-40% depending upon nature of active ingredient(s), and carrier. It depends on factors such as the potency of insecticide, pest to be controlled, release rate of active ingredient, mode and rate of application etc. The lower concentrations are more prone to decomposition whereas the higher ones pose cost and environmental problems.

Granules differ from the powder products mainly with respect to particle size. These are generally expressed as 15/30, 20/35, 20/40, 30/60 etc. implying that at least 90% of the product is within the specified mesh size range and the remaining 10% may be distributed on either side. British standards specifications require 96% of the particles to be within the stated range. The presence of finer particles, which may become air borne by a crosswind during application, is generally considered undesirable. A good attrition quality is, therefore, required to enable granule handling during preparation, packaging etc. In India, 16/44 granules are frequently used.

Whitehead (1976) has classified granules into core and compound. The former consist of a granular base of the required particle size and active material is impregnated or coated on it. The compound granules are created from raw materials of smaller particle size than the finished product. Physical and chemical properties of the active ingredient, the desired concentration, particle size and shape, physical form of technical material, method of application, special features such as slow release, delayed release etc. and economic considerations govern the choice of core or compound granules. Typical properties of blank granules made of some common materials are given in Table 2.15.

TABLE 2.15. TYPICAL PROPERTIES OF BLANK GRANULES FROM COMMON MATERIALS

| Property | Carrier | | | | |
	Sand	Gypsum	Marble*	Bentonite	China clay
Moisture content (max., % m/m)	2	2	2	3	3
Bulk density (g ml^{-1})	1.4-1.7	0.9-1.2	0.8-1.7	0.8-1.2	0.6-1.0
Material passing through:					
2 mm IS sieve	100	100	100	100	100
250 mm IS sieve	5	5	5	5	5
75 mm IS sieve	2	2	2	2	2
Acidity as H_2SO_4 (% m/m)	–	0.5	–	0.5	0.5
Liquid holding capacity (% m/m)	–	15	15	15	15

*Natural calcium carbonate.
IS: 9666-1980.

Granules are dried under ordinary drying conditions or by calcination at high temperatures (around 450°C) to yield regular ordinary dried grade or calcined grade. The former have regular volatile matter (RVM) and the latter, low volatile matter (LVM). With attapulgite, a further stage may consist of adding water and pug milling followed by extrusion prior to drying or calcining. These two grades are designated AA RVM for the pugged, extruded dried grade and AA LVM for the pugged, extruded calcinated grade. The pugging and extrusion operations increase absorptive capacity of granules. Typical properties of RVM and LVM granules are given in Table 2.16.

TABLE 2.16. TYPICAL PROPERTIES OF LOW (LVM) AND REGULAR VOLATILE MATTER (RVM) GRANULES

Attribute	LVM	RVM
Moisture (%)	1	5
Volatile matter at 100°F(%)	5	15
Oil absorption (%)	120	95
Water retention (%)	180	135
Bulk density (lbs ft^{-3})	30-35	40-45
Specific heat	0.25	0.25
Specific gravity	2.46	2.46
pH	7-8	5-6.5
Colour	Buff tan	Off white

Design

Two most important characteristics of granular pesticides are concentration of active ingredient and particle size range. Besides, these should be attrition resistant, flowable and dust free. The maximum concentration of active ingredient is limited by physical properties of the pesticidal chemical and the carrier. For example, only 1.5% endrin is generally coated but upto 40% chlordane has been coated on granular attapulgite. Particle size range is a function of size of the carrier. The active ingredient concentrations and granular size play an important role in regulating release rate of active ingredient.

Several carriers including the sorptive clays (attapulgite/polygorskite, montmorillonite, sepiolite, bentonite, kaolin etc.) and silica (diatomite, perlite, pumice, volcanic cinder, graded low angular sand), and low-sorptive materials (talc, calcite, gypsum, anthracite, graded brick etc.) are employed for formulating granules. The hardness of granules can, in case of clay carriers, be adjusted, as required, by calcination of granules at about 450°C. The active ingredient-carrier compatibility is crucial for good formulation and has to be duly ascertained before hand as in case of powders. Choice of a carrier has to take into account factors such as pH, pKa and others and whenever required, the surface activity has to be deactivated by using deactivators, which of course, will be at the cost of the carrying capacity of the granules for the active ingredient. However, use of deactivators is often not recommended since their ability to percolate to the active sites inside the carrier matrix, is questionable. Toxicant coating, either with liquid toxicants as such or with solution of solid or liquid toxicants in appropriate solvent(s) or melted solids (same as for impregnated dusts) is applied. Choice and amount of solvent is governed by solubility, viscosity and other related considerations.

Design of a granular recipe has to consider the requirements of shelf life and dust formation during handling and application. Another important consideration while designing granules is manner of their use. These considerations hold true for most of the granules. However, in case of water dispersible granules, the instant disintegration and dispersion of granules in water or their ability to form stable suspensions on dilution of slurry prepared in water (as in water dispersible powders) would need consideration. In such cases, recipes similar to water dispersible powders would need to be designed. Similarly, in case of floating granules, their ability to float will have to be catered to. It implies that hydrophobicity will need to be incorporated into the recipe. The last two categories of granules are separately described later.

Specifications and Analysis

The major requirements of granular products are: particle size range; dust formation; number of particles per unit weight; hardness; moisture content; acidity/alkalinity/pKa; flowability; encapsulation; shelf life (storage stability). The particle size must be in the declared range. A minimum of 90-97% is prescribed in different specifications for conformity. Dust (passing through a 75 μm sieve) formed should be less than 1% by mass of the product and that passing, shall not contain more than 8% by mass of the declared nominal active ingredient content. Size and density of the granules will govern the number of particles per unit mass of the material. Other parameters being constant, Whitehead (1976) exemplified the variation in number with the size (Table 2.17). Bioactivity will thus be affected accordingly.

TABLE 2.17. NUMBER OF PARTICLES AS A FUNCTION OF GRANULE SIZE

Size range (mesh)	No. of particles per pound of granules
8/16	1,70,000
16/30	13,00,000
18/35	32,50,000
25/50	87,00,000
30/60	136,00,000

Whitehead, (1976).

Hardness of granules is ascertained through attrition test by rotating them along with small steel balls in a jar for a given time and determining the dust formed by sieve analysis. Hardness, sufficient to withstand the rigours of handling, packaging, transportation and application, is required. The moisture content needs to be within prescribed limit for a given toxicant since hydrolysis (acid or alkaline degradation) can occur. Granules must enable transit and storage without caking or fusing to form agglomerates. Free flow enables uniform application. This, as well as the potential breakdown of the product during handling, transportation etc. are adjudged through water run off, attrition and other tests. Encapsulation enables a safer product handling.

Product after storage at 54 ± 1°C for 14 days should conform to various requirements for a satisfactory shelf life.

Some test methods prescribed for judging conformity to requirements of granules are listed in Table 2.18.

TABLE 2.18. TEST METHODS FOR KEY REQUIREMENTS OF GRANULES

Requirement	Test method	
	CIPAC	*ISI*
Active ingredient	As per specification for a given pesticide	
Acidity/alkalinity	MT 31	11.3 of IS 6940 (1982)
pH range	MT 75	–
Sieve	MT 58.3	12.1.2.2 of IS 6940 (1982)
Dust formation	MT 171	–
Description	–	2.2.1 of IS 11009 (1984)
Flowability	–	–
Storage stability	MT 46	2.2.4 of IS 11009 (1984)
Moisture content	MT 30	4 of IS 6940 (1982)
Bulk/apparent density	MT 58.4	12.2 of IS 6940 (1982)
Encapsulation	–	14.1 of IS 6940(1982)
Attrition resistance	MT 178	–

Note: Rate of release of active ingredient is also prescribed for slow release products.

Preparation

Some typical formulae for making granules are given in Table 2.19.

1. ***Absorbent or Impregnated Granules:*** Solution or suspension of active ingredient is sprayed on carrier granules which have porous molecular structure to absorb liquid. Commonly employed materials are attapulgites, montmorillonites, sepiolites, bentonites and pumices. When the active ingredient is a liquid under normal conditions, it is simply sprayed on granules which may be with or without a deactivating agent. If low concentrations are required, a solvent is added to facilitate even coverage of granules. Low melting pesticides, which are solids at room temperature, can be liquefied by melting and used. The solids which cannot be readily melted are dissolved in a suitable solvent (preferably volatile) and sprayed. Granules are dried or subjected to some solvent removal process(es) like heating, vacuum drying etc. (Fig. 2.5 steps 2, 4; Fig. 2.6).

   Too volatile solvents may pose problem of toxicant crystallisation in spray lines and nozzles. Ribbon blenders, drum blenders, cane and double cane blenders or pan blenders can be used to facilitate blending. Total time for spraying the toxicants for coating be restricted to about 10 minutes.

TABLE 2.19. SOME TYPICAL FORMULAE FOR MAKING GRANULES

Example	Steps/materials	Percent by weight
I. Pesticide "A" granular product		
	Step A. Blend the following:	
	Pesticide "A" (finely powdered)	4.0
	White oil	16.0
	Step B. Impregnate suspension from Step A on:	
	30/60 AA RVM attaclay (attapulgite)	80.0
		Total 100.0
II. 2.5% Pesticide "B" granular product		
	Step A. Blend the following:	
	Pesticide "B"-50% wettable powder	5.0
	Water	25.0
	Step B. Impregnate suspension from Step A on:	
	20/50 A LVM attaclay (attapulgite)	70.0
		Total 100.0
III. 10% 2,4-D granules		
	2,4-D solution (50% acid equivalent)	20.5
	25/50 A LVM Attaclay (attapulgite)	79.5
		Total 100.0
IV. 20% Aldrin granules		
	Aldrin (75% solution in diesel oil)	26.9
	Urea (50% solution in water)	1.5
	30/60 A RVM Florex (attapulgite)	71.6
		Total 100.0
V. 5% DDT granules		
	DDT (technical, 60% solution in xylene)	8.5
	Vermiculite #4 (mica)	91.5
		Total 100.0

Note: I, II are examples of suspension impregnation; III-V solution impregnation.
Polon, (1973)

2. **Non-Absorbent or Coated Granules:** There is little or no absorption of active ingredient into granule matrix and active ingredient remains on surface of the granule. The granule may be coated with a liquid (solution or suspension) or oil (Fig. 2.5, step 1) and then rendered dry and free flowing by treatment with a powdered material. In certain cases, spray of a binding agent is given followed by another layer of active ingredient to increase toxicant strength (Table 2.20). Liquid active ingredient can also be sprayed on granule and coated with a powdered inert. The carriers commonly used include limestones, sands and granites. Drum and cane blenders are the most favoured types. The inert materials used include polyhydric alcohols, polyether compounds (ex. polyethylene glycol), glues, molasses among liquids and talcs, China clay, and other silicates

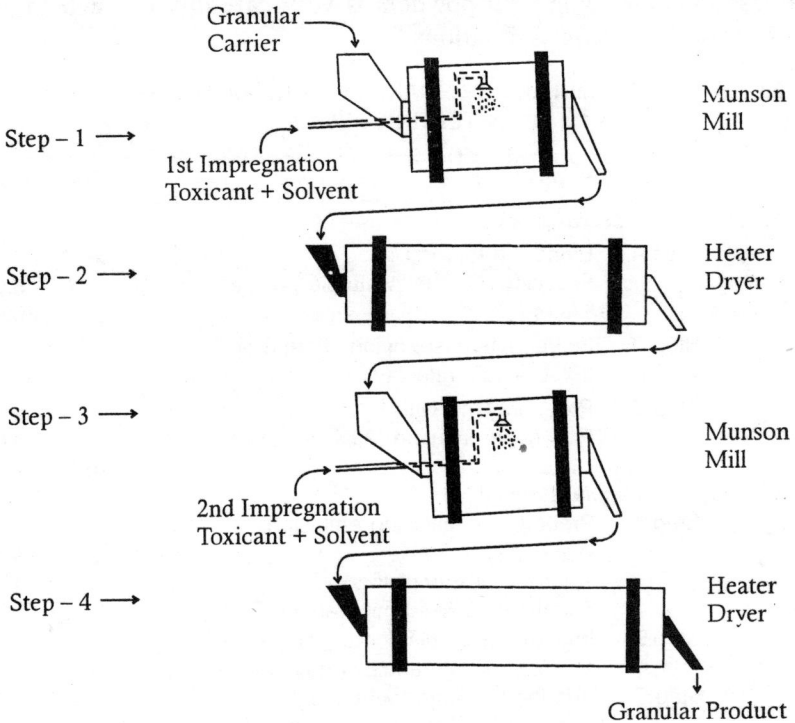

Granular Carrier

Step – 1 →

1st Impregnation Toxicant + Solvent

Munson Mill

Step – 2 →

Heater Dryer

Step – 3 →

Munson Mill

2nd Impregnation Toxicant + Solvent

Step – 4 →

Heater Dryer

Granular Product

Fig. 2.5: Impregnation and Heat Drying of Granules

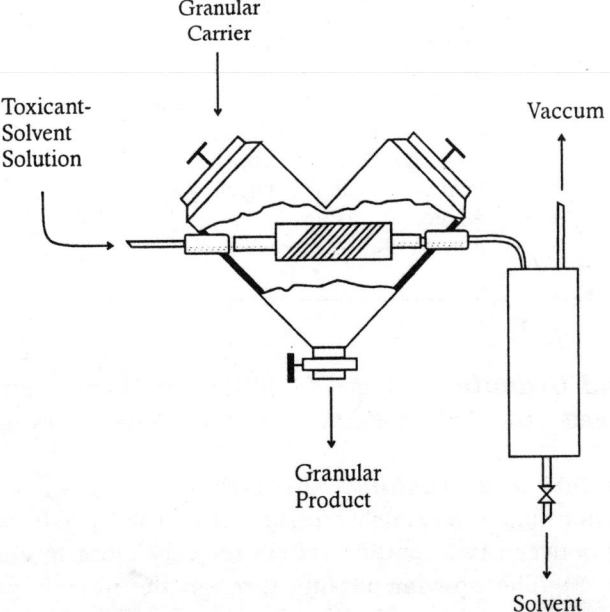

Granular Carrier

Toxicant-Solvent Solution

Vaccum

Granular Product

Solvent

Fig. 2.6: Impregnation and Vacuum Drying of Granules

amongst solids. Drying with powders is very carefully pursued to avoid both excess as well as scanty situations.

TABLE 2. 20. TYPICAL FORMULAE FOR GRANULAR PRODUCTS PREPARED BY STICKING TECHNIQUES

Example	Steps/Materials	Content (w/w)
I. 2.5% *Pesticide "C" granular product*		
Step A.	Blend the follwing	
	Pesticide "C"-50% wettable powder	5.0 parts
	Water	25.0 parts
Step B.	Impregnate suspensions from Step A on:	
	25/50 A LVM attaclay (attapulgite) 65.0 parts.	
Step C.	Post spray mix with:	
	Raw sugar solution (20% in water)	5.0 parts
		Total 100.0 parts
II. 10% *Pesticide "D" granular product*		
Step A.	Prepare the following emulsion:	
	Water	14.8 parts
	Triton X-100 (emulsifier)	0.4 parts
	Atlantic 85 B (light mineral oil)	14.8 parts
Step B.	Impregnate emulsion from step A on:	
	16/30 AA RVM attaclay (attapulgite)	57.5 parts
Step C.	Add the following to the blend:	
	Pesticide "D"-80% wettable powder	12.5 parts
		Total 100.0 parts
III. 10% Pesticide "E" granular product		
Step A.	Prepare the following solution:	
	Acintol R (rosin)	3.0 parts
	Atlantic 85B (light mineral oil)	12.0 parts
Step B.	Impregnate solution from Step A on:	
	30/60 AA RVM Florex (attapulgite)	70.0 parts
Step C.	Add the following to the blend:	
	Pesticide "E" (finely powdered)	10.0 parts
Step D.	Post spray with the following solution:	
	Acintol 10 (rosin)	1.0 parts
	Atlantic 85B (light mineral oil)	4.0 parts
		Total 100.0 parts

Polon (1973).

3. *Compound Granules:* A blend of active ingredient, inert diluent(s) (solid), binding agent(s) and liquid(s) (solvent). It involves the following granulation steps.

Compaction and Crushing: After blending, the powder mix is subjected to compaction under very high pressure. It is usually achieved by pressing the powder between two rotating rollers set very close to each other (Fig. 2.7, Steps 4, 5). The powder passing through the narrow gap is subjected to tremendous pressure; so that it gets compacted and emerges as a thin flat sheet which is broken into smaller particles by crushing or by other means. The blend consists of active ingredient, inert diluent (talc, China clay etc.) and binding agent (molasses or a sulphite lye). The granules produced are irregular in shape and may contain a high proportion of under or over sized granules.

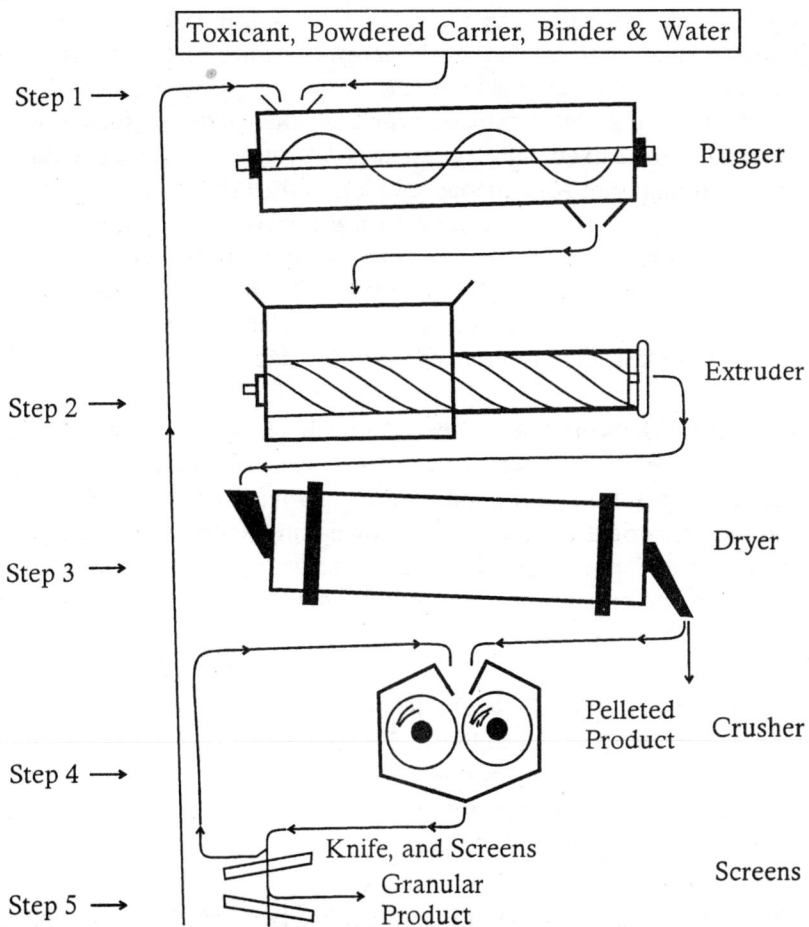

Fig. 2.7: Outline of Extrusion Process of Granulation

Extrusion: In this process, (Fig. 2.7) pressure is used to squeeze the powder blend through holes in a rotating cylindrical die. The powder is converted into cylindrical threads by compaction. These threads are cut to size by meeting knife blades set at the appropriate distance on emergence from the die.

Two systems are used. In one, the powder blend is squeezed between two rollers, one of which is solid and the other perforated. The cylindrical threads of granules are squeezed into the centre of the perforated die and cut to length with the knife blade. In the other system, the powder is fed into the centre of the die and is squeezed from the inside to the outside. The knife blade is arranged externally to the die. Cylindrical granules are produced by both methods. Larger granules (4 mm dia and 6 -9 mm length) are produced in the first system and relatively smaller (2 mm dia and 3-4 mm length) in the latter. The still damp extruded cylinders may be converted into spheres by further rotary action. The granules can be approximately classified to get uniform size range. Capital and energy costs are high and also, the dies and rollers need frequent replacement.

On a small scale in extrusion method, the carrier or solvent and technical materials are mixed and milled to give fine particles to make pulverized technical material. An aqueous solution of the binders such as lignosulphonate and polyvinyl alcohol is added to the above and the mixture kneaded. A suitable granulator then cuts the granules. A band or fluid bed drier at suitable temperature dries these. If necessary, after drying. disintegration is carried out in order to reduce the size of longer extruded granules. These are finally sieved to obtain the granules of required size.

Pan Granulation: The powdered mix is placed in an inclined rotating pan and a fine spray of liquid, usually water, is directed into the tumbling powder. The rolling action produced in the powder alongwith the liquid droplets produce spherical granules by a snowballing action. Great care is necessary to limit the range of particle size. The granulation equipment is cheap to purchase and run, but the operator is often exposed to the powdered product due to the open pan.

Ribbon Blender or Z Blade Mixer: It is a variation of pan granulation. In this, a fine spray of the liquid is directed on to the mixing powders. Starting with a granular core and adding a saturated liquid or a suspension to a mixing core, the granules are obtained. This is extensively used in the large scale production of granular fertilizers.

Fluid Bed Granulation: It is another method based on spraying a fine spray of liquid into the powder mix. Agitation is achieved by means of air passing via porous plate into the body of the powder. The powder becomes airborne and collision between the solid particles and liquid droplets causes granulation (Fig. 2.8).

Prilling Granulation: In this method, production of a molten mix of components or formation of hot highly concentrated liquid system is required. The molten liquid or solution must crystallize to give a hard product on cooling. The granulation is achieved by pumping the liquid into a rapidly rotating disc or perforated cup. The centrifugal action of the disc produces small, spherical droplets. These droplets are cooled to produce solid granules. Air cooling is commonly used in preparation of fertilizer prills. The plant is expensive and can be used only in case of high sale items. Alternatively, droplets are cooled in a liquid, provided the material being prilled does not react or dissolve in the cooling fluid. With most pesticides, the special conditions required for these methods are not easy to achieve.

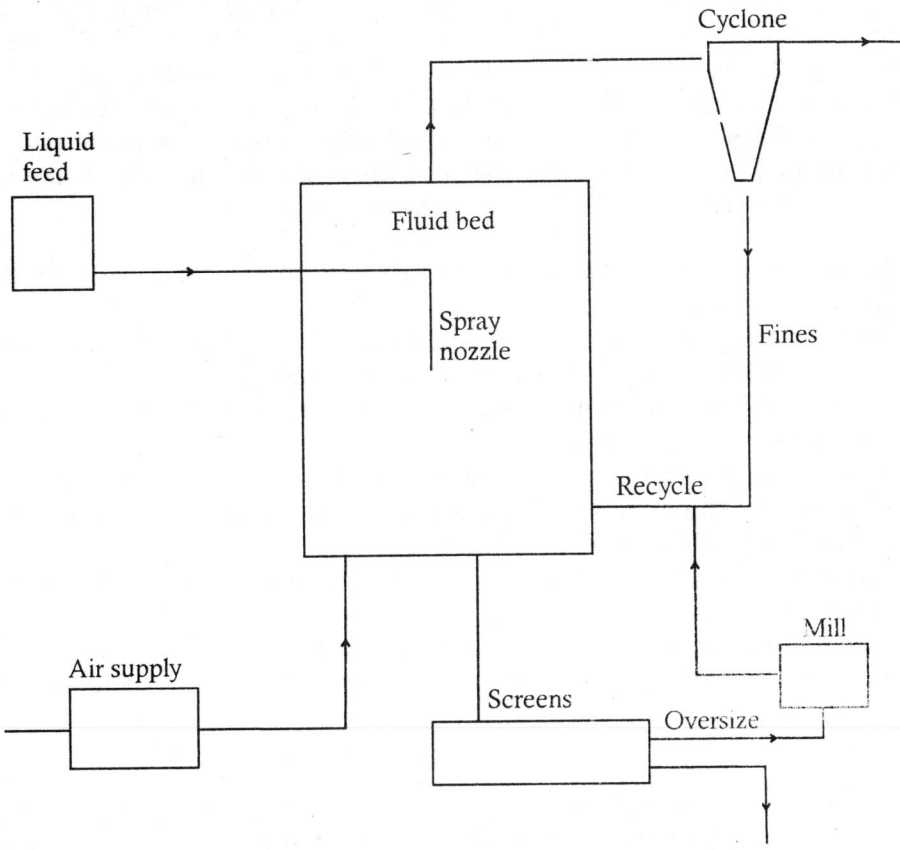

Fig. 2.8: Outline of Fluid Bed Granulation

Innovative Granules

Fine Granule (Ordinary and F.)

Both these formulations are prepared with such active ingredients as may be absorbed and translocated (systemic) in plant tissues. Fine ordinary granules have a particle size range of 105-296 μm (150-48 mesh) whereas 90% of the fine granule (*F*) lie in 62-210 μm or 250-65 mesh size. These are suited for use in situations where the drift of dusts is to be checked. Fine granules of several organo-phosphorus (diazinon, sumithion etc.) and carbamate (bassa, trumacide etc.) pesticides are registered in Japan to control various pests of rice crop (green leaf hopper, planthopper, blast, sheeth blight etc.) (Uejima, 1983). These formulations cause less environmental pollution and fewer hazards to loaders and applicators. The granules providing a good deposit are convenient to handle and apply and are reported to be very efficient in pest control.

Water Dispersible Granule (Dry flowable, WG)

Formulation containing granules which readily disperse in water to form a suspension
—(GIFAP, 1989)

Water dispersible granules consist of hard, uniform sized, free flowing granular particles, containing negligible amount of dust, which disperse or disintegrate readily in water and with minimal agitation, yield a homogeneous sprayable suspension comparable to that from water dispersible powder. Sometimes, such dispersions could be instantly obtained on contact of these granules with water. The formulation is promising for future use since it enables to overcome several problems associated with the use of dust, dispersible powder and other similar products. Some of the advantages pertaining to their safety and convenience are:

(*i*) Being free flowing, the granules can be easily, quickly and completely transferred from the package into the spray tank.

(*ii*) There is insignificant quantity of dust during handling and slurry preparation which avoids hazard to workers and their surroundings.

(*iii*) Spillage does not lead to clinging of the granules to the clothings and the spilled material can be readily swept.

(*iv*) The packed granules can be readily and safely quantified before use.

(*v*) The granules can even be packed in standard water soluble packs, completely avoiding the problem of disposal of container.

(*vi*) The dispersions obtained from such granules are often better than those obtained from water dispersible powder.

(*vii*) The transport and storage is economic and safe.

(*viii*) Problems like settling, thickening etc. associated with wet flowables, are missing in dry flowables.

(*ix*) If required, these granules can also be used like ordinary granules.

In view of various advantages, it has found commercial acceptance in several advanced countries of the world. The formulation has, however, its drawbacks too. It is expensive and is sensitive to variations in the formulation process as well as the formulation raw materials. The capital expenditure is some what high and skilled operators may be required.

A typical water dispersible granule may be described as follows:

(*i*) active ingredient 50-75%, (*ii*) diluent 0-45%, (*iii*) dispersant 1-7% , (*iv*) binder 0-2% and (*v*) wetting agent 0- 2 per cent.

Several products are commercially available. Work on the carrier and surfactant of such recipes has revealed that the requirement for a blank WG could be entirely different from those of a toxicant based WG. The microgranules in the size range of 60 to 100 mesh are particularly suited to obtain instant disintegration and dispersion of such granules. The easiest recipe for their preparation would be to convert water dispersible powder, if available, into such granules. Otherwise, these could be prepared adopting a suitable method from those available for making granules.

FAO (1999) describes the requirements for such granules along with the CIPAC test methods. These include acidity or alkalinity (MT 3) or pH range (MT 75), wettability (MT 53.3), wet sieve test (MT 167), degree of dispersion (MT 174), suspensibility (MT 168), persistent foam (MT 47.2), dustiness (MT 171), flowability (MT 172) and stability at elevated temperature (MT 46). When material is packed in a sealed water soluble bag, dissolution of the bag (MT 176), suspensibility (MT 168) and persistent foam (MT 47.2) are prescribed.

Floating Granule

It is comparatively a recent formulation that employs a carrier and a surfactant, with or without a binder, along with active ingredient. The granules float at the surface of water and release the active ingredient from water surface. The product has been developed to control the situation specific aquatic pests in standing waters. Such granules have been reported for butachlor and methyl parathion (Parmar and Dutta, 1990).

In a parallel development, the Japanese have reported another concept of the floating granule in 1990. These granules first settle towards the bottom of water and then rise towards the surface, and *en route* disintegrate and disperse to release the active ingredient. A part of the active ingredient settles at the bottom and another part floats at the top of the water body.

Emulsifiable Granule (EG)

A granular formulation to be applied as an oil in water emulsion of the active ingredient after disintegration in water which may contain water insoluble formulants

—(GIFAP, 1989)

It is a new innovative way of delivering in a dry form a liquid or a solid active ingredient dissolved in a solvent. It has been described as granule in which droplets of an emulsifiable liquid (EC or a liquid toxicant containing surfactant) have been trapped through a drying process into a polymeric, water-soluble matrix (Chiovato *et al.*, 2002). On addition of EG to water, the matrix dissolves quickly (spontaneous blooming) and regenerates a stable emulsion. The benefits of this technology include:

(*i*) Compared to EC, it has the advantage of a classic WG (easy handling, reduced handling hazard, reduced risk during transportation and storage, easy and risk free measurement etc.).

(*ii*) Compared to WG, its biological activity matches closely to that of EC.

Preparation

Preparation of an intermediate EW (oil in water emulsion), in which aqueous phase is the solution of the encapsulating polymer, is the first step. The EW is then dried following a classic drying method involving either a spray, fluidized bed, Glatt or drum dryer. In laboratory, simple oven drying can be practiced. The technical challenge is to find a polymer that dissolves in water readily and does not coalesce during drying.

Satisfactory compliance to flowability, dispersibility, stability of the emulsion, dust freeness etc. is needed.

Examples: The materials formulated successfully in this form included rapeseed oil (65 %), mineral oil (65 %), methyloleate (50 %), propiconazole (50 %) and metolachlor (50 %).

Slow/Controlled Release Products

Controlled release methods are postulated as means of increasing pest control efficacy while vastly decreasing the quantity of toxic material added to ecology. The methodology aims at a long term, continuous loss of a chemical agent at ultra low concentration for minimal environmental contamination.

Simplest approach to slow release of an active ingredient can be to increase size of particle, thus reducing the specific surface per unit mass and increasing surface per particle. Release of toxicant from a granule will be slower as compared to a powder prepared by using same carrier and other ingredients.

Several advantages are attributed to products possessing slow or controlled release characteristics. These include extended bioactivity, and reduced phytotoxicity, environmental degradation, evaporation, toxicity to non-target organisms, leaching losses and overall pesticide levels in the environment (Lewis and Cowsar, 1971).

In controlled release products, effective duration of an active is increased as explained below:

The first order rate law provides the expression:

$$\frac{dM}{dt} = -k\,M_t$$

where, $\frac{dM}{dt}$ = rate of dissipation or loss or removal in environment (removal rate)

         k = rate constant

         M_t = amount of pesticide present at any time t

The solution of the integrated equation gives

$$\ln\frac{(M_t)}{(M_\alpha)} = -k\,t$$

or $$\ln\frac{(M_\alpha)}{(M_t)} = k\,t \qquad\qquad \ldots(i)$$

where, M_α is the amount present immediately after application ($t = 0$ sec.)

The rate of removal of a pesticide from the environment is expressed in terms of its half life ($t_{1/2}$), which is related to the first order rate constant for removal (K) as follows:

$$k = \frac{0.693}{t_{1/2}}$$

If M_e is the minimum effective level of pesticide, and t_e the time during which an effective level of pesticide is present after a single application, then equation (i) becomes:

$$\ln\frac{(M_e)}{(M_\alpha)} = -k\,t_e$$

It means that to increase the effective duration of action (t_e) of a conventionally applied pesticide, exponentially greater quantities of the pesticide must be applied. In contrast, if the pesticide could be maintained at the minimum effective level (M_e), by continuous supply to restore the fraction dissipated, then the optimum performance of a pesticide would be realized when instantaneous rate of removal equals the instantaneous rate of delivery.

$$\frac{dM}{dt} = -kM_e + k_d M_e = 0$$

where, k_d = rate constant for pesticide delivery.

Here the amount of the active ingredient loss in the environment is controlled without affecting the efficiency of the dose which is being released constantly from within the slow or controlled release formulation.

When a pesticide is formulated for sustained delivery, the duration of action, t_e of the formulation is given by

$$\frac{M_\alpha - M_e}{M_e} = k_d t_e$$

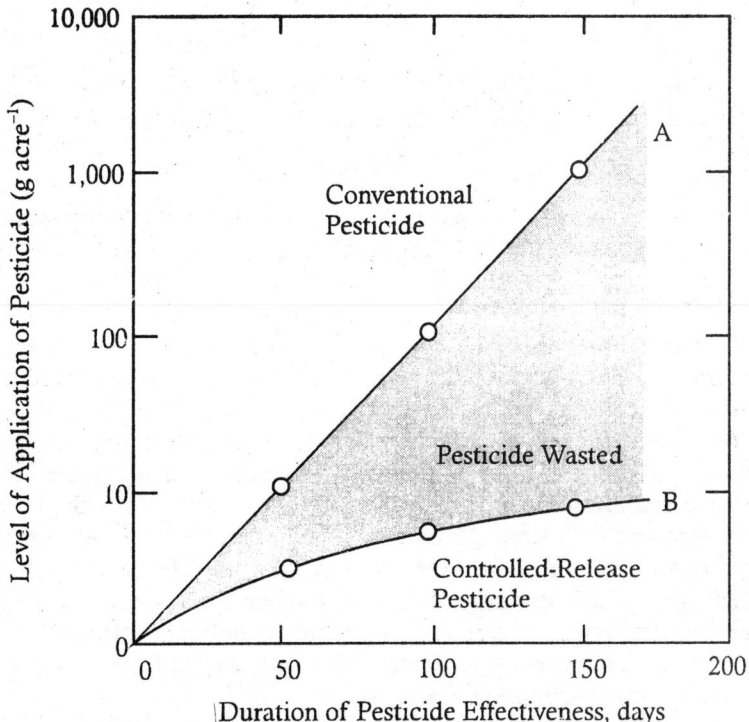

Fig. 2.9: Levels of Application and Duration of Action for Conventional and Controlled Release Formulations

Relationship between level of application and duration of action for conventional and controlled release first order formulations is shown in Fig. 2.9. Assuming

$t_{1/2}$ of 15 days and minimum effective level of 1 g acre^{-1}, the amounts needed for 50, 100, and 150 days of pest control duration are respectively 3.3, 5.6 and 7.9 g in case of controlled release formulation as compared to nearly 10, 100 and 1000 g acre^{-1} in the conventional one. The area between the two curves denotes the amount of pesticide wasted. The magnitude of wastage will increase in pesticides having shorter half lives. Theoretical curves (Fig. 2.10) comparing available pesticide concentration from a conventional *vis á vis* controlled released formulation further highlight the utility of the controlled release formulations (Shasha, 1998).

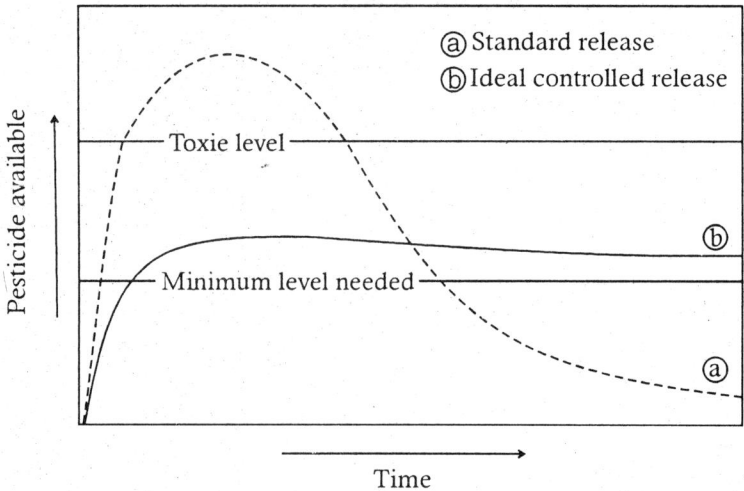

Fig. 2.10: Theoretical Curves Comparing Pesticide Concentrations Available from (a) Standard Spray Application and (b) Ideal Controlled Release Devices (Shasha, 1998)

Approaches to the Development

Concept and practice of controlled release encompasses many mechanisms. The earliest formulations were based on sorption of pesticides on strong sorbents like silica gel, mica and activated charcoal. Most of the current systems are based on more controllable mechanisms such as diffusion through rate-controlling medium, erosion of biodegradable barrier materials and retrograde chemical reactions. Designers of controlled-release formulations or devices usually strive for zero-order (constant) rate of release, but systems with time dependent release kinetics are proving to be useful for pesticides, especially when the rate and duration of release are predictable and well controlled. In practice, the rate of pesticide release may be controlled by several sequential or simultaneous mechanisms, which do not lend themselves to simple analysis, but it is usually possible to determine experimentally an overall "apparent" order of release for these complex systems.

The key approaches to the development of controlled release products are as follows:

Physical Barriers to Toxicant Movement or Release: In the simplest type of product, the solid active ingredient is initially uniformly distributed in matrix of low permeability or is dissolved in a polymer matrix. Regulation of release of active ingredient can also be achieved by physically mixing soluble actives with insoluble filler to form a highly compressed tablet of uniform composition. It has very limited potential because the active ingredient is present only in a small proportion of the filler which must possess high porosity. The pores of a compressed granular structure should be freely permeable.

Compressed wax containing viscous additives such as starch can provide a barrier of very low permeability but it must be used as an outer layer in a composite granule so that a useful loading of active ingredient can be contained in the core. Coating of granules can be carried out by tumbling of the core granules in an oblique rotating drum while alternatively adding filler dust and spray of adhesive solution. At appropriate temperature control, molten wax can build up in solid state by this process. The polymeric coating, like a particulate one, is more efficiently used as an outer layer.

There are three possible ways to stop the active ingredient from spreading from core to coating during manufacture. These include (*i*) the coating polymer may be cross-linked, limiting its swelling capacity (*ii*) the active ingredient may be crystalline solid, with limited solubility and (*iii*) the polymer may be such that the active ingredient, if liquid, swells it to a limited extent only.

An alternate method is to dissolve the toxicant in a liquid in which the polymer is insoluble. This, however, would reintroduce the tailing effect and at best the release would be exponential. The best system for steady state of slow release is to have a crystalline active encapsulated in a corss-linked polymer. Any of the above techniques, alone or in combination can be applied during manufacture.

The combination must retain the properties essential for the desired purpose during storage. Since only a small fraction of the chemical binds on swollen polymers particularly when cross linked, sometimes it is possible that it may react to influence their physical properties disproportionately and the storage stability requirement of the formulation may not be easy to meet.

Physical barriers to toxicant release are also exemplified by hollow fibres, laminated structures (three layers of laminated material, e.g. plastic, with the reservoir in central layer), monolithic dispersal etc. Encapsulation for controlled release has been one of the earliest approaches and involves the spontaneous formation of a coherent envelope around particles dispersed in liquid, usually as aqueous solution. The particles to be coated are usually oil globules although these may contain suspended solid. The material inside the capsule is referred to as core, internal phase or fill whereas coating is known as wall, shell or membrane. In order to apply coating procedures to particles in the range of a few microns, it is necessary to use chemical engineering processes, which can be carried out in bulk.

In instances where toxicant is not soluble in an elastomeric or more commonly a plastic matrix of choice, a third element 'carrier material' which is usually soluble in the matrix, is added. The toxicant is matrix-insoluble but it is soluble and/or dispersible in the carrier. As the carrier moves via solution pressure or thermodynamic instability towards the surface, it carries the pesticide with it. Surface losses through volatilization or other means occur. The system has been refered in literature as three phase carrier system.

The envelope material is almost necessarily one of high molecular weight polymer. The polymer may be pre-formed one, held initially in solution in a continuous phase and slowly precipitated by change of temperature or addition of salts or non-solvents. Suitable envelope substances under suitable conditions form around existing particles at first, a tenuous gel, characteristically thicker at opposite poles under influence of necessary agitation and becoming more compact as conditions of precipitation are slowly made more extreme. After further contraction by action of added tanning agents, coated particles may be washed, filtered off and dried.

A proper choice of encapsulating material, polymer matrix and other components is essential. Some of the common materials are listed in Table 2.21.

TABLE 2.21. COMMON MATERIALS FOR FORMULATING CONTROLLED RELEASE PRODUCTS

Microcapsule	:	Gelatin, gum arabic, polyamide, polyamide-polyurea, polyurea, polyethylene, wax.
Others	:	Acrylic polymer, black cellucon, chlorinated polyethylene, cyclo-dextrin, ethylene-propylene copolymer, ethylene-propylene elastomer, natural rubber, neoprene rubber, nylon reinforced fabric, polyamide, polyester, polyvinyl chloride, poroplastic, styrene-butadiene copolymer, thermoplastic resin, vinyl polymer

For agricultural use, biodegradable polymers derived from natural polymers such as alginate, cellulose, starch (starch xanthate, starch-alkali earth metal adduct, starch borate), sawdust, bark etc. are reported (Shasha *et al.*, 1976; Stout *et al.*, 1979; Hamerstrand, 1981; Shasha *et al.*, 1981; Trimnell *et al.*, 1982; Wing, *et al.*, 1998; Scheiber *et al.*, 1987). Inorganic silica microcapsules prepared by interfacial reactions have also been used as wall materials of microcapsules for pesticides.

Release by Membrane-Moderated Diffusion: Release is based on solubility characters of various elastomeres. Many pesticidal materials are soluble in natural rubber and other polymers possessing thermodynamic characteristics of elastomers. Upon incorporation in elastomeric matrix at levels below the solution threshold, solution equilibrium is established. If the toxic material is volatile, the molecules on or near the surface will desorb into exterior air environment. Likewise, if the toxicant is water soluble, molecules on or near the elastomer surface will, through dissolution, pass into the water interface. In both cases, internal solution pressure will lead to migration of interior solute molecules towards the depleting surface, and a continuous toxicant loss is established at the interface.

Diffusion-controlled membrane devices can be divided into two main categories: 'reservoir systems' in which the pesticide is totally encapsulated within a rate-controlling membrane, and 'monolithic systems' in which the pesticide is dispersed or dissolved in a rate-controlling matrix. These systems are depicted in Fig. 2.11. It has been demonstrated that the diffusion rates from controlled-release systems follow Ficks law of diffusion, which states that the rate of diffusion depends on five factors. Two of the factors involve the geometry or dimensions of the device, and the other three involve pesticide-polymer interactions. Mathematically, the rate of diffusion at a given time t can be given as:

$$\frac{dM_t}{dt} = \frac{2AD}{h}(C_s - C_e k)$$

where,

M = total amount of substance in the matrix after time t

A = surface area of the membrane

h = thickness of diffusion layer

D = diffusion coefficient of pesticide in the polymer

k = partition coefficient of the pesticide between the polymer and the medium which surrounds the device

C_e = concentration of released pesticide in the environment

C_s = saturation solubility of the pesticide in the polymer

A and h are dimensional factors and k, C_e and C_s are the diffusional factors.

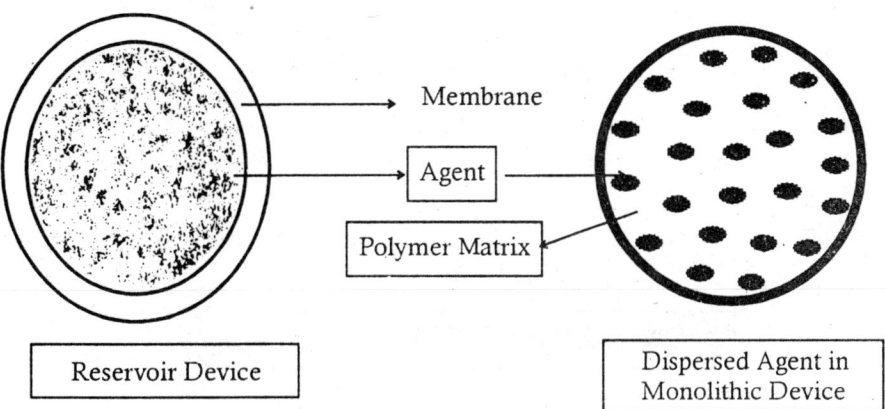

Fig. 2.11: Reservoir and Monolithic Diffusion Controlled Device

The solubility and partition coefficient are susceptible to analysis and prediction in terms of appropriate thermodynamic solution theories. The mobility of penetrant pesticide molecules, as measured by diffusion coefficient, is a kinetic parameter governed by size, shape, and polarity of penetrant and by morphology of diffusion medium.

When applied to reservoir systems, Fick's law predicts that if a pesticide is enclosed within an inert membrane, and if the concentration is maintained constant within the enclosure, then a steady state will be established during which the release rate will be zero order, i.e. constant. Various forms of Fick's law which apply to reservoir system of familiar geometries are given in Fig. 2.12 for the slab or laminate device and for the sphere or microcapsule.

Membrane-moderated monolithic systems in which pesticide is dispersed or dissolved in a rate-controlling polymer matrix are simple to prepare, but they do not have the zero-order release kinetics of the reservoir systems. The pesticide is released from the surface layers of a monolithic device first, and the distance it must diffuse to reach the surface increases with time. Thus, these systems have slowly declining rates of release. The kinetics of release can follow two patterns depending on whether the pesticide is present as a solution or as a dispersion.

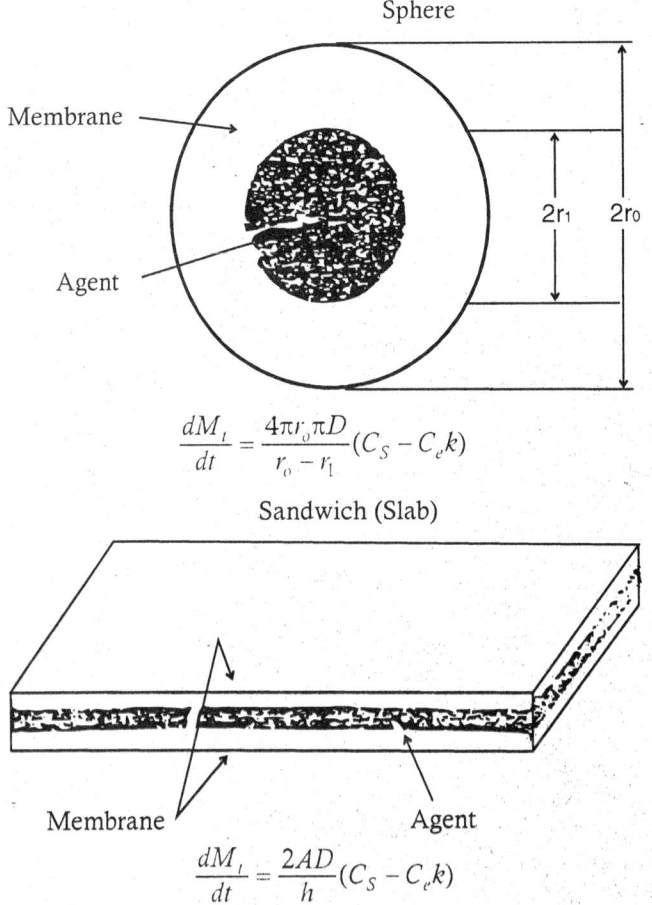

$$\frac{dM_t}{dt} = \frac{4\pi r_0 \pi D}{r_0 - r_1}(C_S - C_e k)$$

$$\frac{dM_t}{dt} = \frac{2AD}{h}(C_S - C_e k)$$

Fig. 2.12: Expression of Fick's Law for Different Reservoir Devices

For delivery of pesticide which has been dissolved in a spherical polymer bead of radius r_0, the expressions of Fick's law which describe the rate of release are given by the equations:

$$\frac{dM_t / M_\alpha}{dt} = 6\left(\frac{Dt}{r_0^2 \pi}\right)^{\frac{1}{2}} - \frac{3Dt}{r_0^2} \qquad \ldots(a)$$

$$\frac{dM_t / M_\alpha}{dt} = 1 - \frac{6}{\pi^2} \times \frac{\pi^2 Dt}{r_0^2} \qquad \ldots(b)$$

Equation (*a*) is called the "early-time approximation" and is valid for $M_t/M_\alpha < 0.5$, and equation (*b*) is called "late-time approximation" and is valid for $M_t/M_\alpha > 0.5$. From these equations it is apparent that about first 50% of the pesticide is released at a rate which decreases as the square root of time, and the rate of release of remaining pesticide follows exponential of first-order kinetics. Equations for the slab and cylindrical geometries have also been derived (Baker and Lonsdale, 1974).

If the pesticide is dispersed in the spherical polymer matrix rather than dissolved, the release kinetics are altered, and the rate of release is described by equation as:

$$\frac{dM_t/M_\alpha}{dt} = \frac{3C_sD}{r_o^2C_o} \times \left[\frac{(1-M_t/M_\alpha)^{1/3}}{1-(1-M_t/M_\alpha)^{1/3}}\right]$$

This equation is valid as long as the total amount of pesticide in the matrix C_o is much larger than the saturation solubility of pesticide in the matrix, C_s.

Fig. 2.13 shows the release profiles of monolithic devices with different initial loadings. It is not possible in this case to express the release rate as a single function of time. If, however, equation is integrated, an expression for the time required for the spherical beads to become totally exhausted of the dispersed pesticide can be derived as:

$$t_\alpha = \frac{r_o^2C_o}{6DC_s}$$

where t_α is time in which the matrix becomes totally exhausted.

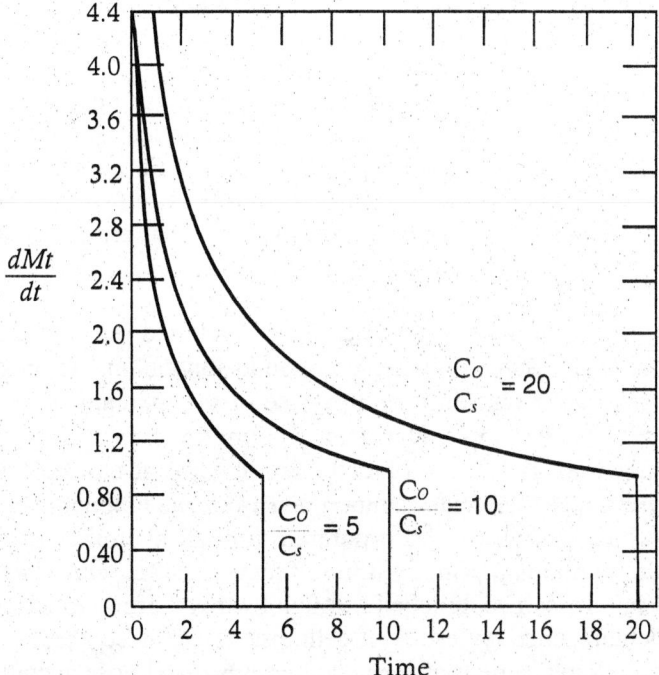

Fig. 2.13: Release Profile of Monolithic Devices with Different Initial Loadings of Dispersed Agent

Release by Erosion: When the polymer (containing dispersed or dissolved pesticide) is water soluble, hydrolytically unstable, or otherwise biodegradable, the active ingredient can be released by diffusion from the polymer or by erosion of the polymer, or by a combination of both diffusion and erosion. Release by erosion is a surface area dependent phenomenon, and if the surface area of a device stays constant while it is eroding, the release of pesticide will follow zero order. The general expression which describes the rate of release by a purely erosion mechanism is given by equation.

$$\frac{dM_t}{dt} = k_E C_o A$$

where, k_E = erosion rate constant

A = surface area exposed to the environment

C_o = amount of pesticide loaded in the erodible matrix.

For the flat film or slab geometry, the surface area does not change as erosion occurs, and consequently, the release of agent is constant until the film ultimately disappears. For an erodible sphere of radius r,

$$\frac{dM_t}{dt} = 4\pi C_o k_e (r_o - k_e t)^2$$

where, dM_t / dt = release rate

C_o = initial concentration of toxicant

t = time

k_e = release constant

It is apparent from this equation that pesticide release decreases slowly with time, and zero-order kinetics are not possible with spherical systems of this type.

If the pesticide is contained as a single reservoir within a spherical bioerodible membrane, as in some microcapsules, the mechanism of release can be the erosion and rupture of the barrier membrane. Various delivery patterns, including constant release, can be achieved by blending microcapsules of appropriate wall thickness.

Chemical Means: Chemical means have also been exploited for retarding the release of active ingredient. An insoluble polymer compound functioning as progenitor of the active ingredient, can cause retardation in release of the toxicant. Esters are widely used as progenitors. If a lower molecular acid is bound by ester linkage to a hydroxyl group-containing polymer as in case of cellulose acetate, it can be trans-esterified with many carboxylic acids. Herbicides containing carboxylic function can be converted to acid chlorides, which can also be incorporated through hydroxyl groups of natural polymers. Herbicides containing primary amine functions react with phosgene to form isocyanates, which in turn can be attached through hydroxyl group to natural polymers. The dichloro or trichloro phenoxy esters of cellulose or cellulose acetate are prepared in a similar manner by replacement of the acetyl groups of celluose acetate by phenoxy acetic acid resulting in water insoluble solids, which slowly release free herbicide.

Herbicide—COOH + $SOCl_2 \rightarrow$ Herbicide—COCl

Herbicide—COCl + Cellulose acetate → Cellulose—OCO—Herbicide

Herbicide—NH_2 + $COCl_2$ → Herbicide—NCO

Herbicide—NCO + Cellulose → Cellulose—OCONH—Herbicide

The link through oxygen atom of a chain of atoms and carboxylic group of ester or amide linkage (by hydrolysis) are the most reactive links in a polymer. More complex cases may undergo undesirable reactions. For example, a carbamate insecticide of general formula RO—CO—NH—CH_3 could be linked through the N-atom (amido nitrogen) to a C atom in a polymer chain carrying a reactive 'pendant' atom or group such as Br—CH_2CO_2H or —COCl. But the linked insecticide could decompose by hydrolysis at linkage 1 or 2 or with assistance or oxidation, even at 3 before the desired liberation occurs at linkage 4 as illustrated below.

$$ArO-\overset{\overset{\textstyle O}{\|}}{C}-NH-CH_3 + RCOCl \xrightarrow{-HCl} ArO-\overset{\overset{\textstyle O}{\|}}{\underset{\downarrow 1}{C}}\overset{2}{\underset{}{\uparrow}} N \overset{3}{\underset{\mapsto 4}{\uparrow}} CH_3$$
$$COR$$

Hydrolytic reactions are often strongly pH dependent. The polymer attached esters of 2, 4-D could be liberated rapidly in an alkaline soil but very slowly in an acidic medium. Cellulose acetate is unaffected by immersion in water under conditions where ethyl acetate or glycol diacetate would be extensively hydrolysed. Cellulose is degraded under very vigorous conditions and then breaks down completely to form glucose. Once the protection of organised crystalline structure is lost, the linkages in the water-soluble oligosaccharides are rapidly broken.

It has been reported that the herbicidal carboxylic acids could be changed to resins by precipitating ferric salts at suitable pH and temperature in presence of aldehydes. These products could result in slow release into soil.

A pesticide could also be linked to yield a monomer which could then be polymerized as given below:

Matrix + Pesticide → Monomer $\xrightarrow{\text{Polymerise}}$ Polymer

$$CH_2 = CH_2 + Herbicide-COOH \xrightarrow{+HgSO_4} \left[\begin{matrix} CH-CH_2{}^{\bullet} \\ | \\ O-\underset{\underset{\textstyle O}{\|}}{C}-Herbicide \end{matrix}\right]$$

$$\left[\begin{matrix} CH-CH_2 \\ | \\ O-\underset{\underset{\textstyle O}{\|}}{C}-Herbicide \end{matrix}\right]_n \xleftarrow{\text{Polymerization}}$$

In the event of an active ingredient and a polymer being incompatible, the compatibility could be improved by using bridging moieties to link e.g.

$$P + B + A \rightarrow P\Big\langle{\begin{matrix} B-A \\ B-A \end{matrix}}$$

where, P is a polymer, A active ingredient and B bridging moiety.

In this way, the various chemical possibilities could be exploited to obtain controlled release products.

Release by Retrograde Chemical Reactions: As opposed to physical combinations of pesticide dissolved or entrapped in polymers, in these formulations, chemical combinations have the pesticide firmly attached to the polymeric substrate by a definite identifiable chemical bond. The active material is released when environmental factors such as water, air, sunlight or microorganisms act to cleave the specific chemical linkages which attach the pesticide to the substrate. The general mechanism for retrograde chemical systems is as follows:

$$\text{Polymer + Pesticide} \xrightleftharpoons[\text{Environment}]{\text{Synthesis}} \text{Polymeric pesticide}$$

Obviously, only pesticides with at least one reactive functional moiety can be utilized in delivery systems of this type. The most common linkages are ester, anhydride and acetal.

The rate of release of pesticide in a retrograde reaction depends on properties of the micromolecule and its surrounding medium. When water present in the environment activates the release of pesticide, the rate of hydrolysis depends on the strength and chemical nature of the polymer–pesticide bond. For example, an anhydride linkage is more susceptible than an ester or amide linkage to hydrolysis. The rate of hydrolysis of a linkage is also dependent on the groups surrounding it. Hydrophobic groups thus offer protection against rapid hydrolysis. An uncrosslinked polymer is much more susceptible to hydrolysis than a highly cross linked one, and a stereoregular or crystalline polymer is less susceptible to hydrolytic attack than an amorphous or atactic polymer.

The kinetic expressions which describe the rate of release of pesticides from retrograde chemical systems vary depending on whether hydrolysis reaction occurs on the surface of an insoluble particle or in solution. For a heterogeneous reaction on the surface of insoluble spherical particles, rate of release of pesticide is given by:

$$\frac{dM_l}{dt} = nk_h 4pr_o^2 C_o$$

where, n = number of particles of average radius r_o

k_h = rate constant for hydrolysis

C_o = concentration of pesticide in polymer linkages.

For water-soluble delivery systems, the rate of release of pendant pesticide groups follows conventional first order kinetics. Finally, retrograde chemical systems which are based on depolymerization reactions may indeed be zero order if the mechanism of release comprises upzipping of the polymer chains.

Categories of Release Mechanism

Controlled release is not synonymous with sustained release, though the two are similar in principle and sometimes overlap. Sustained-release formulations contain several times the normal single application, and they provide for replacement of the agent at same rate which gives a measurable increase in the duration of activity. The rate may decrease due to gradual loss of agent, or increase through a maximum due to breakdown of a protective barrier. A controlled-release formulation, in contrast, may exhibit a fast or a slow release or a constant or a changing release, depending on the design. The principal difference lies not entirely in the profile of release but in the mechanism of release. The distinction is drawn mainly by the degree of control of both the optimum level and the optimum time of availability of the biologically active agent.

Altering the molecular weight or changing the hydrophobicity of the polymer carrier can vary the rate of release. The rate of release also depends upon the degree of substitution of the pesticide moiety within the polymer, the pH of the hydrolytic medium and the size of the particles.

Harris (1974) has categorized the release mechanism on the basis of the release-rate-controlling step into four types.

(*i*) ***Diffusion through a Non-porous Polymer Membrane:*** The molecule moves from a reservoir through enveloping homogeneous polymer membrane by an activated diffusion process. The permeant first dissolves in the surface layers, migrates through the bulk material under a concentration gradient and then evaporates or desorbes from the surface.

(*ii*) ***Diffusion through a Non-Porous, Solid Polymer Matrix:*** A permeant migrates through an enveloping, homogeneous matrix with two variations of the diffusional process. In the first, the permeant is completely dissolved in the surrounding polymer and in the second it is dispersed in the matrix at a concentration greatly exceeding its solubility in the matrix.

(*iii*) ***Diffusion through a Porous, Solid Polymer Matrix:*** Here the active agent is leached out from the matrix by the bathing fluid which enters the matrix through pores. The active agent dissolves slowly into the permeating fluid phase and diffuses from the system along with cracks and capillary channels filled with extracting solvent.

(*iv*) ***Surface Hydrolysis or Dissolution of a Polymer Matrix:*** The rate of release depends on the exposed surface area. A controlled release formulation with a constant output rate, such as water evaporation from a cake in which the rate of evaporation remains constant, exhibits a zero order rate. First order rate occurs when the rate declines with time and is proportional to the amount of active agent still in the reservoir. Release rate follows second order when the output rate is proportional to the square of the amount remaining in the reservoir. These can be represented mathematically as follows:

Zero order: $\dfrac{dM_t}{dt} = k$

First order: $\dfrac{dM_t}{dt} = k \times (M_\alpha - M_t)$

Second order: $\dfrac{dM_t}{dt} = k \times (M_\alpha - M_t)^2$

where, M_a = total amount of active ingredient
             M_t = amount released in time t
      dM_t/dt = release rate
             k = rate constant.

Zero order release is ideal in a formulation, provided the release duration has no deleterious effect on non-target biota (Refer Fig. 2.13 showing ideal controlled release).

Non-matrix soluble materials can be incorporated in elastomers or plastics through mechanical binding. Upon exposure to water, the toxicant molecules near the polymer surface slowly pass into solution. As the surface layers are depleted, a pore structure is formed to yield continuous flow provided the toxicant loading is sufficiently high. Water thus penetrates the pore structure and washes out freshly exposed toxicant molecules. The process is continuous although the rate of loss usually varies with tortuousity of growing pore structure and other factors.

Various elastomers and plastic substances undergo physical, chemical or biological degradation and along with, the toxicant too is lost, if present. Rate of such a release depends upon a number of environmental factors. As a rule, such materials do not provide a continuous constant toxicant loss.

Ingestible baits, which may contain an attractant or attractant and a toxicant bound together in an ingestible matrix—elastomer or another substance are also based on the same principle. The attractive element slowly escapes and affects various chemoreceptors of the target. The pest organism locates the bait, ingests a lethal dose and succumbs. In a contact bait, the matrix element may contain an attractant that is released slowly from the bait facilitating the toxicant to have full contact with the target site. Ordinary non-toxic baits may contain biologically active materials like attractants, repellents etc., which by nature are non-toxic to the target or other species, but alter the target behaviour in such a manner as to reduce the pest population.

Design

Three of the most promising design concepts are (a) capsules of polymeric material filled with a solid or liquid pesticide or with a suspension or solution of the active agent in a fluid, in which the release of pesticide is controlled by diffusion; (b) a heterogeneous dispersion of particles or droplets of pesticides in a solid polymeric biodegradable or non-biodegradable matrix, which controls the release of agent by diffusion through the matrix, by erosion of the matrix, or by a combination of both diffusion and erosion; and (c) chemical bonding of a pesticide to a natural or synthetic polymeric material, as in case of pendant anhydride or ester linkages or formation of micromolecules of pesticides via ionic or covalent linkages, which control the release of the active ingredient agent by hydrolysis, thermodynamic dissociation, microbial degradation, or some other retrograde chemical reaction of the linkages.

The controlled release formulations may be designed based on the principles and approaches described above.

Preparation

Chemically Bound: Such controlled release products can be prepared in following two ways.

(a) By attaching a polymerizable site to the active ingredient (mostly herbicides), followed by polymerization of the new product:

$$\text{Herbicide—COOH} + CH_2{=}CHOCOCH_3 \rightarrow$$

$$\begin{array}{c} \text{OCO—Herbicide} \\ | \\ CH{=}CH_2 \end{array} \xrightarrow{\text{Polymerization}} \left[\begin{array}{c} \text{OCO—Herbicide} \\ | \\ -CH{-}CH_2- \end{array}\right]_n$$

(b) Those prepared by chemically binding derivatives of active ingredients to suitable polymer. For example,

$$\text{Herbicide—COOH} + SOCl_2 \rightarrow \text{Herbicide—COCl} \xrightarrow{\text{Cellulose}}$$
$$\text{Cellulose—OCO—Herbicide}$$

$$\text{Herbicide—NH}_2 + COCl_2 \rightarrow \text{Herbicide—NCO} \xrightarrow{\text{Cellulose}}$$
$$\text{Cellulose—OCONH—Herbicide}$$

The rate of release can be increased by lowering the molecular weight or increasing the hydrophobicity of the polymer carrier. Degree of substitution of the herbicide moiety within the polymer, the pH of the medium and size of the particles influence rate of release.

Microencapsulation: The tiny particles or droplets of the active ingredient are coated to obtain small capsules varying in diameter from a few micrometers to a few millimeters. Microencapsulation enables ease of handling, reduced volatility and toxicity, stability, sprayability and controlled release. This can be achieved by following methods:

Coacervation: This process dispenses the pesticide in a suitable vehicle containing a coating material that will deposit around the pesticide. The coating coacervate is then hardened by chemical cross linking and solvent evaporation to form discrete microcapsules. Himel and Cardarelli (1982) developed a sprayable system based upon coacervation that encapsulated herbicides 'in flight' i.e. encapsulation process takes place as the herbicide disperses within the vehicle and the coat material simultaneously dried.

In coacervation technique, the wettability of the core material with the coating is crucial. Properly wetted solid particles are easier to coat than the liquid cores. If a liquid core material is highly insoluble in the coacervate forming solution, proper wetting may be difficult. It is also critical that the core is insoluble in the polymer solvent and that the polymer does not partition into the liquid core.

Interfacial Polymerization: In a lipophase system containing an organic compound, an aqueous phase emulsifier along with another reactive monomer, the monomer undergoes poly condensation at the interface of the organic compound and aqueous phase to form a polymer that coats the pesticide. Microcapsules so formed contain as much as 90% pesticide, which may be used in sprayable formulations in conjunction with thickening or suspending agents. In interfacial polymerization, solubilities of the reactants in the phases permit a choice of the solvent and polymer.

Air Suspension Coating: It involves coating particles as solutions or melts while suspended in an upward moving air stream. They are supported by a perforated plate having different patterns of holes inside and outside a cylinderical insert. To fluidize the settling particles, air is permitted to rise through the outer annular space. Most of the rising air (usually heated) flows inside the cylinder, causing the particles to rise rapidly. At the top, as the air stream diverges and slows, they settle back at the outer bed and move downward to repeat the cycle. As the particles start flying, they encounter a fine mist of the coating solution either from the top or from the sides. Only a small amount of solution is applied in each passage. Many different coating materials including water-soluble polymers have been used in this process.

Molecular Complex Formation: The process involves complex (coordinated) formation between pesticide and selected organic chemicals like starch and liposomes. By virtue of their spatial geometrical structures, certain chemicals such as Schardinger dextrins, (which originate mostly from starch) are capable of entrapping pesticides at molecular level. These dextrins are 6 to 8 member rings built of D-glucose units with a hydrophobic cavity in which the pesticide can be bound. Liposomes, which are cell-like structures derived from phosphatidyl cholines and phosphatidyl ethanolamines, can also entrap chemicals. These triglycerides are capable of forming closed concentric membranes, that entrap other substances. Triazine herbicides have been encapsulated within such structures.

Centrifugal Extrusion: The process employs a rotating extrusion head containing concentric nozzles used for encapsulating liquids. A jet of core liquid is surrounded by a sheet of wall solution or melt which breaks into droplets as the jet moves through the air. Each droplet gets coated with the wall solution. The molten wall material may be hardened or the solvent may be evaporated from the wall solution while the droplets are in flight. It is an excellent process for forming particles of 400-2000 μm diameter.

Spray Drying: The active ingredient is dissolved or suspended in a melt of the polymer and is entrapped in the dried particles. The viscosity of the solutions to be sprayed can be as high as 300 centipoise. In a modified technique, the core material is incorporated into a low melting fat or wax to form an emulsion or suspension that has to be chilled only below its melting point to form particles. Labile materials can be handled by this technique.

Matrix Encapsulation: Pesticide is dispersed within a polymer and becomes entrapped within many small cells of a continuous matrix. The active ingredient may be dissolved or suspended in various polymers to yield ribbons, sheets or granules. Products obtained here lack a distinctive wall (surrounding each individual particle of active ingredient). Often, an excipient or porosogen is added to such formulations. Such excipients may be inorganic fillers or water-soluble polymers that provide points at which the surrounding medium can penetrate the product in order to regulate the rate of release of active ingredients. Matrix encapsulation can use any of the following methods.

Encapsulation within an Aginate: Alginic acid, a linear polysaccharide composed of polymannuronic and polyglucuronic acid units, is isolated from brown seaweeds. Its sodium salt is water soluble and forms a tough gel like beads of calcium or barium alginate when treated with calcium or barium cations. Alginate beads are made after dispersing the pesticide into a solution of sodium alginate and adding the pesticide into a solution of sodium alginate and then adding the dispersion to an aqueous gellant solution of calcium or barium chloride to entrap the pesticide within them. Beads prepared this way are outwardly similar to many prepared by microencapsulation but differ in that they have a matrix structure rather than having central zone of active ingredient.

Encapsulation within a Lignin Matrix: The process involves emulsification of the active ingredient with an alkaline lignin, followed by acidification to precipitate the composites which are isolated and dried. Upon immersion in water, the composites develop cracks and fissures. The release of active ingredient is mainly by diffusion.

Encapsulation within a Starch Matrix: Starch, an inexpensive polymer, consists of glucose units, and can be easily fractionated to yield amylose, a straight chain polymer, and amylopectin, a branched chain co-polymer. While amylose is capable of forming tough, flexible films, amylopectin does not possess this property. Because of the abundance of free hydroxyls (three hydroxyls for each glucose unit) starch can be easily derivatized (the degree of derivatization or substitution, DS, is defined as the number of substituents per glucose unit) and theoretically, has a degree of substitution upto three.

Simple and economical procedures are described in literature to entrap water insoluble pesticides within starch. The main methods are the xanthate (Parmar *et al.*, 1998; Shasha *et al.*, 1976; Stout *et al.*, 1979), the calcium (Shasha *et al.*, 1981) and the borate (Trimnell *et al.* 1982) procedures, which all involve chemical modification of starch. Another method uses extrusion procedure without chemically modifying the starch (Wing, 1998). The pH of particles formed by different procedures varies. Particles made by the xanthate process have a pH of about 4. With the calcium method, a pH of about 11 is obtained; with the borate, a pH of about 9.5, while with the extrusion process a pH of about 6 is obtained. In both the calcium and borate procedures, modifications can be made to yield final products with pH close to neutral.

Encapsulation within Inorganic Matrices: In this process herbicides have been entrapped in plaster of paris for use with container grown ornamentals. Pesticides are formulated as tablets of various sizes and shapes. Herbicides have also been entrapped within a silicate matrix using sodium silicate and calcium chloride in a manner similar to the alginate method.

Controlled release formulations for delivery of biopesticides to soil have been recently reported for the bacterium *Serratia entomophila* (Enterobacteriaceae), a biocontrol agent for the New Zealand grass grub (*Costelytra zealandica*). The formulation comprises the bacteria stabilized in a biopolymer matrix which is then incorporated as carrier based granule. Granules can be stored for up to 4-6 months at ambient temperature and can be manipulated to obtain fast or slow release products (Johnson *et al.*, 2002). The technology has been successfully used for formulating *Beauveria* sp., *Bacillus* sp. and *Pseudomonas* sp.

In India, controlled release polymeric products of several pesticides have been reported for use in agriculture (Rao *et al.*, 1989; Parmar *et al.*, 1998: Parmar *et al.*, 1999; Kumar *et al.*, 2002 a, b).

OTHER SOLID FORMULATIONS

Powder for Dry Seed Treatment (DS)

A powder for application in the dry state directly to the seed   —(GIFAP, 1989)

DS consists of a homogeneous mixture of active ingredient, filler and other requisite formulants, including colouring matter, to yield a fine, free flowing powder, free from visible extraneous matter and hard lumps. It may be coated on seeds in dry form or the surface on which these have to be applied may be pre-wetted before application. Two important requirements include non-interference with plantability of seed and no undesirable effect on seed viability. Certain seed dressings, used as dry concentrate, are added to seed grains in a planter box to control insects or diseases right from time of planting till seed has germinated. Water dispersible powder type of seed dressing is used for treatment as slurry. The requirements of stability and performance for these formulations are similar to the other dry formulations like dust, depending upon use and application of product.

Key requirements along with CIPAC test methods (FAO 1999) are: acidity or alkalinity (MT 31) or pH range (MT 75), dry sieve test (MT 59.1), adhesion to seeds, and stability at elevated temperature (MT 46).

Tablet for Direct Application (TB)

Pre-formed solids of uniform shape and dimensions, usually circular, with either flat or convex faces, the distance between faces being less than the diameter
—(GIFAP, 1989)

These are intended for direct application in field (e.g. rice paddies) without prior dispersal or dissolution in water. Tablets are pre-formed solids of uniform shape and dimensions, usually circular, with either flat or convex faces, the distance between faces being less than the diameter. The size and weight are determined by manufacturing and/or use requirements. These consist of an active ingredient together with carrier and other necessary formulants. The formulation is dry, unbroken, free flowing tablet, free from extraneous matter.

Key requirements along with CIPAC test methods (FAO, 1999) include water (MT 30), acidity or alkalinity (MT 31) or pH range (MT 75), tablet integrity and stability at elevated temperature (MT 46).

Water Dispersible Powder for Slurry Treatment (WS)

A powder to be dispersed at high concentration in water before application as a slurry to the seed —(GIFAP, 1989)

It consists of active ingredient, carrier and other necessary formulants, including colouring matter. It is a free flowing fine powder, free from visible extraneous matter and hard lumps. Key specifications along with the CIPAC test methods (FAO 1999) include: water (MT 30), acidity or alkalinity (MT 31) or pH range (MT 75), wet sieve test (MT 59.3), persistent foam (MT 47.2), wettability (MT 53.3) and stability at elevated temperature (MT 46).

Water Dispersible Tablet (WT)

Tablets to be used individually to form a dispersion of the active ingredient after disintegration in water —(GIFAP, 1989)

Tablets are pre-formed solids of uniform shape and dimensions, usually circular with either flat or convex faces, the distance between faces being less than the diameter. Tablet technology has been exploited with advantage to develop pesticide products that offer possibility of a unit dose, avoid use of water (preventing degradation of hydrolytically unstable compounds) and overcome physico-chemical incompatibility by the use of multi layer tablet techonology. These can be formed by direct dry compaction or after dry or wet granulation. Appropriate method can be selected based on the nature of the active ingredient to be formulated. Specific requirements that have been prescribed include disintegration time, hardness and friability (Hytte and Pevedic, 2002). The size and weight are governed by manufacturing facilities and use requirements. These are intended for application after disintegration and dispersion in water by conventional spraying equipment.

The key requirements and CIPAC test methods (FAO 1999) include: water (MT 30), acidity or alkalinity (MT 31) or pH range (MT 75), disintegration time, wet sieve test (MT 167), suspensibility (MT 168), persistent foam (MT 47.2), tablet integrity and stability at elevated temperature (MT 46).

Water Soluble Powder (SP)

Powder formulation to be applied as a true solution of active ingredient after dissolution in water, but which may contain insoluble inert ingredients —(GIFAP, 1989)

These comprise of either a water soluble active ingredient alone or formulated with water soluble auxiliaries. Such formulations offer one of the easiest and the cheapest modes of pesticide application. Salts of acid herbicides can be conveniently formulated in this form. Their physico-chemical and chemical requirements, besides the active ingredient, along with CIPAC test methods (FAO 1999) include acidity or alkalinity (MT 31) or pH range (MT 75), wettability (MT 53.3), degree of dissolution and solution stability (MT 179), persistent foam (MT 47.2), stability at elevated temperature (MT 46). When packed in water soluble bag, requirement of dissolution of bag (MT 176) and degree of dissolution and solution stability (MT 179) and persistent foam (MT 47.2) are additionally prescribed.

Water Soluble Powder for Seed Treatment (SS)

A powder to be dissolved in water befor application to the seed   —(GIFAP, 1989)

These comprise of an active ingredient together with any necessary formulant(s) including colouring matter. The product is water soluble, free from extraneous matter and hard lumps. The key requirements along with the CIPAC test methods (FAO 1999) are: acidity or alkalinity (MT 31) or pH range (MT 75), degree of dissolution and solution stability (MT 179), persistent foam (MT 47.2) and stability at elevated temperature (MT 48).

Water Soluble Granule (SG)

A formulation consisting of granules to be applied as a true solution of the active ingredient after dissolution in water. The liquid may contain water insoluble formulants
—(GIFAP, 1989)

These granules consist of active ingredient along with the required carrier and/or other formulants. The formulation is homogeneous, free from visible extraneous matter and/or hard lumps, free flowing and non-dusty. The active ingredient shall be soluble in water. The salient requirements and CIPAC test methods (FAO, 1999) include acidity or alkalinity (MT 31) or pH range (MT 75), degree of dissolution and solution stability (MT 179), persistent foam (MT 47.2), dustiness (MT 171), flowability (MT 172) and stability at elevated temperature (MT 46). When packed in a sealed water soluble pack, the additional requirements are dissolution of the bag (MT 176) and persistent foam (MT 47.2).

Water Soluble Tablet (ST)

Formulation in form of tablets to be used individually to form a solution of the active ingredient after disintegration in water. The formulation may contain water insoluble formulants   —(GIFAP, 1989)

The ST formulation is in the form of tablets for application after disintegration and dissolution in water. It is required to be dry, unbroken, free flowing, free from extraneous matter. The key requirements along with the CIPAC test methods (FAO 1999) are: water (MT 30), acidity or alkalinity (MT 31) or pH range (MT 75), disintegration time, degree of dissolution and solution stability (MT 179), wet sieve test (MT 167), persistent foam (MT 47.2), tablet integrity, stability at elevated temperature (MT 46).

Oil Dispersible Powder (OP)

A powder formulation to be applied as a suspension after dispersion in an organic liquid —(GIFAP, 1989)

The oil dispersible powders disperse readily on addition to an oil or an organic solvent. The dispersions so obtained are applied either with a brush to a specific situation like tree trunks or may be used as per need on wood and wall etc. Special adjuvants that aid in formation of stable dispersions in organic phase constitute an important component of these products. Used engine oils, besides the other oils, are commonly used for the application.

Pellet (PT)

Pesticide pellets are dry formulations with a size larger than 4 mesh. These may or may not be slow release formulations. For example, pesticidal baits for control of rats release the pesticide immediately when taken orally whereas aluminium phosphide pellets used for control of stored grain pests release the active ingredient (phosgene) slowly on contact with environmental moisture. There is no specified limit for the pellet size. It may range from 5 mm to 15 mm. They are generally formed by mixing suitable inert and other appropriate formulants with the toxicant and pan granulating the mix into desired size. Extrusion of the bait is not desired. Concentration of the toxicant is varied as per the requirement and the final product may sometimes contain other adjuvants like an attractant, sticky trapping agents, deodorant etc. Fertilizer pellets are common. Herbicides can also be formulated as pellets. Poison baits are special form of pellets used to control foraging insects and rodents.

Microcapsule Tape

Most microencapsulated pesticide formulations are manufactured and used in the form of aqueous capsule suspensions. Since solid formulations offer benefits in terms of reduced transport costs, ease of handling, reduced operator exposure, simpler pack disposal etc., a novel microcapsule formulation in the form of a flat, flexible sheet has been reported (Landham, 2002a). It is produced by tape casting technology, used extensively in the electro-ceramic and polymer film industries but new to the agrochemical industry. This process enables manufacture of solids with dimensions that are outside the range available to water dispersible granules. Due to this, greater control on solid handling properties such as packaging density and dustiness and wet properties such as dispersion time and sinking characteristics can be requisitely manipulated. Tape properties such as mechanical strength, tackiness in the casting slurry can be selected as per requirement. The flexible sheet format allows the tape to be wound into a continuous roll or post fabricated into configurations such as discs and strips, which can be suitably fragmented for packaging in pre-measured unit doses. The micro granules remain distributed uniformly throughout the length of the tape and are released intact when the solid formulation is dispersed in water.

Solid Formulations with Microbubbles

Solid formulations, particularly the difficultly dispersible, lead to solid sludge formation in spray tank and the amount of this sludge increases with time. It becomes compacted with time, is hard to re-disperse even after effective agitation, and causes uneven distribution with consequent poor biological activity. Even cross contamination of subsequent sprays may result. Mitigation of the sludge formation has been achieved by incorporation of low density solid particulate material with a specific gravity of less than 0.3 and a particle diameter of less than the spray nozzle filter through which it passes (Landham, 2002b).

Low-density solid particulates comprising hollow polymeric micro phases with a density of 0.13 g l^{-1} and a mean particle size of 25 μm when incorporated with a range of solid formulations such as granules, tapes, tablets, powders, gels etc. enhance the spray performance. The modified formulations generate significantly less tank residue and the active ingredient spray concentration displays less of a gradient, compared to the formulations that do not contain micro spheres.

Solid Products with Effervescents

The extrusion process normally produces these granules. Water is added to the solid powder mass to make dough, which is then extruded. A useful option is to carry out extrusion in the absence of water. This will permit addition of acid base couples (effervescing agents) in formulations like water dispersible granules with significant potential benefits for solving spray application issues including excessive dispersion times, large spray tank residues etc. (Formstone, 2002).

Water Dispersible Granule from Microcapsule Suspension

The formulation cobines benefits of encapsulation (toxicant–release control) with those of high loading, free flowing, dust freeness etc. Irreversible adhesion of the microcapsules during drying process results in poor suspensibility and dispersibility, high wet sieve residues and formation of unencapsulated material. A WG/CS formulation offering knockdown efficacy comparable to an EC, using tiny microcapsules with thin capsule walls has been developed (Gimeno et al., 2002). To avoid irreversible aggregation of the microcapsules during spray dry agglomeration, the selection of appropriate protective colloids and the post-encapsulation bridge makers is crucial in such micro encapsulation processes.

CLAY-PESTICIDE INTERACTIONS

Basic Considerations

Pesticide formulation and soil environment in which the pesticide is applied are two major clay-pesticide systems of interest, the former at the stage of formulation and the latter after application. In formulation, the pesticide along with various auxiliaries interacts with the clays or clay minerals such as kaolinite, montmorillonite, China clay etc. used as carrier or diluent. The interaction with soil involves the colloidal matter comprising of clay minerals and organic matter.

Pesticide behaviour in soil is complicated by numerous interactions that occur in solid, liquid and gaseous phases. Pesticide adsorption on solid surfaces in soils depends principally upon pesticide-water, pesticide-colloid and colloid-water interactions operating simultaneously.

Following are the key features of clay-pesticide interaction.

(*i*) Adsorption/desorption of pesticide at clay-water interface which is governed by the physico-chemical factors and controls the quantity of pesticide in soil solution, its mobility and biological activity.

(*ii*) Surface reactions between pesticide and clay, which may catalyze decomposition or protect the adsorbed pesticide from microbial attack.

Depending on structural features, various interactions bind pesticides with clay. The anionic and cationic pesticides interact by ion exchange e.g. diquat and paraquat which yield cations that are exchanged by cation exchange. The non-anionic pesticides with hydroxyl groups or the electronegative groups (carboxyl phosphoryl, thiophosphoryl, amino etc.) undergo coordination with cations. Cations in clay lattice are coordinated with water molecules. 3-Amino-S-triazole (amitrole) forms stable complexes with several transition metals and also with magnesium. Coordinated bonding of carrier cations with the pesticides has been suggested for many pesticides. For example, montmorillonite forms complexes with carbamates (e.g. EPTC), parathion, and fenarimol; whereas bentonite with carbofuran, bendicarb etc.

Interlayer moisture of the clay interacts by hydrogen bonding with electronegative groups of pesticides. Several hydrogen bonds utilizing oxygen atoms on clay surface or edge hydroxyls have been suggested as possibilities for bonding of organic compounds to clay minerals (e.g. R—N—H— ·········· O—clay; C—H---O—clay; R—COOH---O—clay; R—COOR---HO—clay). Parathion is involved in hydrogen bonding through nitro group and water associated with the cation in air dried montmorillonite.

In the absence of any polarisable group, weak van der Waals forces bind the pesticides with the clay.

Soil Clay as Absorbent

Soil generally contains mixture of clay minerals or individual mineral species. These vary considerably in degree of crystallinity and nature of ionic substitution in the lattice. The variability of specific surface and cation exchange capacity (CEC) for individual clay is an expression of variation in particle size and lattice desorption. The number of broken bonds at lattice edges increases as the particle size and the degree of crystallinity decrease. CEC generally increases with an increase in broken bonds which are a major source of CEC in kaolinite and halloysite. The 2 : 1 minerals such as montmorillonite and vermiculite which can expand have very high CEC and surface area. As a result, the extent of adsorption is sufficiently great. The non-expanding 2 : 1 minerals such as illite and chlorite are intermediate in adsorption capacities between 1 : 1 and 2 : 1 clay minerals. The strength follows the order montmorillonite > illite > kaolinite.

In soil, clay minerals may not exist in pure form but as intermediates of various transforming stages. Under alkaline conditions, they may be coated with Al^{3+} and (or) Fe^{+3} hydroxides and perhaps with hydroxide of Mg^{2+}. Since these are positively charged and bound electrostatically to alumino-silicate layers, the properties of soil clays get altered. These positive charges can absorb non-ionic insecticides like DDT more efficiently than other negatively charged clays.

Allophanes or amorphous clays are compounds of randomly arranged silica tetrahedra and iron and alumina octahedra without any symmetry. This amorphous material may constitute a large portion of soil inorganic colloids especially in highly weathered soils like oxisols. The hydroxy compounds of iron and alumina may provide positively charged sites for the adsorption of non-polar pesticides.

Adsorption on clays is also greatly influenced by the ionic composition of the surface. The saturating cations could act in three ways.

(*i*) By competing for adsorption sites with positively charged molecules e.g. cationic and basic herbicides like paraquat, diquat and atrazine.

(*ii*) Acting directly as adsorption site through formation of coordinating bonds; cations can react indirectly via (hydration) water molecules which are responsible for surface acidity.

(Surface $(M)^{n+}$—O—H---B, Hydrogen bond)

(Surface $(M)^{n+}$—O—HB$^+$, Ionic bond)

Under these conditions, electronegativity of cation is an important charcteristic.

(*iii*) Formation of hydroxides on clay surface as in case of Al^{3+} and Fe^{3+} cations. This results in a great increase in adsorption capacity of minerals. Weathering processes modify adsorption capacity of clay minerals.

Sorption

The sorption studies provide information regarding any possible interaction leading to complex formation on a solid surface from gas or liquid phases. In a liquid phase, sorption of toxicant is studied either in pure form or as solution in a suitable solvent. Various mathematical models have been evolved to characterise the type of interaction (Table 2.22) which has been studied by X-ray differaction and infra-red spectral analysis.

Anionic and cationic pesticides interact with cationic and anionic centres respectively in the clay. The non-ionic pesticides with electro-negative groups as carbonyl, phosphoryl, thiophosphoryl and amino undergo coordination with cations. The interlayer water of the clay interacts with electro-negative groups of pesticides forming hydrogen bonds. In the absence of any polarizable group, weak van der Waals forces bind the pesticide with clay.

Cationic pesticides e.g. diquat and paraquat bind to the clay by cation exchange. The non-ionic triazine pesticides get protonated on clay surface or within layer inter lamellar spaces and the resultant cation is held within the clay matrix by Coulombic forces. Propazine, hydroxy propazine, prometone and prometryn undergo probably protonation involving water hydration from clays.

TABLE 2.22. MATHEMATICAL MODELS FOR ADSORPTION

Model	Equation	Application
Langmuir adsorption isotherm	$\dfrac{x}{m} = \dfrac{k_1 k_2 C}{1 + k_1 C}$ x = amount of adsorbate m = amount of adsorbent C = equilibrium concentration of adsorbate in solution k_1 and k_2 = constants	Chemisorption and physisorption
Freundlich adsorption equation	$\dfrac{x}{m} = kC^n$ x = amount of solute adsorbed on m gram of adsorbent k and n = empirical constants C = equilibrium concentration of the solute in solution	Chemisorption and physisorption

The equation can conveniently be written in logarithmic form as follows:

$$\log \frac{x}{m} = n \log C + \log k$$

Model	Equation	Application
Brunauer-Emmett-Teller equation (BET adsorption isotherm)	$\dfrac{P}{V(P_0 - P)} = \dfrac{1}{cV_m} + \dfrac{(c-1)P}{cV_m}$	Multilayer physisorption (gases)

P = equilibrium pressure
P_0 = saturation pressure
V = volume of gas adsorbed when the entire adsorbent surface is covered with a complete unimolecular layer
V_m = volume of gas absorbed when the entire surface is covered by a monolayer
c = constant at a given temperature

Gibbs adsorption isotherm $\Gamma = \dfrac{n}{A} = -\dfrac{C}{RT}\dfrac{d\gamma}{dC} \approx \dfrac{C}{RT}\dfrac{\Delta\gamma}{C} = \dfrac{\Delta\gamma}{RT}$ Physisorption (Liquid-liquid interface)

Γ = Excess concentration of solute per unit surface area (moles cm^{-2})
γ = surface or interfacial tension (dynes cm^{-2})
n = number of moles
A = area of surface
C = activity of the solute
R = gas constant
T = temperature (absolute)

Cations present in clay lattice are coordinated with water molecules. When clay with a certain degree of water content is treated with pesticide, the water associated with the cation will mask the cation and prevent a direct interaction with the pesticide. However, when the clay is dehydrated prior to interaction or if the clay-pesticide complex is dehydrated, the cation can directly interact with the molecule e.g. carboxyl ester of malathion has been confirmed to coordinate with cations (Bowman *et al.*, 1970). Coordinate bonding of cations with pesticide molecules has been suggested to occur in montmorillonite-EPTC, montmorillonite-parathion, bentonite-carbofuran, bentonite-bendiocarb and many others. The frequency changes for carbonyl and nitrile function in IR adsorption of EPTC-montmorillonite complexes (Table 2.23) suggest a coordinated bonding between the carbonyl oxygen and the exchangeable cation.

TABLE 2.23. IR ABSORPTIONS OF EPTC-MONTMORILLONITE COMPLEXES

Exch. cation on clay	C=O Stretching		C—N Stretching	
	Observed	Shift	Observed	Shift
EPTC (No cation)	1655	—	1222	—
Li^+	1593	−62	1227	+5
Ca^{2+}	1594	−61	1232	+10
Al^{3+}	1570	−85	1232	+10

Bailey *et al.* (1968) have suggested that several hydrogen bonds utilize oxygen atoms on the clay surface (edge hydroxyls) as one of the possibilities of binding organic compounds to clay minerals.

$$R—N\text{-}H \cdots O—Clay$$

$$R—CO—OH \cdots O—Clay$$

$$R—CO—\underset{\overset{|}{R}}{O} \cdots HO—Clay$$

$$R—\underset{\overset{|}{R}}{N}—H \cdots O—Clay$$

where, --- = hydrogen bonding

The hydrogen bond of greatest importance is that associated with 'water bridge' between the exchangeable cations and quite a substantial amount of pesticide may be absorbed even though the compound may not be very tightly bound by the polar organic molecule as illustrated by Bailey *et al.*, (1968).

$$\underset{R_2}{\overset{R_1}{>}}C=O \cdots \overset{H}{\underset{H}{>}}O : M^{2+}—Clay$$

The formation of such a bond is indicated in and IR spectrum by a lowering of the OH stretching frequencies of water and has been demonstrated for a number of compounds containing nitrogen and oxygen (Mortland, 1970). Parathion is involved in

H-bonding through nitro group and water associated with the cation in air dried montmorillonite (Saltzman and Yariv, 1976).

When none of the above mechanisms can account for adsorption, it is usually assumed that adsorption is due to van der Waals forces. In solute adsorption, these may be the principal forces causing adsorption of non-ionic, non-polar molecules or portion of molecules that may include short range charge induced forces. The additive nature of van der Walls forces between the atom of absorbate and absorbent may result in considerable attraction for large molecules depending on the extent to which other forces as well as the molecular configuration allow the pesticide molecule to approach the clay surface.

Thermodynamic Functions

A little knowledge of thermodynamics goes a long way in development of proper pest control formulations. Thermodynamic functions are useful in assigning adsorption mechanism and provide a fixed value for a given state of a system. Important functions are (i) Gibb's free energy ($G°$)m (ii) entropy ($S°$) and (iii) enthalpy ($H°$). The free energy changes that may occur when a chemical is absorbed can be used as a measure of extent of driving force of the reaction. A greater magnitude of G value reflects the great extent to which adsorption reaction will take place. However, it is quite possible to have substantial amount absorbed even though the compound is not very tightly bound.

The amount of adsorption at a surface depends upon the amount by which the surface tension changes with increased concentration of the adsorbed material. A generally applicable quantitative thermodynamic expression is the Gibbs equation.

The change in Gibbs free energy of adsorption is given by the equation:

$$\Delta G° = -RT \ln k_o$$

$\Delta G°$ is change in free energy where superscript zero denotes standard state of the formulants and k_o is thermodynamic equilibrium constant. The constant k_o is expressed in terms of several factors other than, and including the stoichiometric concentrations necessary for determination of k, the chemical reaction constant.

The entropy change $\Delta S°$ for adsorption process is also related to free energy change as follows:

$$\Delta G° = \Delta H° - T\Delta S°$$

where, $\Delta H°$ is change in enthalpy.

The standard enthalpy change $DH°$ for the adsorption process can be calculated from van't Hoff equation:

$$\ln \frac{k_2}{k_1} = \frac{\Delta H}{R}\left[\frac{T_2 - T_1}{T_2 T_1}\right]$$

or

$$\log k_2 - \log k_1 = \frac{\Delta H}{(2.303)(1.987)} \times \left[\frac{T_2 - T_1}{T_2 T_1}\right]$$

$$= \frac{\Delta H}{4.576} \times \left[\frac{T_2 - T_1}{T_2 T_1}\right]$$

The thermodynamic equlibrium constant k_o, also called the thermodynamic distribution coefficient, changes with temperature. For adsorption reactions it can be obtained from equation given by Biggar and Cheung (1973).

$$k_o = \frac{a_s}{a_c} = \frac{C_s R_s}{C_c R_c}$$

where,
- a_s = activity of adsorbed solute
- a_c = activity of solute in equilibrium solution
- C_s = concentration of adsorbed solute (micrograms per millilitre)
- C_c = concentration of solute (micrograms per millilitre) of equilibrium solution
- R_s = activity coefficient of adsorbed solute
- R_c = activity coefficient of solute in equilibrium solution

$$C_s = \frac{(\rho / M) A}{S / N(x / M)}$$

where,
- x/M = mole fraction
- ρ = density of solvent
- M = molecular weight of the solvent
- S = surface area in $cm^2 g^{-1}$
- N = Avogadro number (value 6.023×10^{23})
- A = area of cross section of the solvent molecule in cm^2 per molecule

k_o is obtained by plotting ln C_s/C_c versus C_c and extrapolating C_s to zero.

The thermodynamic aspects of adsorption of phosphamidon on monoionic clays as reported by Bansal and Singh (1985) are given in Table 2.24. It would be seen that value for k_o was greater than unity; being higher at 15°C than at 50°C, indicating high preference of dimecron for H- and Na-montmorillonites. The adsorption decreases with a rise in temperature to follow the order: H-montmorillonite > Na-montmorillonite.

TABLE 2.24. THERMODYNAMIC PARAMETERS OF DIMECRON ADSORPTION ON MONTMORILLONITE AT TWO TEMPERATURES

Parameter	H-montmorillonite		Na-montmorillonite	
	15°C	50°C	15°C	50°C
k_o	423.6	38.4	403.4	36.6
G^o (K cal. mole^{-1})	– 3.5	– 2.4	– 3.4	– 2.3
H° (K cal. mole^{-1})	– 12.6	– 12.6	– 12.7	– 12.7
S° (K cal. mole^{-1})	– 31.6	– 31.6	– 32.1	– 31.1

Free energy changes were negative which increased with rise of temperature during the process of adsorption. The reactions were spontaneous at both the temperatures with a high affinity for this organophosphate. The values of $G°$ supported the inference drawn from those of k_o. The negative values of enthalpy reflected the interaction to be exothermic and that the products were energetically stable with the tight binding of pesticide to the clay. The magnitude of change in enthalpy showed that the adsorption process occurs due to protonation and/or coordination.

The negative entropy changes were indicative of a greater order produced during adsorption at both the temperatures in studied clay system. It is suggested that the desorbed water may constitute to entropy gain but S^o is a summation of the gain and loss of entropy value. Immobilization due to complex formation of dimecron molecule with the H- and Na-mortmorillonite results in a decrease in the degree of freedom of the pesticide molecule which contributes towards loss of entropy (Bigger and Cheung, 1973). In case of oxamyl on illites, high preference of adsorption, the decrease of adsorption with a rise in temperature, spontaneity of adsorption, its exothermic nature and an order of adsorption: Na-illite > H-illite > Ca-illite were reported (Bansal et al., 1981).

Practical Aspects

In choosing a suitable inert as a carrier or diluent for pesticide formulation, its compatibility with the toxicant and other formulation auxiliaries must be carefully ascertained. Quite often, the so called inerts interact with the toxicant and other auxiliary materials during storage and cause in them hitherto unforeseen changes, Such interactions are due to the catalytic effects at certain active centres on the surface of the carrier.

There are numerous examples pointing to the harmful effects of clay-pesticide interactions. Endosulfan undergoes slow decomposition on attapulgite. Methyl parathion loses nearly 15% of its activity in 30 days on attapulgite. Parathion breaks easily on kaolin. Such observations have led to a detailed investigation of the pesticide-carrier compatability before formulating a product.

Physical treatments involving heating of carriers like attapulgite to achieve deactivation have been practised earlier with limited success. Subsequently, several hydrophobic substances were coated on carriers to make them suitable for formulation with pesticides. Mixed carrier systems (e.g. diatomaceous earth, calcium silicate and Microcel; pyrophyllite, kaolin and feldspar etc.) have been reported to improve pesticide carrying capacity and stability. Use of deactivators to block the active centres on clays and clay minerals to prevent pesticide decomposition was a consequence to such studies and has been discussed earlier in this chapter (refer surface acidity).

Dry formulations of organophosphates like malathion, parathion, methyl parathion, chlorthion etc. are also quite unstable. It has been found that malathion is highly susceptible to alkaline hydrolysis and decomposition. Thus acidity deactivators such as urea, hexamethylene tetramine, lime, monoethanolamine, sodium carbonate, calcium carbonate and sodium hydroxide which are suitable for chlorinated hydrocarbons like heptachlor and endrin, cannot be used for malathion. The proper method in such cases is to neutralize the alkaline active sites on the carrier with the help of weak acid such as tall oil and rosin acids. Alcohols, ketones, glycols and ethers can be used as stabilizers for dry formulations of parathion and methyl parathion. Diethylene glycol and dipropylene glycol have been suggested for use as deactivators for dry formulations of ethion. It has been found that diazinon decomposes rapidly on all mineral carriers. Walnut shell is a better carrier. Glycols have been recommended as suitable deactivators for phorate (thimet). Phosdrin is found to decompose in contact with highly alkaline carriers with higher moisture content. The types of carriers suitable in this case are talc, pyrophyllite, calcined gypsum and volcanic dust; requiring no deactivators.

Among the carbamate type of toxicants, carbaryl is considered to be the most important. These chemicals are susceptible to alkaline hydrolysis and thus only those carriers which are of acidic nature are suitable for dry formulations of these pesticides. No deactivators have been reported for carbamates yet. Not all the carbamates decompose and 'Sevin' products are stated to be even less degradable.

Nickel, iron, calcium, magnesium and zinc have little or no effect on the rate of hydrolysis while copper has a very pronounced effect on alkaline hydrolysis.

Endosulfan is slowly decomposed by some clays and fillers such as kaolin. Talcs and calcium carbonates are not very active. Dipropylene glycol is a suitable deactivator for the carriers used in preparing dry-stable "Thiodan" formulations. Talc, calcium carbonate, calcium silicates, montmorillonites, pyrophyllites and kaolin type are the best for preparing dry formulations of aramite wherein kaolinites are the most active. Maracarb N (a sodium lignosulphonate) is found to be a good deactivator which masks the active centres of the carrier when used upto 6 percent. Urea is also a good deactivator for acid sites and is used at 3-6 percent.

Park et al., (1967) studied the compatibility of a few insecticides with carriers and found that lindane was influenced by the nature of carrier material. Talc was identified to be the most compatible material with malathion, to which anhydrous maleic acid was added as an anti-decomposition agent.

El-Attal et al., (1975) found that limestone, bentonite and diatomaceous earths of pKa 1.5 can be used to formulate HCH without any appreciable loss of insecticidal activity. Loss of activity of endrin on materials having pKa less than 3.3 has been reported. The materials become suitable as carriers for endrin if deactivated by mixing with required quantity of polyethylene compounds in acetone or with hexamethylene tetramine to pKa value of greater than 3.5.

Attapulgite, a clay mineral used widely in granuler formulations, requires appropriate treatment to ensure adequate stability of the toxicant on storage Captan formulated with untreated attapulgite tended to become phytotoxic consequent on its decomposition on this carrier (Daines et al., 1957). Similar effect has been observed in the case on dichlone. Trademan et al., (1967) found that methyl parathion was unstable on untreated attapulgite with 15% of the pesticide losing activity within 30 days. The decomposition could be prevented by the use of glycol as stabilising agent. Polon and Sawyer (1962) observed that the stability of malathion on attapulgite could be increased by neutralizing the surface activity of the carrier material using substances like tall oil, weak organic acids, glycols etc. Barthel and Lofgren (1964) also found glycols to be effective in stabilising the activity of heptachlor and chlordane on attapulgite. Benesi et al., (1959) used 2% urea, 20% ammonia solution or 2% ammonium bicarbonate to neutralise the acidic centres on attapulgite so as to assure stability of dieldrin.

Parathion is stable when formulated on attapulgite as such. However, it easily breaks down on kaolin. Bell and Kido (1956) could stabilise pyrethrum on attapulgite pre-treated with hydroquinone or its derivatives. However, these were ineffective in case of allethrin.

Transformation of malathion to isomalathion (Halder and Parmar, 1984, Rengasamy and Parmar, 1988) and isomalathion to other products (Rengasamy and Parmar, 1989), release and degradation of carbofuran (Powar and Parmar, 1995), azadirachtin-A (Kumar and Parmar, 1999) are all influenced by the solid pesticide carrier. Carrier cations and relative humidity of the formulation and its storage environment influenced malathion to isomalathion transformation (Rengasamy and Parmar, 1988). No correlation between the physico-chemical characters of carriers and such transformations could be found. However, the liquid carriers have been reported to be relatively inert in bringing about such transformations (Iyer and Parmar, 1987).

Other Carrier Treatments to Improve Stability: A physical process of deactivation is reported by thermal treatment which involves heating attapulgite to high temperature to effect deactivation. Thermally deactivated attapulgite is found to be suitable for formulating diazinon (Schwint, 1966), ronnel (Rosenfield and Van Valkenburg, 1965) and 2,4-D (Galloway, 1962).

LIQUID FORMULATIONS

FACTORS AFFECTING PREPARATION AND PERFORMANCE

The liquids in pesticide formulation play the role of a solvent, diluent, carrier, impregnating agent or other dispersal aids. The classification of different types of formulation solvents may be based on (*a*) composition, (*b*) nature of the chemical and (*c*) structure or function. The most convenient method to classify them is by grouping into polar or non-polar products. The polar solvents include carbonyls or ketones, esters, glycols, glycol ethers, amides etc. The common non-polar solvents are the hydrocarbons and petroleum distillates that may be further classified as aliphatic or aromatic for a functional distinction. The polar solvents may be water miscible or water immiscible, the characters, which along with the economics often govern their choice. Solvency, distillation range or boiling point, specific gravity or density and flash point are some of the other important functional properties of the solvents used in formulating pesticides.

Solvency

Solvency is the basic parameter governing type and the strength of a liquid formulation that can be prepared employing any active ingredient. It is defined as the ability of a solvent to dissolve a material under a given set of conditions. Amongst the solvents commonly used in pesticide formulation, the solvency increases in the rising order from aliphatic to aromatic hydrocarbons and to other polar compounds with different functional groups.

Selection of proper solvent or development of a blend is made easier when interaction between solvent and solute is understood. From the earlier empirical approach in choice of solvents, the trend today is the use of more fundamental methods. The environmental protection and safety regulations have led to the search of newer solvents to replace the old, especially the aromatic ones.

The process of solution/dissolution of a liquid or a solid in a liquid is the result of two dynamic processes. The dispersion of the solute molecules into the liquid and the deposition of dissolved molecules back into the condensed or solid state. Solubility is maximum at equilibrium concentration under the given conditions.

Qualitatively, the solubility of a solute in a solvent may be predicted by considering similarities of their chemical structure, molecular weight in a series of compounds of similar type, dielectric constant etc. Similar chemical structures promote solubility e.g. straight chain hydrocarbons are good solvents for straight chain hydrocarbons, compounds containing hydroxyl, OH, group (alcohol, phenol, organic acid) tend to be more soluble in water or alcohol. In a series of compounds of similar type, the solubility tends to decrease with an increase in molecular weight e.g. methyl and ethyl alcohol are soluble in water in all proportions; amyl alcohol with a 5-carbon chain only 3%, and alcohols of higher molecular weight are practically insoluble. The dielectric constants in a similar range indicate a similar degree of electrical asymmetry in the molecules that promote the solubility of the test components.

The principle of 'like dissolves like' is a useful guide for predicting solubility since polar solvents dissolve the polar materials and vice versa. The principle has many exceptions. For example, polystyrene will be predicted to dissolve in several non-polar solvents yet it is insoluble in hexane but dissolves in diethyl ketone, a polar solvent. The temperature of dissolution must find a mention as it is a critical factor for solubility. Quite often, the compounds dissolving at higher temperature will be thrown out at lower temperature, thus offering a serious limitation in pesticide formulation.

The solubility requirements for different classes of pesticide chemicals vary considerably. Even within a given class of pesticides, such as the chlorinated hydrocarbons, there is a wide variation in the solvency. For example, ordinary kerosene, which is one of the poorest solvents, dissolves a large amount of technical chlordane. On the other hand, endrin has a limited solubility even in the aromatic hydrocarbons.

Physico-chemical Aspects of Solubility

Solution of one compound in another must be spontaneous and is expressed by its relation with the free energy of mixing.

$$\Delta G = \Delta H - T\Delta S$$

where, ΔG = free energy of mixing
ΔH = enthalpy of mixing
ΔS = entropy of mixing

ΔS can be considered positive for the dissolution process. Therefore, the controlling term for the spontaneous process is the enthalpy of mixing. If it is negative or a small positive number, the process could be considered spontaneous.

Single Component Solubility:

Solubility of a single solute can be ascertained from heat of mixing and heat of vaporization. For the heat of mixing, an expression has been developed as:

$$\Delta H_m = \frac{x_1 x_2 V_1 V_2}{x_1 V_1 + x_2 V_2} \left[\frac{a_1^{1/2}}{V_1} - \frac{a_2^{1/2}}{V_2} \right]^2$$

where, ΔH_m = heat of mixing
x_1 = mole fraction of solvent
x_2 = mole fraction of solute
V_1 = volume of solvent before mixing solute
V_2 = volume of solution
a_1 and a_2 = interaction constants

Cohesive energy for a mixture of liquids can be expressed as:

$$\Delta E_m = (x_1 V_1 + x_2 V_2) \left[\left(\frac{\Delta E_1^v}{V_1} \right)^{1/2} - \left(\frac{\Delta E_2^v}{V_2} \right)^{1/2} \right]^2 \phi_1 \phi_2$$

where, ΔE_1^v and ΔE_2^v = energy of vaporization of two liquids
ϕ_1 and ϕ_2 = volume fractions.

This can be rewritten as:

$$\Delta H_m = V_t \left[\left(\frac{\Delta E_1^v}{V_1} \right)^{1/2} - \left(\frac{\Delta E_2^v}{V_2} \right)^{1/2} \right] \phi_1 \phi_2$$

where, V_t = total volume

The term $\dfrac{\Delta E^v}{V}$, the energy of vaporization per unit volume, can be taken as a measure of the internal pressure. It is often called the solubility parameter δ.

$$\frac{(\Delta E^v)^{1/2}}{V} = \delta = \frac{a^{1/2}}{V}$$

In other words δ is the expression most commonly used for the solubility parameter and has the units of $(Jm^{-3})^{1/2}$. Therefore, the free energy of mixing is given by the expression,

$$\Delta G = V_t [\delta_1 - \delta_2] \phi_1 \phi_2 + RT[x_1 \ln x_1 + x_2 \ln x_2]$$

where, V_t = volume at any given temperature
R = gas constant
T = absolute temperature

The solution should be assured as δ_1 approaches δ_2. According to this, two substances should mix when the solubility parameters are equal.

The influence of several important factors has been over simplified by this theory; nevertheless, it is widely used as formulating tool. The solubility parameter δ_b, for a blend is given by:

$$\delta_b = \delta_1 \phi_1 + \delta_2 \phi_2 + \delta_3 \phi_3 \ldots \delta_n \phi_n$$

where, ϕ = the volume fraction.

Two-component Solubility: The above theory remains restricted to non-polar compounds. Polar molecules or molecules containing hydrogen bonds, have interactions influencing both the enthalpy and the entropy. The heats of mixing due to polar effects are corrected by including an additional term ω, the dipole force.

$$\frac{\Delta E}{V} = \delta^2 + \omega^2 ,$$

where, δ = dispersion force

               ω = dipole force

The factor g is introduced to account for the effect of the dipole moment on the nearest neighbours.

$$g = 1 + 2\delta \cos\gamma \cdot \exp\left(\frac{-W}{kt}\right) d\omega ,$$

where, γ = angle between dipole moments

               W = potential of the average torque hindering their rotation.

Values of g are near 1 for dilute solutions but at high concentrations they approach the value of the pure polar solute. The effect on the cohesive energy would be small except for highly polar or hydrogen bonded molecules.

Three-component Solubility: Solubility of three component systems is based on the solubility parameters, polarity and hydrogen bonding in the system. The cohesive energy E, per unit volume is given by the equation.

$$\frac{-E}{V} = \frac{E_d}{V} - \frac{E_p}{V} - \frac{E_h}{V} ,$$

where, E_d = dispersion interaction

               E_p = polar interaction

               E_h = hydrogen-bonding interaction

This gives the following total solubility parameter:

$$\delta_b^2 = \delta_d^2 + \delta_p^2 + \delta_h^2$$

where, δ_b^2 = total solubility parameter for the blend of three component system.

The values of solubility parameters for various solvents are determined from a variety of physical constants. For some common solvents, these are reported in Table 2.25.

TABLE 2.25. VALUES OF SOLUBILITY PARAMETERS FOR SOME SOLVENTS

Solvent	Solubility Parameter (δ) ($cal\,cm^{-3}$)
Butane	6.59
Diethyl ether	7.53
Isopropyl chloride	8.07
p-Xylene	8.83
Benzene	9.16
Acetone	9.62
Dioxane	10.13
Acetonitrile	12.11
Ethanol	12.78
Water	23.53

Kauri-Butanol Value

Kauri butanol (K_b) value is a measure of the solvency of a solvent. The test solvent is titrated into a solution of Kauri wax in butanol at room temperature and the point at which clouding occurs, as indicated by the inability to read a newsprint through the solution, is taken as the K_b value. It is expressed in terms of milliliters of solvent added. With experience, the test is of considerable value and meaningful to the laboratory technicians. It is also expressed as a relative number to that of toluene (105). For most aromatic hydrocarbons used in pesticide formulation, the K_b value is nearly equivalent to the volume per cent of aromatics available in the solvent.

Distillation Range and Boiling Point

The distillation range or boiling point indicates the volatility of the solvent under the given conditions. For a pure solvent, the boiling point is the temperature at the atmospheric pressure where its liquid phase is in equilibrium with the vapour phase. The normal boiling point usually refers to the temperature at which a liquid can be distilled at atmospheric pressure.

Determination

The hydrocarbon solvents and petroleum distillates normally used in pesticide formulation are mixtures of hydrocarbons, each with its own range of boiling point. The boiling point range in such cases is determined by recording the temperatures of the initial drop received in an overhead during fractional distillation, through final cut, until no further material from the sample distills between these points (Engler distillation). The yield of the fractions is expressed as per cent of total volume. Typical hydrocarbon solvents used in pesticide formulation are the xylene-type, which distill over 133 to 165°C. The heavy aromatic naphtha normally distills in the range of 117-287°C. Kerosene type aliphatic hydrocarbons distill in the range of 190-475°C.

Polar solvents are usually of higher purity than the hydrocarbon solvents, and their distillation range is seldom greater than 12°C. The polar solvents with boiling point higher than 90-94°C are preferred in pesticide formulation. Only under pressing circumstances such as problems of solubility or phytotoxicity, the solvents with lower boiling points are used, and that too with caution.

Density

The density (specific gravity) of solvent is the weight of a given volume of the solvent relative to weight of an equal volume of water at a standard temperature. Density of a solvent is its mass per unit volume. It is usually expressed in grams per milliliter (g ml^{-1}). When defining, the temperature at which the density is determined becomes an absolute value. Of the hydrocarbon solvents used in pesticide formulation, the aliphatic type such as kerosene have the lowest density of 0.76-0.79. The xylene type have intermediate values between 0.92-0.97 (Flanagan, 1983).

Determination

The density or relative density of a solvent may be calculated from its weight using a calibration factor proportional to an equal volume of water and the term that corrects for buoyancy of air.

Density, ρ, at test temperature is given by

$$\rho = \frac{W}{V} + C$$

where, W = weight of sample in grams
V = apparent volume in millilitre
C = air buoyancy correction obtained from standard table.

Unit for density is g ml^{-1}

Aromatic Content

The aromatic content of the hydrocarbon solvents is expressed as volume percent. Generally, the solvency and the solvent cost increase with an increase in the aromatic content. The aromatic content of xylene and heavy aromatic naphthas range from 85 to over 95 percent.

Determination

It can be determined directly from the estimation of group molecular weights of aromatics and saturated hydrocarbons from the integral intensities of their respective NMR signals in aliphatic and aromatic regions. Percentages of hydrogen and carbon are calculated from the total integral intensity of the sample and total group molecular weights. The hydrogen content obtained by this method is compared with other standard methods. Results are checked against standard blends of aromatic and paraffinic compounds. The Proton Magnetic Resonance (PMR) is also useful for estimation of aromatic, hydrogen and carbon percentage of the products in the boiling range of 140-300°C. It further enables a quantitative differentiation of aromatic compounds into monocyclic or polycyclic.

Flash Point

The flash point of a solvent is an indication of its flammability. Numerically, it is the temperature at which the vapour forms explosive mixture with air and ignites. Formulation solvent with the higher flash point, consistent with other desirable properties, should be used. For most liquid formulations, the minimum flash point should be above 27°C. The lower flash point solvents need to be given extra care during handling.

Determination

For products having flash points between 17.8 to 48.9°C, Abel closed cup method may be employed. For others with a higher flash point, Pensky Martens closed cup method is used. The test solvent is taken in the cup and is heated at a slow and constant rate with continual stirring. A small flame is directed into the cup at regular intervals with simultaneous interruption of stirring. The flash point is the lowest temperature at which application of test flame causes the vapours above the sample to ignite.

Water Miscibility

Water miscibility of the solvents, where concentrates have to be applied as solution in water, is desirable. It is also desirable when the water insoluble solute contained as solution in a solvent is to be transformed into a suspension in water on dilution. However, in case of emulsifiable concentrates, the solvent and water need not be miscible or only slightly miscible so as to give an emulsion on dilution in water. The aliphatic and aromatic solvents meet this requirement but if their polarity is increased, the problem starts since it increases solubility. Solvents like cyclohexanone and isophorone are slightly soluble in water and can often be effectively used in combination with aromatic hydrocarbons. Solvents of increasing polarity such as glycol ethers and amides may be used, usually only sparingly, in mixtures with hydrocarbons.

Determination

Miscibility is observed by drop wise addition of water to the solvent and shaking. The appearance of cloud shows immiscibility or saturation point.

Viscosity

Viscosity is the resistance that a fluid exhibits during the flow of one layer over another. It is an important factor that governs the change of shape of a liquid on application of force. The coefficient of viscosity of a liquid is the ratio of force per unit area to the velocity gradient i.e. it is the force required per unit area to maintain unit difference of velocity between two parallel planes in the fluid, one centimetre apart. It has the unit $kg\ m^{-1}s^{-1}$ or Pa s. It is also measured in 'poise' (after Poiseuille), which is equal to the viscosity of a fluid in which a force of 1 dyne cm^{-2} causes a flat surface 1 cm^2 in area to move, relative to another such surface underneath it at a distance of 1 cm, with a velocity of 1 cm sec^{-1}. Poise is equal to 0.1 Pa s.

A prediction of whether the flow will be turbulent may be obtained by calculating a dimensionless quantity called the Reynolds number. The Reynolds number for flow in a tube is defined by $dv\rho/\eta$, where, d is the diameter of the tube, v is the average velocity of the fluid along the tube, ρ is the density of the fluid, and η is the coefficient of viscosity. At flow velocities corresponding with values of the Reynolds number of greater than 2×10^3, turbulence is encountered.

For certain colloidal suspensions and solutions of macromolecules, the coefficient of viscosity depends on the rate of shear, and this is referred to as non-Newtonian behaviour. If the shear stress orients or distorts the suspended particles, the coefficient of viscosity may decrease as the shear rate is increased.

When a force F is applied to a particle in a solution-as by applying an electric field, if the particle is charged, or a centrifugal field is formed, the particle will be accelerated. As the velocity of the particle increases, it experiences an increasing frictional force.

For low velocities the frictional force is given by vf, where, v is the velocity and f is the frictional coefficient of the particle. When the velocity is sufficiently high for the frictional force to be equal to the applied force, $vf = F$, the particle will move with constant velocity.

The frictional coefficient f is of interest because it provides some information about the size and shape of the particle. For spherical particles, Stokes showed that for nonturbulent flow,

$$F = 6\pi\eta\, rv$$

where, η = coefficient of viscosity
                               r = radius of the spherical particle

The frictional coefficients of prolate and oblate ellipsoids and long rods may be expressed in terms of the radius of a sphere of equal volume and a factor depending on the ratio of the major axis to the minor axis.

The coefficient of viscosity η of a liquid may be determined by measuring the rate of settling of a sphere of known density, as shown in Fig. 2.14. The force causing the sphere to settle in the fluid is equal to its effective mass times the acceleration of gravity; the effective mass is the mass of the sphere minus the mass of the fluid it displaces. The method is termed as Stokes Falling Sphere Method and uses the equation for calculation of η:

$$\frac{\eta_1}{\eta_2} = \frac{t_1(\rho_s - \rho_{L_1})}{t_2(\rho_s - \rho_{L_2})}$$

where, η_1 = viscosity coefficient of unknown liquid (L_1)
                        η_2 = viscosity coefficient of known liquid (L_2)
                        ρ_1 = density of solid ball
                        t_1 = time for fall in L_1
                        t_2 = time for fall in L_2
                        ρ_{L_1} = density of L_1
                        ρ_{L_2} = density of L_2

The coefficient of viscosity η may also be determined by passing a liquid through a capillary tube and making use of the Poiseuille equation.

$$\eta = \frac{P\pi r^2 t}{8Vt}$$

where, t = time required for volume V of liquid to flow through a capillary tube of length l and radius r under an applied pressure P.

The viscosities of most liquids diminish with increasing temperature. According to the "hole theory", there are vacancies in a liquid, and molecules are continually moving into these vacancies so that the vacancies move around. This process permits flow but requires energy because a molecule must surmount an activation barrier to move into a vacancy. The viscosities of several liquids at different temperatures are shown in Table 2.26.

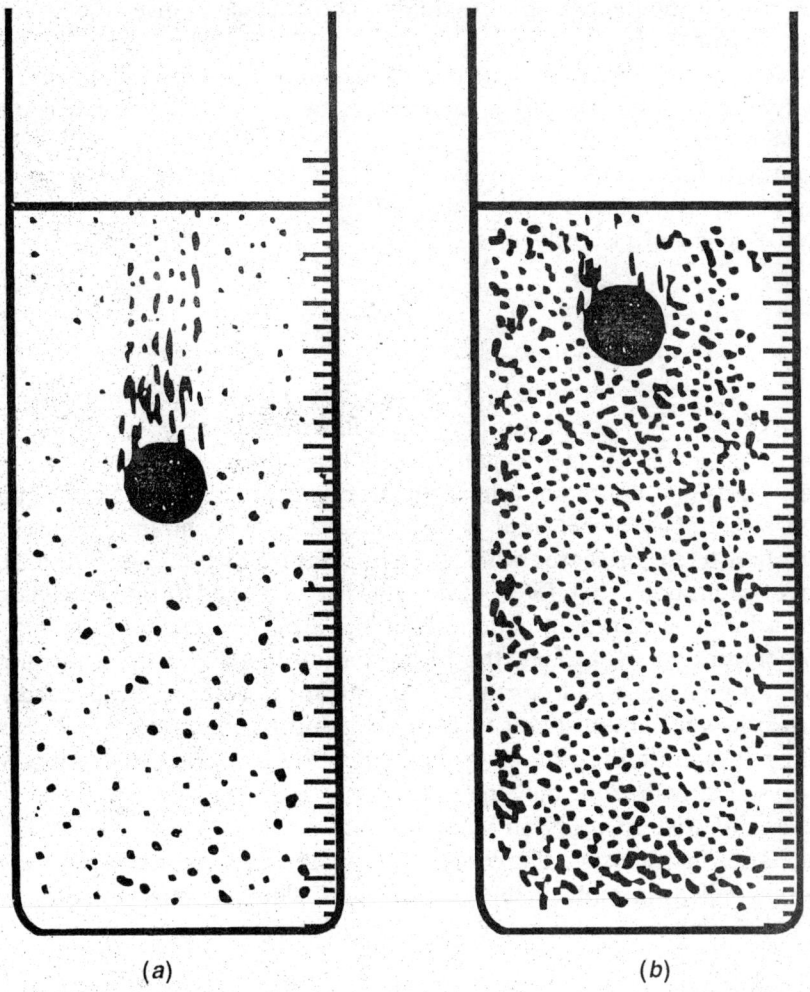

(*a*)   (*b*)

Fig. 2.14: The Spherical Ball Moving in Two Liquids of Different Viscosities (*a*) Less Viscous (*b*) More Viscous

TABLE 2.26. VISCOSITIES OF LIQUIDS IN Pa s (kg m^{-1}s^{-1})

Liquid	*Viscosity* at (°C)			
	0	25	50	75
Water	0.001793	0.000895	0.000549	0.000380
Ethanol	0.00179	0.00109	0.000698	–
Benzene	0.00090	0.00061	0.00044	–
Glycerol	–	0.945	–	–

The variation in the coefficient of viscosity with temperature may be represented by

$$\phi = 1/\eta = Ae^{-E_o/RT}$$

where, E_o = activation energy for the fluidity $(1/\eta)$ which is inverse of η and generally denoted by ϕ.

The viscosity of a liquid increases as the pressure is increased because the number of void spaces is reduced, and it is therefore more difficult for molecules to move around each other.

In contrast, the viscosity of a gas increases as the temperature increases. The viscosity of a perfect gas is independent of pressure.

The co-efficient of viscosity η is given by

$$\eta = \frac{\rho \pi r^2 t}{8VL}$$

where, P = applied pressure
                        r = radius of path or capillary in cm.
                        V = volume of liquid/gas flowing in time t
                        L = length of path or capillary

Viscosity plays an important role in the preparation and use of most liquid products. As the viscosity of a solvent used in an emulsifiable concentrate increases, the rate of crystallization often decreases when the temperature of the concentrate drops below the point of saturation. Therefore, caution must be exercised during cold-stability studies performed during a minimum period of time to avoid determining solubility at a given low temperature when, in fact, the formation of crystals in the concentrate is only delayed. The molecular and crystal alignments are retarded in high viscosity. Seeding (addition of a very small quantity of the crystalline material to the solution) may sometimes accelerate the rate of crystallization by providing nucleating surfaces for further crystal growth from the supersaturated solution.

The ease of dispersibility in water of an emulsifiable concentrate is inversely proportional to its viscosity. The viscosity increases with an increase in concentration of active ingredient in emulsifiable concentrates. The non-viscous concentrate can be best prepared with a suitable solvent of low viscosity.

Determination

Kinematic viscosity is determined by measuring in seconds, the time for a fixed volume of liquid to flow under gravity through the capillary of a calibrated viscometer under a reproducible driving head and a closely controlled temperature. It is a product of measured flow time and the calibration constant of the viscometer and is calculated from the increased flow time t and the instrument constant C by means of the equation $V = Ct$, where, V = kinematic viscosity.

Toxicity

Solvents may either be directly toxic or these may modify the toxicity of the main product against mammals or plants. It is, therefore, essential to ascertain their mammalian and phyto-toxicity as well as that of the formulations prepared by using them. Properties such as acute oral and dermal toxicity, skin penetration and irritation, absorption in the eyes, reproductive toxicity, carcinogenecity, mutagenecity, effect on spray operators, absorption in cuticle of beneficial plants or insects etc. need to be paid a special attention. Each of the toxicological parameters is tested on animals under highly sophisticated laboratory and green house conditions. The readers are advised to refer to specific publications on these topics. The hydrocarbon solvents are generally more phytotoxic to plants than other type of solvents. Higher boiling hydrocarbons are more phytotoxic than the low boiling ones.

Colour

The colour of the solvents meant for application in household and institutional pest control is of great significance because of the likely stains that may left on walls, clothing and other belongings. However, as far as the agricultural application of solvents is concerned, the colour is not of much significance. Special brands of colourless and odourless solvents (e.g.. deodorized kerosene, white mineral oil etc.) are available for such applications.

Determination

A simple method to know the colour of solvents is colour comparison with Munsell Colour Charts. The liquid in a standard reference tube is held behind the aperture separating the closest chip. The Munsell notation for colour consists of hue, value and chroma, which are combined, in that order to designate the colour.

Odour

Like colour, the odour of the solvent is also of minor significance in an agricultural pesticide formulation but this has relevance when used in household or institutional pest control. The changes in the odour are primarily of chemical nature and need to be investigated for their side effects on the target as well as on other physico-chemical characters.

The odour in household or institutional pest control formulations should be kept to a minimum. The deodorized aliphatic solvents are generally satisfactory for this purpose. Often, certain fragrances, perfumes or masking agents are added to formulations to mask their or pesticidal foul odour.

Cost and Availability

The commercial viability of any formulation depends on the abundance and availability of the raw material and other inputs at a reasonably low cost. Economics of entire operation and price of the finished products will depend on these factors in case of solvent also.

Surface Activity

The surface tension of a liquid, γ, is the force per centimeter on the surface of a liquid which opposes the expansion of the surface area. It is due to the unbalanced molecular forces at the surface of a liquid. The molecules at the surface are pulled inward by the other molecules of the liquid, and the liquid tends to adjust itself to acquire the minimum surface area. The spherical shape of the raindrops, the rise of water in a capillary tube, movement of water in blotting paper or soil pores is due to this phenomenon. The surface tension of a liquid decreases with rise in temperature, as well as with the molecular agitation.

Interfacial Tension

It is caused by unequal attractions on the surface molecules in different directions. Defined as the work required for increasing the area of the interface by 1 cm^{-2}, this tension will change if a third material tends to concentrate at the interface, altering its nature. Materials that do this are surface active agents or 'surfactants'. Capillary active or inactive solutes are classified on the basis of their effect on surface tension or interfacial tension i.e. those which can lower the tension are called capillary active while others are capillary inactive. In case of aqueous solution-air interface, inorganic electrolytes, salts of organic acids, bases of low molecular weight, and certain non-volatile nonelectrolytes such as sugar and glycerine are capillary inactive. Organic acids, alcohols, esters, ethers, amines, ketones etc. are capillary active and have a considerable effect in reducing the surface tension.

Gibbs (1876) gave a thermodynamic treatment to this subject.

$$S = \frac{-C}{RT} \frac{d\gamma}{dc}$$

where, S = excess concentration of solute per square cm. of surface
         C = concentration of solute in the solution
         $\frac{d\gamma}{dc}$ = rate of increase or decrease of surface tension with the concentration of solute
         T = absolute temperature
         R = gas constant

If $\frac{d\gamma}{dc}$ is negative, the solute lowers surface tension of the solvent and vice versa.

Processes, which are accompanied by decrease in free energy, tend to occur spontaneously. Thus, the fact that the organic acids can lower the surface tension of water suggests that they will spontaneously be absorbed on the surface interface resulting in a decrease of free energy of the surface.

Measurement

Surface tension and interfacial tension can be measured quantitatively by several methods such as pulling of a wire ring from the surface of a liquid, weighing of drops which fall from a special glass tip, determination of shape of a hanging drop, reflection of light from ripples on the surface of a liquid, and the pressure required to blow gas bubbles in the liquid. When the liquid is in contact with an immiscible liquid but not with air or its own vapour, the interfacial tension may be quite different.

The most accurate method of determining surface tension of a liquid consists in measuring the height to which it rises in a capillary tube.

A capillary tube of radius r is immersed in a vessel containing liquid of density d. The liquid wets the tube and rises in it till the force of gravity pulling it downwards counterbalances the surface tension tending to pull it upwards. Let the height to which the meniscus rises at this equilibrium point be h. The meniscus is curved, and the average height of the column of the liquid is the significant quantity. For ordinary work, the height to bottom of the meniscus may be taken but for more precise work a correction factor equal to $1/3^{rd}$ of the radius is added.

Then, downward force $= \pi r^2 h d g$

where, g = acceleration due to gravity.

If, the surface makes an angle θ with the walls, and only the vertical component of the force is effective in pulling the liquid upward in the capillary and acts along the whole length of the surface,

$$\text{Upward force} = 2\pi r \gamma \cos\theta$$
$$\text{At equilibrium } 2\pi r \gamma \cos\theta = \pi r^2 h d g$$
$$\gamma = h d g r / 2 \cos\theta$$

For many liquids, including water, the angle of contact is very small i.e. θ is nearly zero, and cosine zero being 1, surface tension $\gamma = \frac{1}{2} h d g r$.

Wetting

It is the spontaneous formation of an area of intimate contact between a liquid and another surface. Wetting is due to lowering of interfacial tension either by some degree of attraction between the molecules of two materials, or by the presence of a film of a third material in the interface which is attracted towards both the other surfaces. Wetting is caused by splashing and spreading of a liquid and hence depends on all factors responsible for spreading.

Spreading

Subsequent to wetting, the spreading occurs and the latter depends on both interfacial tension and other forces in the surface. A contact angle of 90° or less wets a surface followed by its spreading which depends on its various characters.

When a drop falls on a solid surface it experiences three types of interfacial tension, that is towards solid-air interphase γ_{SA}, towards air-liquid interphase γ_{LA} and towards solid-liquid interphase γ_{SL} (Fig. 2.15).

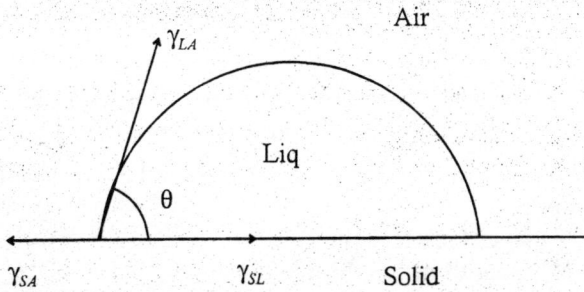

Fig. 2.15: Contact Angle of, and Various Forces Acting on a Drop in Contact with Solid Surface

Spreading of the drop depends on spreading coefficients, and is given by

$$S = \gamma_{SA} - (\gamma_{LA} + \gamma_{SL}) = \gamma_{SA} - \gamma_{LA} - \gamma_{SL}$$

The quantity S is the Harkin's spreading coefficient. If the value of Harkin's coefficient is zero or positive, the force γ_{SA} can balance or overcome the sum of the forces preventing the spreading and the liquid can spread completely over the surface. If the value is negative (as it usually is), the liquid will fall short of complete spread.

The γ_{SA} and γ_{SL}, however, cannot be measured directly on a solid surface, which is rigid and not free to contact. For comparative purposes, this can be avoided by using a second liquid instead of a solid. For example, the Harkin's coefficient of aqueous solutions of surfactants against a mineral oil may be used. This evaluation may not hold well against totally dissimilar surfaces.

A more general method of overcoming the difficulty in the measurement of γ_{SL} and γ_{SA} is to take into account the contact angle θ. If spreading is incomplete, the forces on a drop edge will be as shown in Fig. 2.15.

The component of the forces along the solid surface is the only one effective in moving the droplet edge. Not all of γ_{LA} is available in the right direction, only a part of it is available for opposing spreading and hence modified Harkins coefficient can be written as follows (Adamson, 1976):

$$S = \gamma_{LA}(\cos\theta - 1)$$

where, γ_{LA} = surface tension of the liquid

$\cos\theta$ = contact angle of surface of the drop with solid surface

Both γ_{SA} and γ_{SL} are thus eliminated and the spreading coefficient expressed in terms of γ_{LA} the surface tension of the liquid and θ, the contact angle on solids or crop, both of which can be measured. This coefficient is always −ve or zero because $\cos\theta$ cannot exceed one. It can, therefore, give a measure of incomplete or barely complete spreading, but not an excess of spreading ability as can the original Harkin's coefficient.

The different tensions on the liquids, can be measured with comparative ease to give the Harkin's coefficient. On solids, the measurement may be difficult. The advancing angle may differ radically from the receding angle, due to a change in the surface after wetting, e.g. by displacement of an adsorbed film of air, moisture, grease etc. The 'equilibrium' angle is sometimes calculated arbitrarily as the angle whose cosine is the average of the cosines of the advancing and receding angles. Either the advancing or the equilibrium angle may be used, depending upon the phenomenon of interest.

Spreading coefficients are reliable only when the same criteria are used and the same surface is involved. The latter restriction lessens but does not destroy the value of spreading measurements on standard surfaces.

Run off

Furmidge (1962) has established that run off of the drop will occur if,

$$mg \sin\alpha = \gamma \times W (\cos\theta_R - \cos\theta_A)$$

where, m, g and α denote mass of the drop, gravity and angle of fall of the drop with surface respectively,

γ = surface tension

W = width of the drop

θ_R and θ_A = receding and advancing contact angles respectively

The retention factor 'F' which is a function of number of drops retained, is:

$$F = \theta_M [\gamma(\cos\theta_R - \cos\theta_A)/\rho^{1/2}$$

where, θ_M = mean of the contact angles

                 ρ = density of liquid

Wetting of a surface occurs subsequent to spreading. Spreading involves liquid-solid interphase but wetting is taken only after the excess liquid is drained off.

Applications of Surface Activity

Appropriate moderation of the surface active materials is often necessary for effective use of formulations. The suspensibility of suspensions, stable emulsion formation in emulsifiable concentrates, appropriate wetting and spreading characters in various products, dispersions of powders or granules in water etc. will all be influenced by surface activity.

A proper choice of materials based on surface activity considerations cannot be made by theoretical considerations alone. These require a considerable trial and error, particularly in view of the multiplicity of the factors involved. For example, when the wetting of surfaces is under consideration, it may imply wetting of surface of the solid being suspended, leaf, insect, or other target. So the nature of both the material and the surface will govern the optimum. Likewise, suspension or emulsification depends on the interfacial tension and film formation, droplet size, on dynamic surface tension, wetting and spreading on surface and interfacial tensions all around, all the values of which are seldom available. Also, exact relations between measurable values and the final behaviour are seldom known. Frequently, they depend upon methods and mechanical details of use e.g. emulsification, deposition etc.

A material used for a purpose affects many other characters too. For example, the emulsifiers or suspending agents effect droplet size, wetting, spreading, deposition etc. A good material for one purpose may lead to trouble in another area of use. The overall behaviour finally desired is often not clear and is poorly defined. For practical reasons, a single formulation must be made to cover a variety of conditions. It is seldom possible commercially to put out a product which is designed for a single purpose, except under special circumstances.

The actual choice of materials is narrowed down considerably and is based on solubility, availability, compatibility, cost and other factors discussed before. The final product is, thus, a compromise between the requirements and possibilities.

CARRIERS AND DILUENTS

The solvents which are used to dissolve the active ingredients and other auxiliaries are the major carriers and diluents in liquid formulations. These can be divided into two broad chemical classes, polar solvents such as water, alcohols etc. and the nonpolar solvents like paraffins, benzene etc. (Refer Table 2.27. for eluotropic series of solvents). The dissolution of substances in various solvents often follows the principle of 'like dissolves like' though there are many exceptions to the rule e.g. polystyrene, which otherwise should dissolve in nonpolar solvents, is insoluble in nonpolar solvents e.g. hexane or light petroleum ether. On the other hand, it is soluble in diethyl ketone, a polar solvent.

TABLE 2.27. TWO ELUOTROPIC SERIES OF SOLVENTS IN ORDER OF INCREASING ELUTIVE POWER

Series 1	Series 2
n-Pentane	Petroleum ether (b.p. 30-50°C)
Petroleum ether	Petroleum ether (b.p. 50-70°C)
n-Hexane	Petroleum ether (b.p. 70-100°C)
n-Heptane	Carbon tetrachloride
Cyclohexane	Cyclohexane
Carbon tetrachloride	Carbon disulfide
Trichloroethylene	Diethyl ether
Benzene	Acetone
Methylene dichloride	Benzene
Chloroform	Toluene
Diethyl ether	Esters of organic acids
Ethyl acetate	Ethylene dichloride, chloroform,
Pyridine	and methylene dichloride
Acetone	Alcohols
n-Propanol	Water
Ethanol	Organic acids
Methanol	Mixtures of acids with bases, water,
Water	alcohols or pyridine
*Strain N.H. (1942) Chromatographic adsorption analysis. Interscience Publishers Inc. New York	G. Wohleben in Handbuch der Lehensmittelchemie Vol. II, Part I, Springer Verlag, Berlin-Gollingen-Heidelberg

General Solvents and their Properties

Physical characters of some important solvents are listed in Table 2.28.

Ketones

The simplest ketone is acetone or propan-2-one. It is a colourless, volatile liquid, with a characteristic odour. It is completely miscible with water, alcohols, ethers, hydrocarbons and their halogen derivatives, other esters, ketones, fatty acids, vegetable and animal oils, and is an excellent solvent for a large number of substances and resins.

Methyl ethyl ketone (butan-2-one, MEK), another important ketone is an excellent solvent for organic materials and resins. It is a colourless, flammable and volatile liquid, somewhat more toxic than acetone. Isomeric pentanones, diethyl ketone (pentan-3-one) and methyl n-propyl ketone (pentan-2-one) are the other important ketones having similar properties.

TABLE 2.28. PHYSICAL CHARACTERISTICS OF SOME IMPORTANT SOLVENTS

Solvent	Boiling Point(°C)	Melting Point(°C)	Density (g ml^{-1})	Vapour pressure (mm Hg)	Solubility parameter (δ cal ml^{-1})
Acetone	56	− 64	0.791	231	9.62
Methylethyl ketone	80	− 87	0.806	90	9.48
Diethyl ketone	103	− 40	0.816	35	9.06
Methyl isobutyl ketone	115	− 80	0.802	16	8.88
Methyl n-propyl ketone	102	− 86	0.810	35	8.99
Methanol	64	− 98	0.796*	127	14.50
Ethanol	78	− 117	0.789	59	12.78
Isopropanol	82	− 89	0.785	44	11.44
n-Propanol	97	− 127	0.804	20	12.18
n-Butanol	118	− 90	0.810	6	11.60
Isobutanol	108	− 108	0.806*	9 (20°)	11.24
Ehtylene glycol	198	− 17	1.116	0.06	17.05
Diethylene glycol	246	− 10	1.118	0.00	14.24
Propylene glycol	187	−	1.040	0.07	14.99
Furfuryl alcohol	170	− 15	1.219	5.5(55°)	--
Benzene	80	5	0.879**	95	9.16
Tuluene	111	− 95	0.862**	28	8.93
m-Xylene	144	− 25	1.381	7	9.06
n-Pentane	36	−131	0.626	513	7.02
Cyclohexane	81	7	0.799	98	8.19
Diethyl ether	35	− 116	0.713	530	7.53
1,4-Dioxane	102	12	1.035	35	10.13
Di-isopropyl ether	69	− 60	0.726	148	7.06
Di-phenyloxide	259	28	1.073	0.01	10.10
Di-chlorodi-isopropyl ether	187	−	1.114	0.60	9.08
Ethylacetate	77	− 84	0.901	97	8.91
Amylacetate	148	− 71	0.879	--	8.50
Methylacetate	57	−98	0.927	217	9.46
n-Butylacetate	126	− 77	0.882	11	8.69
Di-ethyl carbonate	126	− 79	0.982*	10(20°)	8.80
Methylene chloride	40	− 97	1.336	437	9.88
Carbon tetrachloride	77	− 23	1.595	115	8.55
Chloroform	62	− 64	1.498*	194	9.16
1,1,1-Trichloroethane	74	− 31	1.325	100(20°)	8.64
1,1, 2-Trichloroethene	87	− 73	1.456	74	9.16
Perchloroethylene	121	− 22	1.631	18	9.28
Arcton 113	48	− 35	1.600*	335	7.25
Dimethysulfoxide	100d	− 6	1.100	0.6	12.00
Dimethylformamide	148	− 60	0.944	4	11.79
Acetonitrile	82	− 43	0.783	86	12.11
Furfural	162	− 37	1.160	--	11.80
Nitromethane	101	− 29	1.130	84	12.90
Pyridine	115	− 42	0.982	21	10.62
Diethylamine	56	− 50	0.711	241	8.04
Triethylamine	90	− 115	0.723*	68	7.12

Specific gravities: *15°, **25°, other values at 20°C.

Vapour pressure: All values at 25°C, unless otherwise mentioned.

       d = decomposition

Alcohols

Monohydric Alcohols: Alcohols of low molecular weight are liquids and those of the higher molecular weight, are solids. Methanol is the first member of the homologous series, followed by ethanol, propanol, butanol, etc. Methanol (b.p. 65°C) is colourless volatile liquid and is completely miscible with water and a number of organic solvents. It is an excellent solvent. The vapours of methanol are poisonous with a TLV of 200 ppm. It is flammable and the vapour can form an explosive mixture with air or oxygen. Ethanol, (b.p. 78°C) generally used as rectified spirit, is also a volatile, colourless, flammable liquid, with a TLV of 1000 ppm. Like methanol, it is also an excellent solvent for most pesticides.

Isopropanol, or propan-2-ol (b.p. 82°C) is another widely used solvent, derived from propylene. It is a clear, flammable, colourless liquid, miscible with water and most organic solvents such as esters, ketones, hydrocarbons, alcohols, vegetable oils such as castor oil, linseed oil and many essential oils. It is a solvent for natural gums, resins, oils, waxes, shellac and copal. It forms azeotropes with many solvents e.g. benzene, toluene, ethyl acetate, methyl ethyl ketone, carbon tetrachloride and water. It is sometimes used as a binary azeotrope with water. Water also improves its solvent power. Vapours of isopropanol are mild irritant of eyes, nose and throat.

Propyl alcohol or n-propanol, (b.p. 97°C), often finds use in various applications where ethanol or n-butanol are used. It is a good solvent for gums, resins, oils etc.

n-Butanol, b.p. 117°C, is only partly miscible with water but is completely miscible with other alcohols, ketones, esters and most other organic solvents. It is a colourless, mobile, flammable liquid with a mild odour. It is used as a solvent for many natural products including castor oil, linseed oil, gums and waxes and several other products. It is potentially more toxic with TLV 100 ppm, and above concentration of 50 ppm in air, causes discomfort to nose, throat and eyes. Isobutanol, (b.p. 108°C), butanol (b.p. 100°C) and tert-butanol b.p. 83°C too resemble n-butanol in most traits. Their narcotic effect is less severe than n-butanol, and hence are less hazardous to mammals.

Polyhydric Alcohols: The common polyhydric alcohols include ethylene glycol (1, 2 dihydroxyethane), diethylene glycol (2, 2-dihydroxydiethyl ether) and propylene glycol (1,2-dihydroxy propane) Ethylene glycol is a colourless, odourless, slightly viscous, hygroscopic liquid (b.p. 197°C). It is miscible in all proportions with water, glycerol, acetic acid, most alcohols, pyridine, furfural and acetone, but not appreciably soluble in benzene, toluene, xylene, chloroform, trichloroethylene, carbon tetrachloride and ether. No special precautions are necessary for handling ethylene glycol. Its lethal dose for adults being large, dangers can be easily prevented.

Diethylene glycol is a colourless, almost odourless, viscous and hygroscopic liquid (b.p. 244°C). It is miscible in all proportions with water, methanol, ethanol, ethylene glycol, acetic acid, acetone, furfural, pyridine, chloroform, nitrobenzene and aniline. Aromatic compounds generally have high solubility in diethylene glycol whereas saturated hydrocarbon oils are almost insoluble. Diethylene glycol should never be taken internally or used where it might come in contact with food. There is a little fire hazard from it.

Propylene glycol or 1,2-dihydroxypropane is somewhat viscous, colourless, almost odourless, non-flammable liquid with a sweet taste (b.p. 187°C). It is miscible with water and several organic solvents. Besides a solvent and adjuvant in perfumes and flavouring extracts, it is also used as antifreeze and mould growth inhibitor. Its toxicity is considerably less than ethylene glycol. It has no effect on skin but may cause irritation to eyes on contact.

Hydrocarbons

Fractional distillation of petroleum yields several hydrocarbon fractions of commercial importance. The term crude petroleum (mineral oil) usually encompasses the gases occurring naturally in the oilfields and liquid petrol from the wells along with dissolved solids, which can be separated by fractional distillation. It varies in composition with the locality of its occurrence, but all commercial products contain paraffins ($C_1 - C_{40}$), cycloparaffins and other aromatic hydrocarbons. Almost all low boiling petroleum fractions are composed of paraffins. The composition of the higher boiling fractions differs from source to source. Besides the hydrocarbons, compounds containing oxygen, nitrogen, sulphur and metallic constituents are also present in it.

The important components of crude from formulation point of view may be: light petrol (C_5H_{12}—C_7H_{16}), a solvent, b.p. 20-100°C; benzene (C_6H_6), solvent for dry cleaning, b.p. 70-90°C; ligroin (C_6—C_8), solvent, b.p. 80-120°C; petroleum (gasoline, C_6—C_{11}), motor fuel, b.p. 70-200°C; kerosene (paraffin oil, C_{12}—C_{16}), used for lighting, b.p. 200-300°C; gas oil (heavy oil, C_{13}—C_{18}) and fuel oil. Fractions having b.p. above 300°C, e.g. lubricating oil (mineral oil, C_{16}—C_{20}), are used as lubricants, greases; vaseline (C_{18}—C_{22}) and paraffin wax or hard wax (C_{20}—C_{30}) for candles, waxed paper etc. The residue is asphaltic bitumen (C_{30}—C_{40}), asphalt tar and petroleum coke.

Another major source of hydrocarbons is the natural material, coal. On destructive distillation it yields gases, tar, coke and ammoniacal gas liquor. The coal tar is a thick viscous liquid, black in colour, and with an obnoxious smell. Its fractional distillation yields a number of liquid products widely used in liquid formulations.

The composition of coal tar is governed by the method of carbonisation of coal. The tar produced by low temperature carbonisation is richer in aliphatic compounds. Carbonisation of coal in vacuum at 250-270°C yields a product containing upto 45% of unsaturated hydrocarbons and about 40% of the liquid paraffins. If the carbonisation is carried out at progressively higher temperatures, the proportions of these substances decrease with a corresponding increase in the amount of free hydrogen. It is due to the fact that aliphatic compounds are converted into aromatic substances and hydrogen is eliminated.

During distillation, tar is typically separated into five fractions, as follows:

Fraction	Name	Distillation Temp. (°C)
I	Light oil or crude naphtha	Upto 170
II	Middle oil or carbolic oil	170-230
III	Heavy oil or creosote oil	230-270
IV	Anthracene oil or green oil	270-360
V	Pitch	Residue

Light oil principally consists of hydrocarbons, benzene, toluene and xylene with small amounts of pseudocumene and mysitylene. In addition, there are other basic substances such as pyridine and acidic substances such as phenol.

Middle oil on cooling deposits crystals of naphthalene and from the residue, phenol (carbolic acid) is separated.

Heavy or creosote oil has a greenish fluorescence and contains phenol, cresols, naphthalene and anthracene. It is employed for preserving timber. Anthracene or green oil contains solid hydrocarbons such as anthracene and phenanthrene, both of which are separated and used for the preparation of dyes.

Petroleum ethers boiling between 40-80°C are the most volatile, consisting of low molecular weight aliphatic hydrocarbons like hexane. These are soluble in many organic solvents such as chloroform, benzene, alcohol, ether etc. These can be used as solvents for some oils, waxes and resins and pesticides. Of the other grades and ranges of hydrocarbon solvents, benzenes, ligroin, white spirit, naphthas need mention as these are extensively used in formulation industry. Their solvency power generally increases with an increase in the aromatic content of the solvent.

Benzene, C_6H_6, is the simplest of the aromatic type. It is used as a solvent for oils, ester, gum, polystyrene and several resins. Benzene is colourless liquid with a characteristic smell, immiscible with water and glycols. It has a TLV of 25 ppm and is highly flammable. Lately, its possible carcinogenicity has come in its way for a large scale application in various fields.

Toluene or methyl benzene, an aromatic solvent, is finding wide application in the lacquer and adhesive industry. The toxicity of toluene is somewhat lower than that of benzene (TLV 200 ppm).

Xylene, dimethyl benzene, has three isomers, all present in the commercial product. The *meta* isomer is in the highest amount. The boiling points are all within 6°C *i.e.* o-xylene, b.p. 144°C, m-xylene b.p. 139 °C and p-xylene b.p. 138 °C. Xylene is a solvent for oils, fats, waxes, pitches, gums and resins. Its TLV is tentatively assigned as 100 ppm.

Cyclohexane, b.p. 81°C, is a colourless, pleasant smelling liquid similar to benzene in solvency but is less toxic than benzene. It is miscible with other organic solvents and is a good solvent for paraffin wax, shellac, bitumen and rubber.

Terpentine oil obtained from pine trees, is also commonly used as a solvent. It is a mixture of cyclic terpene hydrocarbons, mainly α-pinene. It is a colourless, pleasant smelling liquid boiling at 160°C and above. The TLV of turpentine is 100 ppm.

Ethers

Aliphatic ethers are alkoxy alkanes. Their boiling points are in close proximity of the hydrocarbons which can be theoretically obtained by substituting a CH_2 group in place of the oxygen atom. For example, n-pentane boils at 36°C and diethyl ether at 35°C.

Ethers and polyethers are extremely good solvents for natural gums, resins, oils, waxes and fats. Ethers can form explosive peroxides in presence of oxygen during storage and, therefore, precautions should be observed.

Diethyl ether is a highly volatile, colourless liquid, widely used for extraction of natural products from plant materials. It is sparingly soluble in water, but miscible with ethanol in all proportions.

1.4-Dioxane, $(CH_2)_4(O_2)$, a cyclic ether is a colourless, stable liquid with pleasant odour. It is miscible with water as well as most organic solvents and can dissolve most oils, waxes and resins. It is used as a stabilizer in the chlorinated solvents. The TLV of dioxane is 100 ppm.

Diphenyl ether (diphenyl oxide or phenoxy benzene) is practically colourless, cystalline solid with geranium like odour. It is insoluble in water but dissolves in most organic solvents and is thermally stable up to 400°C. It is used as a component of heat transfer media.

Dichloro-di-isopropyl ether is a colourless, non-flammable liquid with a mild odour. It has low volatility and is used as a high boiling solvent for oils, fats, waxes, greases and many other organic compounds. It is a skin irritant and its vapours cause irritation in the eyes.

Esters

Esters or alkyl/aryl alkanoates are in general susceptible to hydrolysis, yielding parent alcohol and acid. One of the commonest esters is ethyl acetate, $CH_3COOC_2H_5$. It is a colourless, volatile, flammable liquid with pleasant fruity odour. Generally miscible with organic solvents such as other esters, ketones, halogenated hydrocarbons and alcohols, ethyl acetate is a good solvent for oils, fats, waxes, resins etc. Its TLV is 400 ppm.

Diethyl carbonate is miscible with most common organic solvents like petroleum ether, carbon tetrachloride, ether, alcohol, acetone, benzene. Addition of 5-10% alcohol can increase its solvency and improve range of solutes, which can dissolve in it.

Halogenated Hydrocarbons

The greatest advantage of halogenated hydrocarbons is their non-flammability. The chlorinated hydrocarbons are strong solvents for oils, fats, plastic, rubber and resins. Their low surface tension ensures good penetrating power.

Methylene dichloride, (dichloromethane), b.p. 40°C, is the most volatile chlorinated solvent used to dissolve a number of materials. It is a colourless liquid with a characteristic ethereal odour. It is a fairly stable chlorinated hydrocarbon, in which, but for stringent applications, the addition of stabilizers is necessary. Its TLV is 500 ppm.

Chloroform or trichloromethane (b.p. 60-61°C) is a heavy, colourless, volatile liquid with a rather pleasant odour. It is a powerful solvent for many substances. It is readily miscible with most organic liquids. Its TLV is 50 ppm.

Trichloroethylene, 1, 1, 2-trichloroethene, is a highly volatile, colourless liquid, which is widely used as solvent for oils, fats, greases, plastics, resins etc. It is miscible with a number of other organic solvents. It may decompose in presence of oxygen and UV light. Its TLV is 100 ppm. It forms explosive mixture with alkalies.

Perchloroethylene or 1,1,2,2-tetrachloroethene may sometimes replace trichloroethylene in applications where a more dense and less volatile solvent is required. It is miscible with most organic solvents and can dissolves waxes, oils, fats, resins etc. It is extensively used for dry cleaning. Its TLV is 100 ppm.

1,1,1-Trichloroethane, is a non-flammable solvent with TLV of 350 ppm. To check its instability in presence of moisture and light, cyclic ethers, nitro compounds, epoxides, nitrites, alcohols, and ketones are generally mixed with it.

Freons or chlorofluroethanes have also been extensively used for various applications. 1,1,2-Trichloro-1,2,2-trifluoroethane is most widely used as a solvent. It is a colourless, pleasant smelling liquid with a TLV of 1000 ppm.

The halogenated solvents are, a subject of considerable criticism because these are suspected to act as ozone depletors in the environment. These are being replaced by dimethyl ether as aerosol propellant. The liquified petroleum gas (LPG) and compressed carbon dioxide or nitrous oxide are the other likely propellants.

Solvent Mixtures

The solvent miscibility data given in Fig. 2.16 may be used as a broad guideline to know the compatibility of common solvents. However, precise miscibility has to be ascertained as per the requirement of the formulation of pesticides. It is difficult to predict precisely the properties of the solvent mixtures based on the properties of components and their proportion in a mixture. Generally, the mixture shows properties closer to those of the poorer solvent. If the two solvents are quite dissimilar in nature, then addition of a third material may cause actual separation into two liquid layers.

While choosing an "intermediate" solvent for keeping a third component such as a toxicant in solution in a poor solvent, the following considerations should be kept in mind.

(*i*) The intermediate solvent should have an extremely high solubility for the toxicant so that its minimum quantity dissolves in it.

(*ii*) It should be highly miscible with other solvent to prevent a separation of two liquid layers.

(*iii*) Normally, quantity of intermediary solvent required will be more than that predicted from its solvency alone.

General Observations *vis à vis* Commercial Pesticide Formulations

Many a modern pesticides consist of moderately polar molecules for which polar solvents are optimal.Ketones are usually much better solvents than hydrocarbons for such molecules. Out of these, acetone is completely miscible with water whereas methyl ethyl ketone and cyclohexanone have limited (~10%) solubility. Chloroform is an excellent solvent for many pesticides but only sparingly water soluble (0.5%). It is costly, and its density is very high. Thus, usually mixture of crude hydrocarbons from coal tar or petroleum sources etc. are employed for preparation of ECs. For example, methyl naphthalene (b.p. 230-250°C) is much used in formulation of organochlorine insecticides.

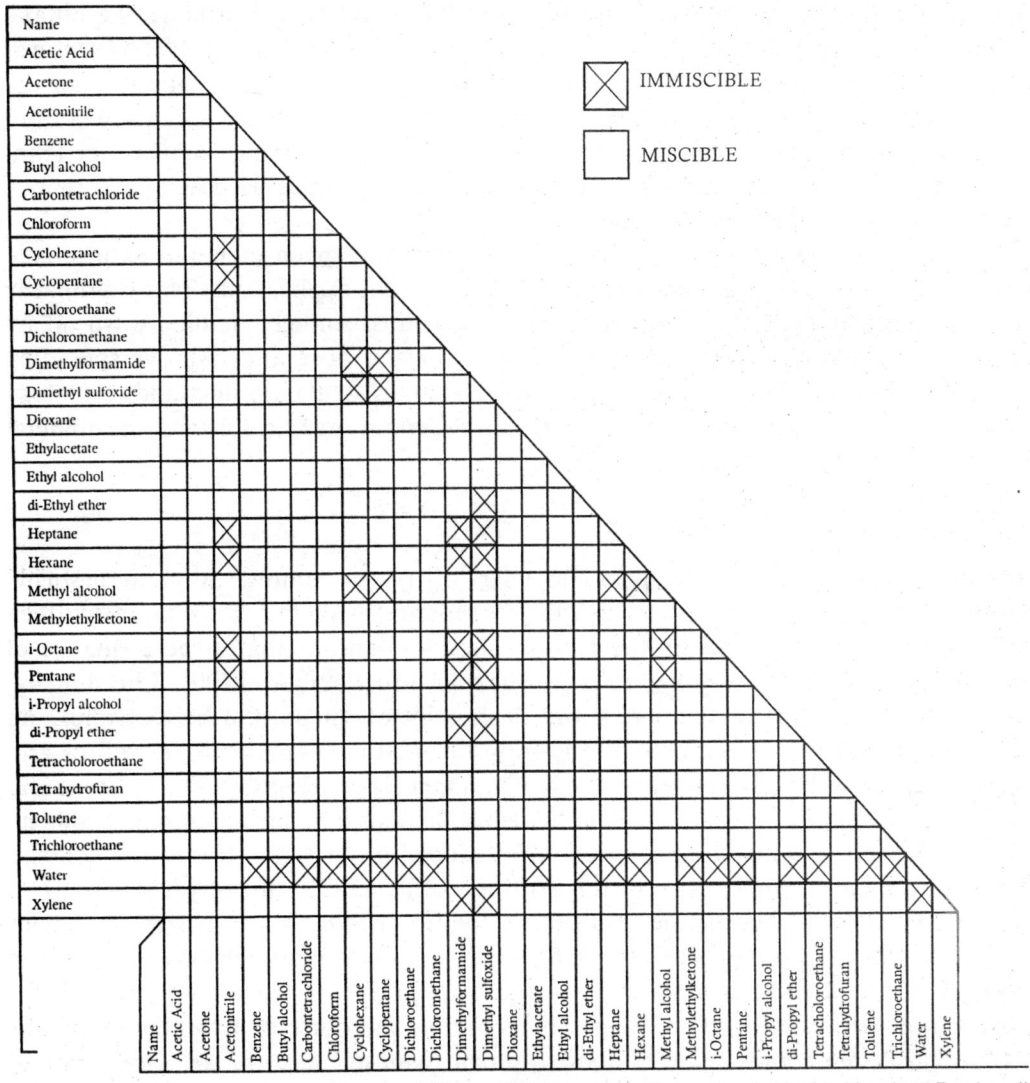

(The HPLC Applications Book Vol. I, Hewlett Packard)

Fig. 2.16: Solvent Miscibility Chart

The solubility of a polar solute may be increased by adding polar solvents, taking care that crystallization does not occur either during storage (shelf life) of the product or on its dilution in spray tank during application. Thus, it may be permissible to add a ketonic solvent to an aromatic solvent to keep the product in solution at low temperature e.g. –5°C, provided, of course, the product remains in solution at 10°C without the addition of the ketone.

Solvents such as fluorocarbons (ex. 1,1 and 1,2-dichlorotetrafluoroethanes, b.p. – 2°C and 4°C respectively, difluorodichloromethane, b.p. – 28°C) are most commonly used propellants for domestic dispensers. These possess slight odour and low toxicity and are non-flammable. These are better than the flammable hydrocarbons like butane which finds use in foam formulations. These in combination with a better solvent can be recommended for use in aerosol formulations for outdoor application. With the development of container technology and protective coatings, potentially more corrosive but cheaper solvents (e.g. methyl chloride) are becoming popular. For foam applications, CO_2 dissolved in water based liquid is sometimes used as an alternative to the more general butane emulsions.

Vegetable oils such as soybean and cotton seed have been explored as pesticide diluents to replace water in ultra low volume (ULV) foliar applications. This is primarily to avoid pesticide hydrolysis, protect chemicals against sunlight, reduce wash off by rain, reduce volatility, drift, adjuvant requirement and dose and improve spreading and penetration in insects and plants. However, stability of the product and toxicology will have to be worked out in oil combinations. Certain organic solvents and mineral oils also find use as carriers in ULV products.

Solvent(s)/Fractions

The physico-chemical and chemical characters of sixteen commercial grade pesticide formulation solvents/fractions used in India are reported in Table 2.29. Of these, fourteen were investigated for effect on the physico-chemical and biological performance of fenvalerate emulsifiable concentrates (Sharma and Parmar, 1990). The density, kinematic viscosity, refractive index and flash point of the solvent fractions bore no relation with their hydrocarbon composition (aromatics, naphthenes, paraffins and olefins) but were directly related to the boiling range of the fraction. The toxicant and surfactant solubilites were often high in solvents/fractions with higher aromatic content. The surfactant solubility was apparently influenced by its polyoxyethylene content. Except in case of light aluminium rolling oil (LARO), the physico-chemical characteristics of the fenvalerate emulsifiable concentrates in various solvents/fractions complied with the Indian Standard: 1997 (1987). Except the emulsions from ECs in aromax, C IX, iomax, aromax HAE and extract II, which were mildly phytotoxic against *Lagenaria siceraria*, with the plants recovering in course of time, the remaining 0.02% fenvalerate emulsions were non-phytotoxic against *Pisum sativum*, *Cicer arietinum* and *Lagenaria siceraria*. The concentrates prepared in certain fractions of aromax, CIX and extract I showed significantly superior bioactivity against *Tribolium castaneum* Herbst. The composition of different lots of the same solvent from the same source could influence the pesticidal performance.

MAJOR FORMULATIONS
Solution
Emulsifiable Concentrate (EC)
A liquid, homogeneous formulation to be applied as an emulsion after dilution in water   —GIFAP (1989)

TABLE 2.29. PHYSICO-CHEMICAL CHARACTERS OF SOME FORMULATION SOLVENTS USED IN INDIA

S. No.	Solvent	Boiling range (°C)	Density (g ml⁻¹ at 20°C)	Refractive index (20°C)	Kinematic viscosity (cs at 20°C)	Flash point (°C)	Hydrocarbon composition (w%)		
							Aromatics	Naphthenes	Paraffins
1.	Aromax	167-305	0.91	1.5228	1.90	70	62.5	8.6	28.9
2.	Naphtha	152-225	0.84	1.4733	0.91	46	52.5	10.5	37.0
3.	C-IX	158-282	0.88	1.5054	0.82	45	96.0	0.0	4.0
4.	Xylene	134-137							
5.	Cyclohexanone	152							
6.	C_1 Top, fraction of Heavy Kerosene	160-240	0.79	1.4413	0.86	44	17.1	0.0	82.9
7.	LARO (Light Aluminium Rolling Oil, fraction of Heavy Kerosene)	240-270	0.82	1.4632	2.39	115	71.3	11.1	17.6
8.	C_2 bottom fraction of Heavy Kerosene	270-320	0.83	1.4638	3.34	160	18.9	16.1	65.0
9.	8/19 Iomax HAE	181-362	0.90	1.5321	2.19	72	70.0	10.8	19.2
10.	8/20 Iomax	155-280	0.93	1.5088	1.22	50	68.8	11.5	19.7
11.	7/112, Extract-I	142-190	0.84	1.4793	0.79	37	65.3	10.5	24.2
12.	7/113, Extract-II	170-240	0.88	1.5035	1.18	61	62.8	8.5	28.7
13.	7/114, Composite	153-230	0.88	1.4955	1.01	48	64.4	12.7	22.9
14.	8/109, Heavy Aromatic Extract-I	175-250	0.88	1.5009	1.26	63	69.0	11.5	29.5
15.	8/110, Heavy Aromatic Extract-II	195-280	0.92	1.5227	2.32	82	61.0	7.6	31.4
16.	8/111, Heavy Aromatic Extract-III	151-240	0.82	1.4573	1.33	46	23.4	22.2	54.4

Sharma (1988), Sharma and Parmar (1990)

Emulsifiable concentrate is a solution of active ingredient and surfactant in water immiscible solvent(s), which on addition to water forms usually an oil in water emulsion. The emulsion may be formed with or without agitation. These concentrates may, at times, contain other auxiliaries such as stabilisers, stickers, defoamers, dyes etc. The concentration of active ingredient depends on the potency of the active ingredient and its attainable solubility in the solvent system. Emulsifiable concentrates, widely ranging in active ingredient strength (2.5 to over 50%) are commercially available.

At the time of dilution with water, if the ratio of oil to water is less than 1:10, a water in oil emulsion results. In it, the oil portion constitutes the continuous or external phase and water, the discontinuous or the internal phase. Such emulsions are termed 'invert emulsions' and the EC employed to obtain these, the invert EC. Since the oil constituting the external phase is of relatively low vapour pressure, the invert emulsions form significantly larger droplets and evaporation of the continuous phase is thus, minimized. Due to this, an almost constant size of the droplets is maintained from time emulsion emerges from application equipment till its impingement on the target. This also reduces the drift of particles in air. The invert ECs are prepared with soluble esters of herbicidal acids.

When the volume of water employed to dilute an EC is large, an oil in water emulsion results. In it, the continuous phase of emulsion is water and the discontinuous phase, the oil. The size of droplet is considerably smaller and reduces further during move-ment through air. Such emulsions are characterized by a low phytotoxicity as compared to invert emulsions. Most pesticidal ECs yield oil in water emulsions during use.

Ever increasing cost of the petrol based solvents coupled with toxicological and phyto-compatibility constraints, have led to a search for alternate products to replace ECs. Water based liquid formulations are being projected as potential viable options that have replaced some of the ECs in certain advanced countries of the world. However, it is still the most popular and extensively used formulation.

In india, several innovative studies on emulsifiable concentrates are reported (Parmar, 1985, Kumar and Parmar, 2000).

Design

Compatibility of active ingredient with solvent and surfactant is of paramount importance in choice of the ingredients of an EC. The most commonly used solvents are the xylene type, the heavy aromatic naphtha type, aliphatics of kerosene range. Quite often, a mixed solvent system is employed to overcome the solubility problem of the active ingredients. Polar solvents are frequently employed as co-solvents to improve the solubility. The choice of ingredients in such mixtures and their amounts are governed by a successful compliance of physico-chemical requirements of the resultant products. The solvency, physico-chemical and bio-compatibility, high flash point, low volatility, low cost, easy availability etc. are the important considerations in choice of a solvent. Usually, a blend of 2-5 anionic and non-ionic surfactants is used at a level of 2 to over 10 per cent. Calcium salt of alkylbenzene sulphonic acid is the anionic emulsifier while the non-ionics are alcohol ethoxylates, alkyl phenol ethoxylates, fatty acid ethoxylates, sorbitol ester ethoxylates, triglycerides, etc.

Specifications and Analysis

Emulsifiable concentrates are required to comply with specifications relating to the content of active ingredient, acidity or alkalinity, low temperature stability, emulsion stability, heat stability and the flash point. Depending upon strength of the concentrate, the content of active ingredient must be within range of the prescribed limits. For a nominal value of up to 10.0%, the permitted variation is +10 and – 5% of the nominal content; above 10 and up to 50%, it is ± 5% and above 50, + 5 and – 3 per cent. The acidity or alkalinity must also remain within the prescribed limits for a product, both before and after the heat stability test (54 ± 1°C, 14 days). Similarly, the cold test takes care of low temperature stability of the product. No turbidity or separation of solid and/or oily matter are desirable in either of the pre and post heat treated samples. The product must yield a stable emulsion on dilution with normal and/or the standard hard water. In either case, the creamed layer, oil layer and the sediment settling at bottom, should not exceed 2 ml in 100 ml at 30°C. The flash point of the product should be below 24.5°C. The key requirements of emulsifiable concentrates along with recommended test methods for their determination (ISI, 1982, FAO, 1999, CIPAC, 1970) are summarized in Table 2.30.

TABLE 2.30. KEY REQUIREMENTS OF EMULSIFIABLE CONCENTRATES AND THEIR TEST METHODS

Requirement	Test method FAO/CIPAC	ISI
Active ingredient	As per specification	
Acidity/alkalinity	MT 31	13.5
Low temp. stability	MT 39.3	13.14
Persistant foam	MT 47.2	—
Emulsion stability	MT 36.11, MT36.2or MT 173	13.3
Flash point	—	13.2
Heat stability	MT 46	13.4

FAO (1999), ISI (1982)

Preparation

Commercially, emulsifiable concentrates may be prepared by taking individual constituents of the recipe in a dissolving/mixing tank. Various proportionating devices, such as, metering pumps, balances etc. are used to avoid contact with workers. Normally, a stirrer at ambient temperature is good enough for mixing. However, need based heating may be required to dissolve and mix the ingredients. The product is tested for quality after thorough homogenization and filtered before packing. The mixing vessels and most parts of the equipment with which the material has to come in contact, are made of stainless steel. Similarly, flame proof pumps, motor and other electrical fittings are recommended.

Suspension and Emulsion

Aqueous/non-aqueous flowables

Dispersion Science in Pesticide Formulation

The incorporation of insoluble, powdered solids into liquid media is vital for manufacture of chemical coatings, printing inks, pigment dispersions etc. and likewise, the dispersion concentrates of pesticides. The principles employed in paints and inks in achieving complete wetting of powders and separation of particles of microsize, possessing dispersion stability to avoid settling and caking, are very much applicable to aqueous and non-aqueous flowable pesticide formulations.

Pesticide Dispersions: Now a days water based suspensions are most commonly used along with other flowables. Flowable formulations are predispersed systems which enable good dispersibility upon dilution, are dust free and easy to measure. In addition to the active ingredient, the surfactants in such suspensions are used to overcome handling and shelf life problems. Pesticidal suspensions of solids are more efficient if they can be produced with the a.i. in a finely divided state with no agglomerates and with sufficient dispersion stability to prevent settling and compaction during storage.

Suspension concentrate (SC) formulation of a pesticide is conventionally prepared by milling or comminution process using a wetting agent and/or a dispersant to effect the production of a colloidally stable suspension of the active ingredient in a continuous aqueous phase. In oil based flowables, a liquid pesticide or inert oil comprises the continuous phase and an insoluble solid pesticide comprises the dispersed phase. Slow release formulations employing micro-encapsulations also apply dispersion methods. Microencapsulations are produced by the formation of a polymer skin around the liquid pesticide which has been dispersed into fine droplets. The procedure involves dispersion of a liquid pesticide containing a dissolved monomer in an aqueous medium. Additives are added to various dispersions for performance of different functions. Surfactants may act as emulsifier in ECs and microemulsions and as wetting and dispersing agent in wettable powders, water dispersible granules (WG) and other flowables. Polymers also serve as protective colloids and dispersing agents in concentrated emulsions and dispersions. Pesticide can be physically dissolved, adsorbed or dispersed in a polymer matrix, which controls the release by diffusion or erosion. Additives like water soluble polymers can alter properties of dispersions and modify the release response.

Molecular Interactions in Colloidal Dispersions: The possible inter-particle forces in colloidal systems are:

(i) Those arising from surface potential and subsequent counter ion cloud i.e. electrostatic forces.

(ii) Those arising from electromagnetic fluctuations, i.e., van der Waals forces or dispersion forces/London forces.

(*iii*) 'Born' forces, usually of very short range and repulsive.

(*iv*) 'Steric' forces dependent on the geometry and conformation of molecules (frequently adsorbed) in an interfacial region adjacent to the particle.

(*v*) 'Solvation' forces arising from the removal, displacement or rearrangement of solvent molecules in an interfacial region.

(*vi*) 'Magnetic/induction' forces.

(*vii*) 'Exclusion' forces.

The interaction forces can be physically measured by balancing the internal force of interaction by an externally applied force (creating an equilibrium). Dispersion interactions such as London-van der Waals forces are negligibly small and hence are not the forces causing coagulation/instability in the system. Interactions between polymers are responsible for stability as well as flocculation. Frequent encounters between particles occur due to Brownian movement, gravity and convection. A dispersion is colloidally stable when its particles remain permanently free.

Stability of Dispersions: In general the balance of forces of interaction between colloidal particles determines, whether a colloidal dispersion remains stable or undergoes coagulation or flocculation. A colloidal system is conventionally considered to be stable when the rate of attainment of minimum free energy is so slow that no significant flocculation or coagulation occurs. However, the existence of surface charge on the particles alone cannot give stability to the dispersion.

Colloidal stability is also governed by nature of pesticide and dispersing agent used as well as the conditions such as particle size, electrolyte concentration, temperature, type of polymer, etc. encountered by the system.

Particle Size : Physical stability of a dispersion of particles depends on their size. Micro particles have strong cohesive forces which promote reagglomeration in 30-50 µm clumps in dry state. During milling, high shear wet grinding equipment can reduce particle size of the active ingredient to needed size range and also activate clays used in the formulation and stabilize suspensions. Excessive particle size reduction of the active ingredient will increase surface area and surfactant demand. Combination of shear and impact action will give desired particle size and activate suspension agents.

Electrolyte Concentration: At low electrolyte concentration, many colloidal particles remain in a dispersed state while at higher concentrations, they lump together and destabilize the system. Colloidal stability is governed by nature of the pesticide and electrolyte concentration. However, polyelectrolytes can also yield a deflocculated suspension. Addition of electrolytes to the medium for dispersion reduces the electrostatic repulsions by compression of double layer surrounding the drops, thus allowing closer approach before electrostatic repulsion becomes significant. Addition of a salt to dispersions causes enhancement of coalescing efficiency.

Viscosity: Excessive viscosity during milling may result in incomplete dispersion and poor stability. Very low viscosity causes wear and tear of the mill and unwanted particle size reductions. Flocculation produces a rapid increase in viscosity thus causing problems during preparation. Strong flocculation may cause severe thickening on storage and reduce the spontaneity of dispersion. A rapid increase in viscosity occurs above a critical volume fraction of the dispersed phase.

Temperature: Inter-particle interactions are strongly temperature dependent. Attraction increases with a reduction in temperature. During milling process, temperature has to be regulated to avoid damage by heat. High temperatures may also cause flocculation. When the dispersion is cooled, distinct phase separation occurs.

Polymers: Colloidal dispersions in presence of polymers exhibit charge stabilization/ steric stabilization. The molecules of the polymer can be very effective in inducing instability in the dispersed systems. The flocculation caused by polymers is due to the so-called depletion effect/volume restriction effect. Polymers can keep the dispersion sterically stabilised.

Aqueous concentrated suspension formulation of pesticide requires use of powerful dispersing agents which keep the particles in a deflocculated state. Several agents like ionic and non-ionic surfactants, non-ionic macromolecules and polyelectrolytes can be used to get a deflocculated suspension. The best agents for producing deflocculated suspensions are polymers of block and graft type (Refer to surfactants, Chapter 3). These not only function as binders but also as dispersants, which require adsorption of macromolecules on to powder particle surface to give steric stabilization. Non-ionic polymers can be added to the dispersion medium and stability imparted by free polymers in solution. This is called electrostatic stabilization which is in contrast to steric stabilization where adsorbed or attached polymer is responsible for repulsion between colloidal particles. Phase separation into a disordered dilute dispersion and an ordered concentrated dispersion occurs when the concentration of free polymer exceeds a certain limiting value.

Block polymers can be effective dispersants when the polymer is soluble and part of the molecule can interact with particle surface of the powder. The highly soluble outer segment can remain in solution while inner segments interact with surface of the particles. When the dispersed phase of pesticides undergoes sedimentation, water-soluble polymers can be added to redisperse them. The polymers function by inducing a controlled level of flocculation between the suspended particles of the system by a bridging mechanism. Hydroxy ethyl cellulose (HEC) is one such polymer. Methoxy polyethylene oxide-methacrylate, a copolymer of polymethyl methacrylate-methacrylic acid, is a graft polymer used as a dispersing agent in pesticidal suspensions.

The block and graft polymers are designed to have two main groups, one with strong affinity to particle surface and the other to provide steric stabilization. These are sufficiently solvated by the continuous medium and have sufficient length (molecular weight) to provide a steric barrier. The dispersion is colloidally stable up to an optimum level of polymer. Addition of a free polymer such as ethylene oxide to sterically stabilised dispersions induces weak flocculation above a certain critical concentration of free polymer which is dependent on the molecular weight of the chain.

Surface Charges: The dispersibility of a colloid is also a function of its charge. Attractive inter-molecular forces lead to a reduction in the free energy of the system which results in phase separation. Flocculation occurs only when there is an attractive part in the interaction potential (opposite charges) and the attraction must be strong enough to exceed the entropic energy of mixing. The magnitudes and signs of surface charges with two types of particles in dispersion medium will be important if they are oppositely charged. In all dispersions, attractive forces are similar while repulsive forces may be electrostatic or steric.

Rheology of Pesticidal Dispersions, Sedimentation and Redispersion: Pesticide dispersions/suspensions are thermodynamically unstable systems and will undergo spontaneous sedimentation of the dispersed phase under gravity forming dilatant sediments/clays. The sediment does not re-disperse readily to give a homogeneous dispersion and tends to thicken, increasing the viscosity. Stabilising agents, usually water soluble polymers like hydroxyethyl cellulose, are added to such suspensions to inhibit or slow down the sedimentation rate (flocculating clays such as bentonite can also be used). The particles which settle down under gravity are able to move past each other (as a result of the strong repulsive force between the particles) forming a closely packed sediment requiring a large number of revolutions for re-suspension. Addition of polymers induces weak flocculation after which it is easy to re-disperse the sediment. Otherwise, strong flocculation may cause severe thickening on storage and reduce spontaneity of dispersion.

The nature and level of gel forming agent required to prevent the formation of hard cakes/clays depends on the density difference between disperse phase and medium, volume fraction of the dispersed phase and interaction of the antisettling system with the pesticide particles. In general, addition of a second disperse phase to the continuous medium that is capable of forming a 3-dimensional 'gel' network in the continuous medium will be useful in controlling sedimentation.

Rheological Measurements: Structural suspensions exhibit visco-elasticity i.e. their response to strain is partly of a viscous and partly of an elastic nature. Measurement of viscosity and elasticity provides quantitative information on the extent of structure formation in the system. These visco-eleastic properties can be correlated with its long term stability. So, optimum conditions needed for the prevention of claying of suspensions are arrived at in the laboratory by rheological studies and measurements are taken at small and large deformations. Dispersing power of a dispersing agent can be evaluated from measurement of viscosity as a function of volume fraction of the settling particles. The properties of pesticidal dispersions such as viscosity, density etc., which determine their response to settling mechanical force give an idea of the deformation and flow of matter under stress.

Wet Flowable (Suspension Concentrate, SC)

A stable suspension of active ingredient(s) in a fluid, which may contain other dissolved active ingredients(s) intended for dilution with water before use   —GIFAP (1989)

A suspension concentrate is a physical system, containing finely divided solid particles in a liquid dispersing medium. The proportion of the solid usually ranges from 5 to 60 per cent. In the pesticidal suspension concentrates, the solid may be a single active ingredient or a mixture of several active ingredients with or without a carrier. The liquid is often water, perhaps with an addition of water-miscible solvents, although anhydrous organic solvents such as mineral or fatty oils may also be used. Furthermore, the liquid itself may be a true solution which contains, for example, one or several active ingredients or formulating aids, or it may be an emulsion.

Several wet flowable formulations have been developed during the last two to three decades. They are becoming much more important because of increasing solvent costs and environmental restrictions on pesticide auxiliary materials. As the flowable formulation is a liquid, it possesses the following characteristics which show its superiority as compared with the wettable powder.

(*a*) Easy to measure and convenient to handle.
(*b*) Less nuisance and hazards than those involved with handling of pesticide dusts.
(*c*) No nozzle blockage.
(*d*) Easier ULV application.
(*e*) No stains on the sprayed surface.
(*f*) Being an aqueous flowable, it does not contain organic solvents, so offering following additional advantages:

  (*i*) Cheaper auxiliary materials
  (*ii*) Less toxic and less irritant to mammals
  (*iii*) Non-inflammable and easy to pack, transport and store
  (*iv*) Less phytotoxic.

Suspension concentrates are also comparable with emulsifiable concentrates and emulsions. The viscosity of the spray is not very high, it is usually less than 50 centipoise. The worker can easily measure required volume of concentrate and mix it with other materials.

When poured in water they disperse spontaneously, matching the spontaneous emulsification of emulsifiable concentrates, thus ensuring at the outset an even distribution of the active ingredient in spray, which also avoids over or under dosing.

Suspension concentrates must display good suspension stability and active ingredient should not settle. The dispersed phase should be stable during storage. Sediment, if formed, should be easily dispersed on shaking to form a homogenous suspension.

Suspension and chemical stability are therefore, important properties, particularly because pesticides are often required to withstand lengthy storage at high and low temperatures without any loss of biological activity. The major breakdown processes in suspension concentrates are the irreversible flocculation, settling i.e.claying or caking, crystal growth (Ostwald ripening) and non-spontaneous dispersion on dilution.

Design

Active Ingredients for Wet Flowable: Nearly all active ingredients used in crop protection can be processed into dispersible powders. However, an active ingredient from which a suspension concentrate is to be made must possess a number of essential properties to satisfy the set requirements.

(*a*) **Melting Point:** In principle, all compounds that are crystalline at room temperature can be processed. However, as pesticides are required to possess storage stability at elevated temperatures, the melting point of an active ingredient to be processed to form a suspension concentrate, should normally be above 60°C.

(*b*) **Solubility of the Active Ingredient:** Solubility of the active ingredient in the dispersing medium is an extremely important requirement. The solubility at room temperature and at 40°C should not exceed 1 : 10,000 (100 ppm) if Ostwald ripening is to be avoided. If solubility is greater, the physical system will not display adequate stability.

(*c*) **Chemical Stability:** It is obvious that only compounds which exhibit sufficient chemical stability in the dispersing medium can be processed to a suspension concentrate. Some indication as to whether a compound will satisfy this requirement may be found in its structural formula or derived from certain of its structural features. Simple experiments can also be made to clarify this question. For example, if water is used as dispersing medium, the resistance of the compound to alkalies or acids can be determined.

The chemical stability of an active ingredient is often influenced by reaction between the active ingredient and inert materials. In solid products or liquid products with more than one phase (ex. aqueous suspension concentrate), such reaction can, under certain circumstances, lead to change in the physical properties of the formulated product. This needs to be kept in mind while choosing the ingredients.

Inspite of various advantages with flowable formulations, problems in the area of preparation and stabilization can arise. Extensive testing must, therefore, be carried out to ensure high formulation quality. The type of dispersing and milling equipments and chemical nature and concentration of wetting, dispersing and suspending agents must be chosen carefully to optimise biological and physical stability.

Wet grinding is the only suitable method for manufacture of a suspension concentrate formulation, and the mill which produces a fine and narrow particle-size spectrum, is the sand mill. For a proper milling effect, the grinding elements must be harder (sp. gr. > 1.5) than the product to be milled. Materials suitable for use as grinding elements largely depend upon the initial particle size and desired ultimate particle size of grinding stock (the product to be milled) as well as upon the type of mill employed. Usually, the size of the grinding elements ranges from 0.3 to 10 millimetres. The initial particle size of grinding stock should always be between 30 and 200 microns. If grinding stock is coarser, it may be pre-ground using ball or colloid mills.

If dispersing agents are needed to keep grinding stock dispersion in suspended form, their amount should preferably be expressed as a percentage of grinding stock. It normally ranges from 0.5 to 10 per cent. Practically all commonly used surfactants may be used as dispersing agents in suspension concentrates. These must strongly adsorb on particle surface to avoid displacement during particle collision and provide a strong barrier against close contact of the particles (the sphere of strong van der Waals attraction). This barrier may be created as a result of electrostatic double layer repulsion, steric interaction or a combination of both. Lignosulphonates being polyelectrolytes provide a powerful barrier against flocculation but are not able to be strongly adsorbed on the hydrophobic pesticide particles due to lack of hydrophobic groups in their molecular skeleton. Hydrophobic groups have been successfully introduced in some newer lignosulphonates for this purpose.

It depends upon grinding stock as to whether typical dispersing agents or typical emulsifiers alone are suitable or whether a combination of both is needed. Another point, that must definitely be examined, is that it might not be expedient, for reasons pertaining to stability or application techniques, to add thickening agents, substances that cause thixotropic effects, wetting agents, defoamers, buffer substances and others.

The main problem associated with suspension concentrates is that while they are relatively easy to produce as colloidally stable systems, they have an inherent tendency to sediment on storage to yield, what is refereed to as dilatant clay sediments. The rheological properties of these 'clays' mean that they are not readily re-dispersed by inversions since they tend to thicken, or the viscosity increases with the increasing shear rate. The sediment may pack to the close packed limit if sedimentation is not controlled.

It is a common practice to add 'structure agents' to the formulation in order to retard the sedimentation rate. Such agents should have sufficient (low shear) viscosity to prevent their individual settling. These are either water soluble polymers (thickening agents ex. hydroxy ethyl cellulose) or flocculated clays like sodium montmorillonite or bentonite. Water soluble polymers usually function by inducing a controlled level of flocculation between suspended particles of the system but have a tendency to increase the (high shear) viscosity, requiring strong agitation to diperse the particles on dilution with water. The rheological behaviour of sodium montmorillonite or oxides such as silica is sensitive to small variations in pH and/or electrolyte concentrations. There are at least two principal mechanisms proposed by which a water soluble polymer may induce flocculation of suspended particles, and hence retard sedimentation and reduce formation of dilatant clays. These are:

(*i*) The bridging mechanism by which an adsorbing polymer may simultaneously be attached to more than one particle leading to a network-like formation in the suspension.

(*ii*) The depletion mechanism, where a non-adsorbing polymer produces a depletion layer or region of negative adsorption around the particle, which may be due to attractive interactions between particles resulting from osmotic forces of the excluded polymer.

Structured suspension concentrates are generally pseudoplastics with little or no thixotropy.

The assessment and prediction of long term stability of suspension concentrates is apparently not an easy task, requiring careful rheological investigations at low and high deformations and not the latter alone, for which sufficient data is not available.

Assessment and Prediction of Physical Stability of Suspension Concentrates:

Buscall *et al.* (1982) gave following equation for predicting the physical stability of polystyrene suspension:

$$U/U_o = [1 - (\phi/p)]^{kp}$$

where, U = settling rate of concentrated model suspensions of polystyrene of volume fraction ϕ in a Newtonian liquid

U_o = settling rate of an infinitely dilute suspension (predicted from Stokes law)

p = volume fraction at which settling ceases to occur

k = constant (~ 5.4 for polystrene).

In a non-Newtonian medium e.g. ethyl hydroxy cellulose solution, the settling rate was found to depend on zero shear viscosity η_o. Under ideal conditions, settling may be governed by the zero shear viscosity. However, in practical systems more complex structures are formed, which show visco-elastic behaviour and the settling rate may be governed by more than one rheological parameters.

Flocculation of the particles may be monitored by a number of techniques, such as sediment volume measurements, turbidity, flocculation, electron paramagnetic resonance (EPR) studies of adsorbed polymer and ultra-velocity studies of suspensions.

Preparation

Grind time plays an important role in preparation of flowables and depends upon the type of mill used.

Grind time = kd/n

where, k = a constant which factors in the effect of the material being milled, type of media and mill

d = diameter of medium

n = speed of agitator, in revolutions per minute.

It is evident that time for grinding can be minimised by reducing the medium size, which essentially increases the number of molecules per unit volume and, therefore, number of impacts per unit time. Also, increasing mill speed has the effect of reducing time for grinding by increasing medium momentum as well as the number of impacts per unit time.

Another important factor when attempting to prepare flowables of specific sample sizes with small mills is medium charge. The milling equipment is selected to achieve precise and reproducible samples of a specific size. Both wet and dry mills are used.

An easy method of preparing a suspension concentrate is to stir dispersible powder in water. However, dispersible powders are made with the aid of mechanical mills or air mills which yield relatively coarse and broad particle size spectrum, leading to sediment that cannot be redispersed by shaking or stirring. The absence of interfacial force, which prevents close packing, is also responsible for such settling.

Specifications and Analysis

The parameters that best describe the performance characteristics of SC are pourability, water dispersibility, suspensibility, re-suspensibility, wet sieve and persistent foam. The particle size range and viscosity are excluded from the specifications for need of internationally accepted, simple methods of determination. The particle size range is described but is limited by a number of indirect quantifiable parameters such as wet sieve analysis, suspensibility, pourability and water dispersibility which are influenced by it. Since most SCs show non-Newtonian flow characters, viscosity is only one part of a much more complex rheology. Pourability and water dispersibility adequately describe the flow (rheological) properties (FAO, 1999). The key requirements and test methods are given in Table. 2.31.

TABELE 2.31. KEY REQUIREMENTS AND TEST METHODS FOR SUSPENSION CONCENTRATES

Parameter	CIPAC Test method
Mass per ml at 20°C	MT 3.3
Acidity/alkalinity	MT 31
Pourability	MT 148
Spontaniety of dispersion	MT 160
Suspensibility	MT 161
Wet sieve test	MT 59.3
Persistent foam	MT 47.2
Storage stability (0°C)	MT 39.3
(54 ± 2°C)	MT 46

Emulsion Concentrates (EW, Emulsion, Oil in Water)

A fluid heterogeneous formulation consisting of a solution of pesticide in an organic liquid dispersed as fine globules in a continuous water phase —GIFAP (1989)

Micro and macro emulsion formulations hold promise as replacements for the emulsifiable concentrate by reducing or displacing organic solvents. The abuse and various drawbacks such as scarcity, phytotoxicity, dermal absorption, foul smell, low flash point etc. of the solvents are reduced. These employ a solid or a liquid active ingredient with or without solvent, surfactant (emulsifier) and water. The solvents, whenever used, are added in very small amount and those preferred, include the ketones like cyclohexanone and acetophenone, alcohols like cyclohexanol; esters like cyclohexylacetate and diethyl phthalate, alkylated aromatics like xylene, methyl naphthalene etc. The emulsion concentrates are more economic in view of the ever rising solvent costs, reduced operator hazard and phytotoxicity and enhanced bio-activity due to finer size. Besides, these offer advantages in packaging, transportation etc.

The preparation of emulsion concentrates has been particularly facilitated by the new type of phosphorylated emulsifiers. The emulsions are both macro (particle size 500-1000 μm) and micro (particle size 3-10 μm).

The process of formation and breakdown (flocculation and coalescence) of emulsions can be thermodynamically analysed in terms of the energetics as shown in Fig. 2.17

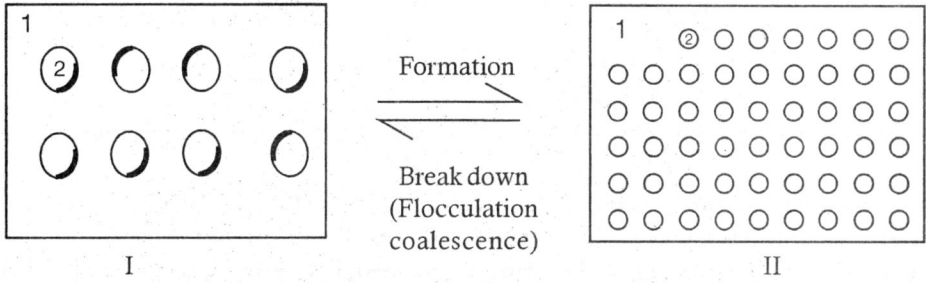

Fig. 2.17: Formation and Breakdown of Emulsion

The free energy change ΔG for the formation of emulsion concentrate consists of an interfacial energy $\Delta A\ \gamma_{1,2}$ due to an increase in the surface area, ΔA, as it goes from state I to II ($\gamma_{1,2}$ is interfacial tension) and an entropy term, $T\Delta S$ (config)., associated with increase in number of configuration of the large number of droplets (T is the absolute temperature). Thus

$$\Delta G\ (\text{form}) = \Delta A\ \gamma_{1,2} - T\Delta S\ (\text{config.})$$

where, $\gamma_{1,2}$ is interfacial tension for the particles in the final liquid (the emulsion).

In case of macromolecules, $\Delta A\gamma_{1,2}$ is much more than change in entropy $T\Delta S$ (config.) ($\gamma_{1,2}$ can be reduced to few milli Newton per meter (mNm^{-1}) by use of surfactants) and, therefore, energy must be expended in the non-spontaneous emulsification process. The addition of surfactants lowers the value of $\gamma_{1,2}$ and hence reduces the energy for emulsification. Since ΔG is positive, there is always a tendency for the emulsion to go from state II to I. This tendency can be reduced (or eliminated) within the time scale required by creating an energy barrier between the droplets which prevents their close approach and if such an approach is inevitable, as in concentrated emulsions, thinning and disruption of the thin film between the droplets is prevented. This is certainly essential to prevent droplet coalescence. This can be prevented if sum of change in repulsive force, π_R, as a result of the infinitesimal reduction in the film thickness δ and the change in surface energy π_r due to film expansion, exceeds the increase in attractive (van der Waals or dispersion) force π_A, as δ is reduced, i.e.

$$\frac{d\pi_R}{d\delta} + \frac{d\pi_r}{d\delta} > \frac{d\pi_A}{d\delta}$$

Film instability usually commences at specified spots in the lamellae due to local thinning in that region. The latter may result from spontaneous growth of surface fluctuations whereby the decrease of free energy due to van der Waals force exceeds the increase due to an increase in surface area. Alternatively, film thinning may be the consequence of fluctuations in the thickness of the whole film.

The forces that oppose thinning can be considered in terms of Deryaguin model where the interaction can be regarded as arising from an additional or disjoining pressure acting at right angles to the plane of the film; implying that the surface tension of a thin film between two (flattened) droplets would vary with its thickness. The disjoining pressure, π is the change of surface energy with distance. The total specific surface energy of a thin parallel lamellae of a liquid is given by,

$$\Delta G\, \text{surf.} = \gamma + \int_{o}^{\delta} \pi\, d\delta$$

where, γ = surface tension of liquid.

The equation shows that π is a function of thickness, δ, whereas γ is independent of thickness.

For non-contact between emulsion droplets, π should have a positive value. It can be negative as in cases where the attractive forces are dominant. Several forces are operate on liquid films, the relative magnitudes of which determine whether π is positive or negative. These forces may be 'long range' or 'short range' surface forces. The former comprise of van der Waals attractive and double layer repulsive forces and the latter are due to chemical, dipole interaction, hydrogen bonding, hydrophobic bonding etc.

Design

In case of liquid active ingredient, the technical material as such or its solution in minimum quantity of a solvent, constitute the oil phase. Solid active ingredients are dissolved in a solvent to form the oil phase.

No rule of thumb can be prescribed in choice of emulsifiers. Both ionic and the non-ionic as well their blends have been reported to work. The most appropriate one has to be arrived at through screening trials.

Rate of addition of one liquid phase to the other and blending speed need to be optimised by monitoring the physico-chemical parameters of the resultant emulsions. To ensure conformity of shelf life, viscosity of the final product too needs optimization. Quite often, a thorough mixing of the formulation through a vigorous shaking before use is recommended.

The emulsifier choice is also governed by the desired type of emulsion i.e. water in oil (EO) or oil in water (EW). Accordingly, the emulsifier is dissolved in appropriate liquid phase. For water/oil systems, water is added to oil-emulsifier mix and for oil/water systems, this mix is added to water. Even the addition of water-surfactant mix to the continuous oil phase and oil phase to the continuous water phase in water-surfactant mix can be tried.

Specifications and Analysis

Emulsions, like suspension concentrates, are metastable systems, requiring re-homogenization after a period of time. Emulsions being often non-Newtonian liquids, show a complex rheology which cannot easily be described by simple measurements of viscosity. Rheology can influence the dilution characteristics which are checked by the emulsion stability test. The key requirements and the test methods are summarized in Table. 2.32.

TABLE 2.32. REQUIREMENTS AND TEST METHODS OF EMULSION CONCENTRATES

Requirement	CIPAC test method
Mass per ml at 20°C	MT 3.3
Acidity or alkalinity or pH range	MT 31
pH range	MT 75
Pourability	MT 148
Emulsion stability and re-emulsification	MT 36.2, 36.1.1 or 173
Persistent foam	MT 47.2
Storage stability (0°C)	MT 39.3
(54 ± 2°C)	MT 46

Microemulsion (ME)

A clear to opalescent, oil and water containing liquid, to be applied directly or after dilution in water, when it may form a diluted microemulsion or a conventional emulsion
—GIFAP (1989)

A microemulsion is a thermodynamically stable (syn. soluble concentrate), isotropically clear dispersion of two immiscible liquids, consisting of microdomains of one or both liquids stabilized by an interfacial film of surface active molecules. One or more active ingredients may be present in either the aqueous phase or, the non-aqueous phase or in both phases. A variety of microemulsion formulations may be prepared in which the aqueous phase can be considered the dispersed phase or the continuous phase or alternatively, where the two phases are considered to be bicontinuous. In all cases, the microemulsion will disperse into water to form either conventional emulsions or dilute microemulsions. It may be noted that micro-emulsions are often stable only within a given temperature range. Their storage in that range is, thus, essential.

In manipulation of free energy change of emulsion formation (refer emulsion concentrate), if $\gamma_{1,2}$ is made very small, $T\Delta S$ (configuration) may balance or exceed $\gamma_{1,2}$ and ΔG (formation) becomes zero or negetive, allowing spontaneous emulsification. Clearly, if $\gamma_{1,2}$ is zero or negative, ΔG (formation) will be negative and the interface will spontaneously expand, adsorbing the surfactant molecules until their activities are lowered to such volume that $\gamma = 0$. Thus, one feature of microemulsion formation is zero interfacial tension between the two phases or rather a negative value of $\gamma_{1,2}$, at the original composition of the two phases. Clearly, such zero or negative interfacial tensions need not be maintained at equilibrium and in the latter case, small positive values are usually reached. Such low interfacial tensions are rarely achieved using one surfactant only. The γ reaches a limiting value at critical micelle concentration (CMC), at which the interface is saturated with surfactant molecules. Above this concentration, micellization starts and the activity of surfactants remains nearly constant. Two or more surfactants may have cumulative effect if they do not form mixed micelles e.g. water soluble (potassium oleate or sodium dodecylbenzene-sulphonate) and oil soluble (hexamol or pentamol). Their combined effect may be large enough to reduce the interfacial tension to zero or at finite concentration, below zero. The oil soluble surfactant is referred to as co-surfactant.

The main contributories in free energy of formation of microemulsions are: (*i*) the interfacial tension, can split into two terms, δ uncharged (the interfacial tension obtained if no electrical double layer was formed) and γ charged (related to the surface potential and surface charge density); (*ii*) free energy of mixing of droplets into the continuous medium, in osmotic term ΔG osmotic, which is determined by interaction between the droplets and their volume fraction.

Generally, ΔG osmotic is negative and its magnitude increases with the decrease in ϕ (values as low as -1.0 m Nm^{-1} may be reached at very low ϕ values of the order of 10^{-5}). On the other hand, change in value of γ is large and positive even for modestly charged droplets. This requires γ uncharged to have a high negative value in order for $\Delta G = 0$. So, theoretically, a potentially negative but ultimately positive small interfacial tension is required for the thermodynamic stability of microemulsions.

Design

A specific interaction (yet incompletely understood) amongst the molecules of oil, emulsifiers and water appears to yield a microemulsion. Without it, the work input e.g. by high shear in mixing or ultrasonics, or increasing emulsifier content, will not produce the desired product. The water/oil (w/o) systems are yielded by adding water to an oil-emulsifier mix and the oil/water (o/w) emulsions by adding this mix to water with mild stirring. The former are relatively easier to formulate. Sometimes, the water soluble surfactants, in water phase and the oil soluble surfactants in oil phase are used. For microemulsions, a high level of surfactant 10-30% of oil weight is required.

There are no guidelines for selection of a surfactant, but it is a crucial ingredient. Some idea may be had from the hydrophilic-lipophilic balance (HLB) values e.g. a low HLB (4-7) for w/o and a higher HLB for o/w microemulsions. Of the two, one surfactant has to be water soluble or dispersible (ionic or high HLB non-ionic) and the other oil soluble (alcohol or low HLB non- ionic). The temperature of phase inversion (PIT, i.e. the temperature at which an o/w emulsion inverts to a w/o emulsion) also provides information concerning the type of oil, phase volume relationships and concentration of emulsifier. It provides information about the requisite chemical type of emulsifier. It is based on the observation that HLB of non-ionic surfactant changes with temperature and that the inversion of the emulsion type takes place when the hydrophilic and lipophilic tendencies of the emulsifier just balance. No emulsion is formed at the PIT. Thus, the emulsion stabilized with non-ionic surfactants are o/w at low temperature and change to w/o at elevated temperature.

The cohesive energy ratio (CER) is expressed by an equation based on thermodynamic parameters to match an emulsifier for a given oil. The CER combines the ratio of dispersing tendencies (R) with the London cohesive energies (solubility parameter) and hydrogen bonding cohesive energies at the water side of the interface. Thus, the CER equals the ratio of head volume to tail volume multiplied by the square of the ratio of the solubility parameter.

Co-surfactant partitioning $(\gamma_o/w)a$, where, a is alcohol, is another parameter employed. It is known that one of the conditions for spontaneous formation and stability of dispersion of small droplets is zero interfacial tension. This is brought about as a result of the dimensional surface pressure in the monolayer of adsorbed species, i.e.,

$$\gamma_i = \gamma_o/w - \pi$$

where, γ_o/w = interfacial tension before adding emulsifier.

Since γ_o/w is of the order of 30-50 mNm^{-1}, π should reach the same value for γ_i to reach zero. Such high π values are achieved on penetration of molecules from the oil phase into the interface. Such high π values will eventually lead to the ejection of hydrocarbon molecules from the interface. It has, therefore, been suggested that the initial negative value of γ_i was the result of not a high value of π but a large depression of γ_o/w as a result of the partition of the cosurfactant between the oil phase and the interface, so that its interfacial tension against water is now reduced to $(\gamma_o/w)_a$. The latter approach ~15 mNm^{-1} irrespective of the original value of γ_o/w . This makes it possible to retain at least certain molecules of oil at the interface, so that π exceeds $(\gamma_o/w)_a$. The lower the percent of cosurfactant needed to depress γ_o/w to say 15 mNm^{-1}, the better the candidate.

Preventing Coalescence

Several methods can be used to prevent coalscence.

Mixed Surfactant Films: These account for enhanced emulsion stability due to the following reasons:

(*i*) Increased dilational elasticity, which increases resistance to lateral displacement of the film as the droplets collide.

(*ii*) Increased dilational viscosity.

(*iii*) Hinderance to diffusion of the surfactant molecules from the condensed film.

(*iv*) Formation of liquid crystalline phases at the o/w interface.

Use of Macromolecules: Gums and proteins stabilise emulsions against coalescence. The stabilization is assumed to be owing to formation of 'thick' and/or 'tough skins' at the oil/aqueous solution interface. Recent use of synthetic macromolecules, explains the enhanced stability against flocculation and coalescence in terms of the theories of steric stabilisation. Moreover, the macromolecular film formed at the oil/water interface can, in view of its visco-elastic properties, in certain circumstances, provide a mechanical barrier to coalescence. This also explains that the adsorbed protein molecules which are denatured on adsorption, collapse at the interfaces, forming a fairly rigid film. Careful studies have shown some correlation between the visco-elasticity and creep stress relaxation curves of these films and stability against coalescence.

Surfactant-polymer Mixtures: These form a very useful combination; the surfactant due to its rapid diffusion to the interface because of low molecular weight and its effect on lowering the interfacial tension, is more suitable for emulsification and the adsorbed polymer provides a more effective steric and/or mechanical barrier against coalescence. An example of such a mixture is sodium dodecylbenzene sulphonate or cetyl trimethyl ammonium bromide and polyvinyl alcohol. It appears that the polymer and the surfactant complex at the interface, in view of its charge, provides a more powerful barrier against coalescence as compared with the polymer alone.

Stabilization by Solid Particles: Very effective stabilization against coalescence can be attained by using very finely divided solids as emulsifying agents. Such emulsions are sometimes referred to as 'Pickering' emulsions. The type of emulsion produced depends on which phase preferentially wets the solid particles. Under thermodynamic equilibrium conditions, the adsorption of solid particles at the liquid/liquid interface is governed by their contact angles against the solid, an obtuse angle against the solid phase facilitating the stabilization. The presence of solid particles at the liquid-liquid interface plays an important role in preventing the thinning of the liquid film between the droplets. For the solid particles to be effective, they should form a continuous monoparticulate film. Contact angle hysteresis is probably important in preventing displacement of the meniscus. For that purpose, rough asymmetric particles such as bentonite clays are more effective than smooth spherical particles. The particles should also form a coherent film at the interface. This is achieved by capillary forces which tend to bring the particles close together at the interface. The smaller the radius of curvature of the meniscus between them, the larger is the force of attraction. This explains the need for very fine solids.

Specifications and Analysis

In view of their change into emulsion or dilute microemulsion on dilution with water, these are treated like emulsifiable concentrates with some additional modifications to take account of potential use problems relating to storage and use at low temperature FAO(1999). The key requirements are given in Table. 2.33.

TABLE 2.33. REQUIREMENTS AND TEST METHODS FOR MICROEMULSIONS

Requirement	CIPAC test method
Mass per ml at 20°C	MT 3.1
Acidity or alkalinity or	MT 31
pH range	MT 75
Persistent foam	MT 47.2
Emulsion stability and re-emusification	MT 36.1, 36.2.2, 173
Storage stability	
0°C	MT 39.3
54 ± 2°C	MT 46

Preparation

The formulation of microemulsions as that of conventional macroemulsions, is an art. Inspite of precise theories, which explain the physics and chemistry of their formation, the current advancement in the science of microemulsions does not enable a prediction with accuracy of the product to be obtained with mixtures of ingredients. The usual surfactant selection norms based on hydrophile-lipophile balance, the phase inversion temperature and the cohesive energy ratio system, apply. Once the recipe is finalized, the preparation is a routine blending of the ingredients.

Aerosol (AE)

A self contained sprayable product —CSMA(1955)

An aerosol is stated to be a "a suspension, of solid or liquid particles with a diameter less than 50 µm, in air or gas". The preferred particle size range has been stated differently in different publications including less than 30 µm, 12-20 µm or even less than 4 µm. American Mosquito Control Association has described particles in the size range of 0.1-50 µm as aerosols and fogs. Such fine particle size is obtained in practice by employing a small orifice, the thin issuing stream, low viscosity liquid and its actual boiling on entering the air.

The Chemical Specialities Manufacturing Association (CSMA) in 1955 defined an aerosol as a self-contained sprayable product in which the propellant force is supplied by a liquefied gas. This includes products used for space application, residual surface coating, foams generation etc. and excludes others like the whipping cream. The spray of aerosols (mist formation) constitutes their most important character. The mist is formed when an aerosol is discharged and the propellant changes from a liquid to a gas at the atmospheric pressure. Some flash evaporation occurs in the expansion chamber of the valve, which is at lower pressure than the container. Additional flash evaporation occurs when the vapour-liquid phase exits the outer orifice. This is followed by evaporation in the environment. The consecutive evaporations govern the final spray pattern.

An aerosol is just not any specific product but the whole package comprising of container with a valve, a liquefied gas propellant, solvent, active ingredient and other auxiliaries packaged under pressure. The package is safe and convenient for the users, has a long shelf life (5-10 years) and dispenses the product in a desired fine or coarse form. These employ homogeneous (consisting of single liquid phase and a vapour phase) as well as heterogeneous (three phased, two liquids or a liquid and a solid phase in combination with a vapour phase) systems. The latter may be emulsion or suspension. However, these suffer from their high cost, invisibility of contents and the use of an actuator function. Flammability of the hydrocarbon propelled systems and ozone depletion by the fluorocarbon propelled systems are the other drawbacks.

Design

Aerosol systems comprise of the aerosol intermediates, the propellant and the valves and actuator. For designing aerosols, all these components have to be clearly understood.

Aerosol Intermediates: These include the active ingredient(s), solvent(s), water, emulsifier(s), inhibitor(s) and fragrance(s). The term active ingredient implies the spectrum of products as per the FAO definition of 'pesticide' (**FAO** 1999) including the conventional insecticides, synergists, insect growth regulators, attractants, repellents etc. A classification of active ingredients by function as well as by aerosol product application is reported by Flashinski (1998, Table 2.34).

A wide array of hydrocarbons, alcohols and chlorinated solvents find use in aerosols. The solvency, flash point, toxicity, rate of evaporation, distillation range, specific gravity/density, miscibility with other solvents, cost etc. are the key considerations in selecting a solvent. Water is a good solvent and has several advantages such as non-inflammabilty, ready availability, low cost, low toxicity etc. but its use is contained by its immiscibility with liquefied gaseous propellants. Use of alcohols as co-solvents and dimethyl ether propellant system can overcome these problems. With the development of container technology and protective coatings, corrosive solvents such as methyl chloride are also finding extensive application in pesticidal aerosols.

The emulsifiers are used in water based aerosol system to produce stable water (foam) and/or oil out (spray) emulsions. The inhibitors help to prevent corrosion of the metal containers used for aqueous based formulation.

TABLE 2.34. CLASSIFICATION OF INSECTICIDAL AEROSOL FORMULATIONS

Product type	Functions of active ingredients					
	knock down agent	Synergist	Kill agent	Residual kill agent	Insect growth regulator	Repellent
Space spray aerosol	Pyrethrin	PBO	Sumithrin	—	—	—
	Tetramethrin	—	Resmethrin	—	—	—
	Pynamin forte	—	Permethrin	—	—	—
	Bioallethrin	—	DDVP	—	—	—
	Neopynamin forte	—	—	—	—	—

Surface aerosol	Pyrethrin	PBO	—	Permethrin	Hydroprene	MGK
	Tetramethrin	MGK 264	—	Cypermethrin	Fenoxycarb	—
	Neopynamin forte	—	—	Deltamethrin	Dimilin	
	DDVP	—	—	Cyfluthrin	—	—
		—	—	Propoxur	—	—
		—	—	Dursban	—	—
		—	—	Malathion	—	—
		—	—	Sumithion	—	—
Multi-purpose aerosol	Pyrethrin		DDVP	Propoxur	—	—
	Tetramethrin	—	—	Cypermethrin	—	—
	Neopynamin forte	—	—	Permethrin	—	—
	Pynamin forte	—	—	—	—	—
	Bioallethrin	—	—	—	—	—
	Esbiothrin	—	—	—	—	—
	Prallethrin	—	—	—	—	—
Aerosol fogger	Pyrethrin	—	DDVP	Permethrin	Hydroprene	—
	Tetramethrin	—	—	Cyphenothrin	Fenoxycarb	—
Aerosol repellent	—	—	—	—	Deet	—
	—	—	—	—	—	—
Aerosol mothproofer	—	—	Permethrin	—	—	—
	—	—	Vaporthrin	—	—	—

Flashinski (1998).

Aerosol Propellants: The propellant expels the contents of an aerosol container. Organo fluorine compounds (flurocarbons) have been used most widely ex. Freons [flurocarbons such as difluorodichloromethane bp. −28°C], and [1,1 and 1,2-dichlorotetrafluoroethanes (b.p. − 2°C and 4°C)]. They have been the common propellants for domestic dispensers. These have the advantages in being almost odourless, non-toxic and non-inflammable. These are also better solvents than the flammable butanes which are used in foam formulations, and hence are widely used along with the additional solvents, in aerosol formulations for outdoor application. The use of fluorocarbons is now being discouraged because of their role in the depletion of the ozone layer in the upper atmosphere. Currently, hydrocarbon propellants such as propane, isobutane, n-butane, dimethyl ether etc. either alone or as suitable blends, are coming to the fore. Compressed gases (high pressure propellant) such as carbon dioxide, nitrous oxide, nitrogen etc. also find use in aerosols. Solubility of compressed gas in liquid concentrates is an important consideration since the gas alone is not preferred in liquid form because it generates more pressure and requires very thick containers. Carbon dioxide scores on this point and is widely used in insecticidal aerosols. The compressed gases offer advantages of being colourless, odourless, non-toxic, non-inflammable, inexpensive etc. and are environmentally benign. Moreover, the temperature variations do not result in significant pressure changes as compared to the liquefied gaseous propellants. The major disadvantage of carbon dioxide as propellant is in production of an inferior spray (more wet and less forceful) which becomes coarser with fall in pressure. It also results in formation of carbonic acid, which may lead to corrosion. Filling of a container is also slower with carbon dioxide than with other propellants.

Valves and Actuators: Valve controls the product flow from dispenser and governs the quality of the dispensed spray or foam. Mounting cups and corresponding gaskets are used to clamp the valve stem, stem gasket, spring and body together and to provide a seal to the opening of the container. The stem gasket provides the seal to prevent leakage of product and weight loss. The gasket needs to be compatible with the product and also provide proper swell to close the opening in the stem when the valve is not in operation. The valve stem may contain the metering orifices to measure the spray rate and control the particle size, if so desired. The upward and downward movement of the stem is checked by springs, made of stainless steel. The valve body provides a base for the spring and also makes a connection with the dip tube. If the body orifice is less than that of the stem, the valve body orifice becomes the metering orifice. The valve body can also contain a vapour trap orifice which permits additional vapour phase propellant to mix with the produce to provide a finer particle size. Dip tubes transfer the product to valve body and also serve as a metering device in certain systems. These also serve as product reservoir for spraying of products such as insect repellent sprays etc. Valve actuators may be simple spray buttons or an integral part of various spouts or spray domes. These control the operation of valve and determine the nature of the product (foam, spray, stream etc.). For uniform particle size of the spray, the actuator orifice should be larger than the metering orifice.

Some examples

The compositions of some of the earlier aerosol insecticide formulations are exemplified below:

	Composition	Parts by weight
Formula 1	Pyrethrum (20 per cent solution)	2
	Cyclohexanone	5
	DDT	3
	Lubricating oil	5
	Freon-12	85
Formula 2		
	DDT	5
	Cyclohexanone	5
	Dichlorodifluoromethane	90
Formula 3		
	Pyrethrum Extract (20 per cent)	2
	DDT	3
	Petroleum solvent	12
	Freon-12	83

Preparation

Vapour pressure of the propellent and concentration of the active ingredient in the product are most important variables that affect a spray. An increase in concentration and vapour pressure of the propellant decreases particle size. An optimum for bioactivity may be 60% propellant. The vapour pressure of total formulation is important since even a high vapour pressure propellant may produce a coarser spray than a low vapour pressure propellant, if a pressure depressing solvent is included in the formulation. The valve parameters are also significant, particularly while using a vapour tap valve in combination with a low vapour pressure propellant. It results in a decrease in vapour pressure and discharge rate, resulting in a change in spray during the discharge. Proper construction of the actuator and its orifice is important to have the desired type of spray. A typical aerosol valve is made of a series of channels, orifices and expansion chambers, all of which influence delivery rate. The prime metering points available for manipulation of the discharge rate, in order of flow, are the dip tube, valve body orifice, vapour tap orifice, stem orifice and actuator orifice. The other key parameters to be kept in mind are the vapour pressure, density and fill weight. Vapour pressure can be manipulated through the application of Rault's and Dalton's laws. Density determinations are simple and from these values, the container fill weight can be ascertained.

OTHER LIQUID FORMULATIONS

Soluble Concentrate (SL)

A clear to opalescent liquid to be applied as a solution of the active ingredient after diluting in water. The liquid may contain water insoluble formulants —GIFAP(1989).

These comprise of solution of the active ingredient in suitable solvent(s), alongwith other necessary formulants. The formulation is a clear or opalescent liquid, free from visible suspended matter and sediment, to be applied as a true solution of the active ingredient in water. Wherever solubility consideration permits, aqueous concentrates/ solutions (AS) are prepared.

AS are water solutions of pesticides which are diluted with water before use. Water is used as solvent but sometimes it is necessary to add a water miscible solvent in order to reach an economically acceptable concentration of the active compound. For this purpose, alcohols, such as butyl cellosolve (butoxyethanol) or diethylene glycol ($\beta\beta'$-dihydroxyethyl ether) are preferred to methanol or acetone as these have higher flash-points. It is important, where possible, that the concentrate should not incur the higher freight charges imposed on flammable products.

The number of pesticides that can be formulated in this simple way is limited by solubility and/or hydrolytic stability. With the exception of herbicide amitrol (and the insecticides octamethyl phosphoramide and dimefox, which now have only very limited use), the solubility factor limits aqueous formulations to salts of active acids or, in the case of diquat and paraquat, of active bases. The active ingredients from acids include 2,4-D and 2,4,5-T as their dimethylamine salts, MCPA and some hormone herbicides. The inactive ion of the salt may have to be chosen for solubility as well as the cost. Dimethylamine salts give adequate cold stability and good field performance. At times, salts of sodium, potassium or lithium are also used.

When the water used for dilution contains traces of magnesium, calcium or iron, insoluble precipitate may be formed. Addition of sequestering agents such as sodium polyphosphate, ethylenediamine tetracetic acid and its derivatives; citric acid salts and some lignosulphonates, to the spray tank can overcome the problem.

The aqueous concentrates of the salts of herbicidal acids are chemically stable. However, for new products, the stability should be verified. The packaging of such concentrates in metallic containers may cause corrosion. Pigmented, high baked phenolic linings and epoxy linings in the pails and drums may help to reduce the problem. High-density polyethylene bottles are particularly useful in small packing.

A peculiar problem with the soluble/aqueous concentrates is the precipitation of the tarry matter during dilution. The excessive precipitation of the tarry matter may lead to filter or nozzle blockage, therefore filterability test after dilution should always be carried out in the formulation laboratory. Such concentrates may or may not contain surfactants. Though concentrated solvent water solutions are easily diluted with water, it does not mean that stirring is not at all needed. Stirring is a must for proper spraying of the diluted formulation.

Besides the active ingredient content, the acidity or alkalinity, persistent foam and storage stability at low and elevated temperature are the key requirements prescribed for this formulation (FAO, 1999).

Oil Miscible Liquid (OL)/Oil Concentrate

A liquid, homogeneous formulation to be applied as a homogeneous liquid after dilution in an organic liquid   —GIFAP (1989)

Oil concentrate or oil solution concentrate or oil miscible liquid is a homogeneous formulation to be applied as a homogeneous liquid after dilution in an organic liquid. The solutions in an organic solvent contain about 25 to 75% of the toxicant. These are usually diluted with a suitable cheap solvent to the desired concentration before application. The solvents to be employed for making these concentrates should not only sustain high concentrations of the active material but also be easily miscible with cheap diluting organic solvents. Besides, the active ingredient content, flash point, acidity or alkalinity and heat and cold stability are the important requirements (FAO, 1999). No solid precipitation or layer separation should be observed, on dilution in another solvent.

Ready to use oil solutions at concentrations below five per cent find extensive use as sprays against household pests. These are normally prepared in aliphatic hydrocarbon solvents such as mineral oil, white mineral oil, deodorized mineral oil, kerosene, etc. For household applications, the solvents to be used, besides meeting above stated requirements, must be colourless and odourless.

Oil concentrates are usually prepared in xylene, heavy aromatic naphtha and similar other aromatic hydrocarbons. When solubility is a problem, more powerful solvents such as isopropanol or cyclohexanone may be added as co-solvents. These however, must be miscible with the primary solvent. For dilution, inexpensive hydrocarbon solvents like fuel oil, diesel oil etc. are usually employed.

Ultra Low Volume (ULV) Liquid (UL)

A homogeneous liquid ready for use through ULV equipment   —GIFAP (1989)

Ultra low oil concentrates are characterized by their volume of spray, usually 1-4 litre ha^{-1}. Even aqueous concentrates may be used in this form. A liquid of moderate viscosity is recommended to obtain the optimum 100 μm droplets which check the effect of drift and provide the best coverage. The volatility and particle size distribution are of primary importance. These parameters are affected by volatility of the solvent, viscosity and surface tension of the formulation, application equipment, distance of the nozzle from the spray target and climatic conditions at the time of application. The ground based sprayers as well as low flying aircraft are used for their application. The prescribed requirements include the active ingredient, acidity or alkalinity, kinematic viscosity, volatility and low and elevated temperature stability (FAO, 1999).

Liquid Seed Dressing

Liquid seed dressing may be formulated in any of the liquid forms depending upon the active ingredient and its compatibility and stability in that form. Thus, these formulations may be as soluble concentrate (SL, AS, LS etc.), emulsifiable concentrate (EC), emulsions (EW, EO, ES), suspension concentrate (SC), or the like. To distinguish the treated seeds from the untreated ones, a dye may often be incorporated in these products. Compliance to the prescribed physico-chemical requirements is essential depending on the form to be used (Refer FAO, 1999). An important requirement is their non-interference with the plantability and viability of the seeds.

Water-soluble powder for seed treatment needs to be water soluble and free from visible extraneous matter and hard lumps. Its key requirements, besides the active ingredient, are acidity or alkalinity, degree of dissolution and solution stability, persistent foam and stability at elevated temperature (FAO, 1999).

Suspo-Emulsion (SE)

A liquid heterogeneous formulation consisting of a stable dispersion of active ingredients in the form of solid particles and fine globules in a continuous water phase
—GIFAP (1999)

It is a multiphase formulation that consists of a suspension, side by side with an emulsion, with water as dispersion medium. It enables combination products from water insoluble solid and liquid active ingredients simultaneously, where one or more of the active ingredients is in emulsion form. It is intended for dilution with water before application. Like other aqueous liquid formulations, it is easy to handle and measure, dust free, non-inflammable and water miscible. The formulation is thermodynamically unstable and therefore, its pourability (ensure that it can be poured from the container), water dispersibility, suspensibility, wet sieve and persistent foam tests are relevant. Like SC and EW, particle size distribution and viscosity are excluded from its specification for the same reasons. The preparation and stabilisation of such a multiple phase system is difficult.

Key FAO (1999) specifications with CIPAC test method (MT) include mass per ml at 20°C (MT 3.3), acidity or alkalinity (MT 31) or pH range (MT 75), dispersion stability (MT 180), wet sieve test (MT 59.3), persistent foam (MT 47.2), stability at 0°C (MT 39.3) and stability at elevated temperature (MT 46.1).

A suspo-emulsion containing chloridazon as solid and dillate as liquid pesticide has been reportedly stabilized in a minimum quantity of a common ionic dispersing agent and a non-ionic water soluble polymeric auxiliary agent. First, a formulation of chloridazon was formulated as a common suspension concentrate and then dillate was emulsified appropriately in the suspension. It was observed that greater the share of suspension i.e. solid particles, the finer were the droplets to which dillate was distributed, due to relative small energy expenditure.

Aqueous Capsule Suspension (CS)

A stable suspension of capsules in a fluid, normally intended for dilution with water before use   **—GIFAP (1989)**

In CS formulations, the active ingredient is present inside discrete, inert, polymeric microcapsules suspended in an aqueous medium. They are diluted with water before application. Their total as well as free active ingredient is of interest. When in equilibrium with the medium, the pH of the formulation has to be in a range that is devoid of any undesirable effect on the active ingredient. Viscosity too is important but as with conventional suspension concentrates, it cannot be described in a simple way. It is one part of the complex rheology of these non-Newtonian fluids. The rinsability and water dispersibility are considered to adequately describe the rheological properties of a CS formulation. Since these formulations are sprayed through conventional application equipment, both the discrete particle size and the agglomerates must be controlled. Currently, no internationally accepted methods for determining a range with particles of this size is available. Wet sieve tests have, therefore, been used to ensure the sprayability of the diluted formulations. Another important consideration is freeze/thaw stability. Freezing may result in capsule failure through crystallization or by other mechanisms, resulting in a change in the properties of the formulation including the release of the active ingredient into the aqueous medium. Therefore, surviving the freeze-thaw cycle is a vital parameter for performance of the formulation. Similarly, information on rate of release of active ingredient in such formulations is an index of their performance. In instances where this information is vital, appropriate test methods and values have to be prescribed.

The major requirements along with the CIPAC test methods as per FAO (1999) are: mass per ml at 20°C (MT 3.3), acidity or alkalinity (MT 31) or pH range (MT 75), pourability (MT 148), spontaneity of dispersion (MT 160), suspensibility (MT 161), wet sieve test (59.3), persistent foam (MT 47.2), storage stability including freeze/thaw stability and stability at elevated temperature (MT 46).

Reverse Phase Microcapsule Suspension

The formulation has been reported specifically with glyphosate isopropylamine salt— a herbicide (Guyomar, 2002). An aqueous solution of the herbicide has been encapsulated in hydrocarbon fluids forming a reverse phase microcapsule suspension. The formulation showed significantly superior bioefficacy as compared to glyphosate salt SL formulation, revealed no compatibility with herbicidal partners in tank mix and exhibited excellent physical stability for three months at room temperature.

Foam

If in an aerosol pack, the pressure is less, the liquid more viscous and the orifice big, it results in foam—a system of air dispersed in a liquid. Quite often, foam forming and stabilizing agents are used. Application as foam has an advantage that a uniform application of the material can be visibly verified. Foams so formed should collapse after a lapse of time to enable better contact. If containers can be suitably protected, then methyl chloride can be used as a solvent. Foams find some application in weed control.

Smoke Generator

These are liquid/solid formulations which generate smoke or smog on ignition . The smoke contains adequate amount of toxicant or insect repellent to control or drive away the pest.

The most common smoke generators are the coils e.g. the mosquito repellent or killing coils. These employ the toxicant or repellent material in a fine powder form, a filler particularly organic, capable of smouldering easily, a binder of the water soluble swelling gum type and other additives like slow combustion promoting agents, perfumes etc. An example of a solid carbohydrate in combination with sodium chlorate will explain the working as follows:

$$2C_6H_{10}O_5 + 7NaClO_3 \rightarrow 10H_2O + 9CO_2 + 3CO + 7NaCl$$

If rate of burning and size of orifice are regulated, a lot of smoke will be produced.

Several organic fillers can be used in such smoke generating coils. Finely ground wood, leaves, stalks, bark of several trees etc. are used. Plants from the family Lauraceae are frequently used in mosquito coils since these burn easily and impart a pleasant odour to the smoke. A widely used binder for mosquito coils is the leaf powder or bark of *Machilus thunbergii*, a tree indigenous to South East Asia. The powder is known as 'Tabu powder' and contains mucilaginous gum. Common starch is also widely used. Synthetic water soluble gums such as methyl cellulose, hydroxyethyl cellulose, sodium carboxymethyl cellulose etc. also find extensive application, often in combination with cheaper binders. Common dyes that can be incorporated include malachite green, brilliant green YS, methylene blue etc. The combustion promoting agents are sodium or potassium nitrate or sodium chlorate. The sodium salts tend to attract moisture but the potassium salts are devoid of this tendency. Some perfumes of local interest are added to give a fragrance to the smoke. Incense additives such as gum benzoin, are often used.

A typical example of liquid smoke generators is in solution formulation. Malathion solutions are used to generate smoke/vapours in various mosquito (malaria) management programmes. Such solutions are mostly prepared in petroleum based hydrocarbon solvents like kerosene, white mineral oil etc. The other ingredients are selected, as per need.

Cream and Ointment

Typical products in 'creams' and 'ointments' classes are the personal and veterinary insect repellent creams. Natural pyrethrum based creams and ointments are extensively used. The creams for personal use employ besides the active ingredient, a wax (Pollwax), lactic acid, glycerine or sorbitol and water. The aerosol dispersed personal creams contain the active ingredients, lanolin, acetyl alcohol, stearic acid, triethanolamine, glycerine, water, perfume with or without a propellant. Grease for cattle against ticks may contain the active ingredient, heavy oil solvent and the commercial petroleum jelly. In most pyrethrum based products, piperonyl butoxide can be used as a synergist. Lately, creams and ointments based on neem oil and other neem ingredients are also picking up.

REFERENCES

Adamson, A. W. (1976). *Physical Chemistry of Surfaces*. Wiley, New York.

Almassy, G., Antal, J., Bohanszy, Laszlo, Dienes, L. (1973). Absorbents containing silicic acid and natural silicates as carriers for plant protectives. *Magy. Kem. Lapja.*, **28**(2): 83-89.

ASTM (1966) Oil absorption of pigments by Spatula Rub Out, Part 20, American Society for Testing and Materials, Philadelphia, Pa. ASTM-D-281-31, p. 161.

Bailey, G. W., White, J. L. and Rothberg, T. (1968). Adsorption of organic herbicides by montmorillonite. Role of pH and chemical character of the absorbance. *Proc. Soil Sci. Soc. Am.*, **32**: 222-234.

Baker, R. W. and Lonsdale, H. K. (1974). In *Controlled Release of Biologically Active Agents*. (eds. A. C. Tanquary and R. E. Lacey). pp. 15-61. Plenum Press, New York.

Bansal, O. P. and Singh, N. (1985) A thermodynamic approach to the adsorption of phosphamidon with H and Na montmorillonites. *J. Indian Soc. Soil. Sci.*, **33**: 513-519

Barthel, W. F. and Lofgren, C. S. (1964). A comparison of some granular carriers for chlordane and heptachlor. *J. Agric. Food Chem.*, **4**: 341.

Bartlett, B. R. (1951). The action of certain inert dust materials on parasitic Hymnoptera. *J. Economic Ent.*, **44**: 891-896.

Bell, A. Kido, G. S. (1956). Hydroquinone and its derivatives as stabilizers for pyrethrin and allethrin. *J. Agric. Food Chem.*, **4**: 341.

Benesi, H. A. (1956). Acidity of catalyst surfaces I. Acid strength from colors of absorbed indicators. *J. Am. Chem. Soc.*, **78**: 5490-5494.

Benesi, H. A. (1957). Acidity of catalyst services II. Amine titration using Hammet indicators. *J. Physic. Chem.*, **61**: 970-973.

Benesi, H. A., Sun, Y. P. and Detung, K. D. (1959). Stabilization of pesticidal compositions. U.S. Patent 2, 868, 688.

Biggar, J. W. and Cheung, M. W. (1973). Adsorption of Picloram (4-amino-3, 5, 7-trichloropicolinic acid) on Panoche epharata and Palouse soils. A thrmodynamic approach to the adsorption mechanism. *Proc. Soil Sci. Soc. Am.*, **37**: 863-867.

Bowman, B. T., Adams Jr., R. S. and Fenton, S. W. (1970). Effect of water upon malathion adsorption into five montmorillonite systems. *J. Agric. Food Chem.*, **18**: 723-727.

Buscall, R., Goodwin, J. W., Hawkins, M. W. and Ottewill, R. H. (1982). Viscoelastic properties of concentrated lattices, Part I. Methods of examination. *J. Chem. Soc. Faraday. Trans.,* **78**: 2873-2887.

Chiovato, A., Hecaen, J., Coret, J. and Douglass, A. G. (2002). Emulsifiable granule (EG): An innovative formulation technology for crop protection. *Book of Abstracts*. Topics 1-4, Vol. I. 10th IUPAC International Congress on the Chemistry of Crop Protection. Basel, Switzerland, p. 387.

CIPAC (1970). *CIPAC Handbook*. Vol. I. *Analysis of Technical and Formulated Pesticides*, (ed. G. R. Raw), Collaborative International Pesticides Analytical Council (CIPAC). Heffer and Sons Ltd., London.

CSMA (1955). *Glossary of Terms Used in Aerosol Industry*. Chemical Specialties Manufacturing Association, (CSMA), Washington D.C.

Daines, R. H., Luckens, R. J., Brennan, E. and Leone, I. A. (1957). Phytotoxicity of captan as influenced by formulation, environment and plant factors. *Phytopath,* **47**: 56.

David, W. A. L. and Gardiner, B. O. C. (1950). Factors influencing the action of dust insecticides. *Bull. Ent. Res.,* **41**(1): 1-61.

El-Attal, El. Sowy, M. S. and Said, A. A. A. (1975). Laboratory studies on some Egyptian granular carriers for lindane and endrin. *Bull. Ent. Soc. Egypt., Econ. Ser.,* **8**: 7-11.

FAO (1999). FAO Specification for Plant Protection Products. AGP: CP/30- 306.

Flanagan, John (1983). Principles of pesticide formulation. In: *Formulation of Pesticides in Developing Countries*. United Nations Sales No. E 83 II B.3. pp. 13-64.

Flashinski, S. J. (1998). Aerosol formulations. In: *Pesticide Formulation, Recent Developments and their Applications in Developing Countries*, (eds.) Wade Van Valkenburg, B. Sugavanam and S. K. Khaitan, UNIDO Publication, New Age International (P) Ltd. Publishers, New Delhi, India. pp. 264-292.

Formstone, C. A. (2002). Formulating solid products—incorporating effervescent agents. Book of Abstracts. Topics 1-4, Vol. I, 10th IUPAC International Congress on the Chemistry of Crop Protection, Basel, Switzerland. p. 403.

Fowkes, F. M., Benesi, H. A., Ryland, L. B., Sawer, W. A., Delling, K. D., Loeffler, K. S., Folokemer, F. B., Johnson, M. and Sun, Y. P. (1960). Insecticide decomposition, clay-catalysed decomposition of insecticides. *J. Agric. Food Chem.,* **8**(3): 203-210.

Furmidge, G. G. L. (1962). Physico-chemical studies on agricultural sprays 4. The retention of spray liquid on leaf surfaces. *J. Sc. Food. Agric.,* **13**: 127-140.

Galloway, A. L. (1962). Method of preparing pelleted pesticidal compositions. *U.S. Patent,* 3, 056,723.

Gibbs, J. W. (1876). *The Collected Works of J. Willard Gibbs*. Yale University Press, New Haven, Conn, reprinted in 1948.

GIFAP (1989). Catalogue of pesticide formulation types and international coding system. Technical monograph No. 2, International Group of National Asociation of Manufacturers of Agrochemical Product (GIFAP), Brussels.

Gimeno, M., Gams, K., Liljegren, K. and Moser, M. (2002). Water dispersible granules from microcapsule suspension. Book of Abstracts. Topics 1-4, Vol. I, 10th IUPAC International Congress on the Chemistry of Crop Protection, Basel, Switzerland. p. 400.

Goel, R. K., Gupta, M. E. and Jones, B. A. (1975). Interaction between clay minerals and bipyridilium herbicides. *Res. Rev.,* **57**: 1-23.

Gooden, E. L. (1944). Size specifications for fine particles. *J. Econ. Ent.,* 37: 204-208.

Gould, E. S. (1962). *Mechanism and Structure in Organic Chemistry*. Verlag Chemie. Berlin.

Gunther, F. and Gunther, J. D. (1971). Pesticide formulation. *Residue Rev.,* 36: 12-26.

Guyomar, P. Y. (2002). Reverse phase microcapsule suspension of glyphosate isopropylamine salt in hydrocarbon fluids in combination with herbicidal partners. Book of Abstracts. Topics 1-4, Vol. I, 10th IUPAC International Congress on the Chemistry of Crop Protection, Basel, Switzerland. p. 405.

Halder, A. K. (1982). Effect of carrier on isomalathion formation in solid malathion formulations. M.Sc. Thesis, Post Graduate School, Indian Agricultural Research Institute, New Delhi.

Halder, A. K. and Parmar, B. S. (1984). Effect of carrier on isomalathion formation in malathion powders. *J. Pesticide Sci.,* **8**: 147-150.

Hamerstand, G. E. (1981). Starch encapsulated pesticides : a preliminary cost estimate. 8th. Intern. Symp. Controlled Release of Bioactive Materials.

Harlow, F. A. (1957). The toxicity of DDT in abrasive and non-abrasive dusts to the rice weevil, *Calandra oryzae* L. (Coleoptera: Curculionidae). *Ann. Appl. Biol.,* **45**(1): 90-113.

Harris, F. W. (1974). Proc. Controlled Release Pesticide Symposium, Univ. of Akron, Akron, Ohio, USA c.f. Shasha, B.S. (1998) Economics and technologies of controlled release formulations of pesticides. In: *Pesticide formulation: Recent Developments and their Applications in Developing Countries*.

(eds Wade Van Valkenburg, B. Sugavanam and Sushil K. Khaitan), UNIDO, New Age International (P) Ltd., Publishers, New Delhi, India.

Hayes, M. H. B., Pick, M. E. and Jones, B. A. (1975). Interaction between clay minerals and bipyridilium herbicides. *Res. Rev.* **57**: 1-23.

Hemmett, L. P. and Deyrup, A. J. (1932). A series of simple basic indicators, I. The acidity functions of mixtures of sulphuric and perchloric acids with water. *J. Am. Chem. Soc.*, **54**: 2721-2739.

Henrict, J. J. (1972). Some considerations on the particle size determination in pesticide formulations. In: *Pesticide Chemistry Vol. V: Herbicide Fungicide and Formulation Chemistry.* Proc. 2nd IUPAC Cong. of Pesticide Chemistry (ed.) A. S. Tahori, Gordon Beach Science Publishers, London. pp. 471-484.

Himmel. C. F. and Cardarelli, N. F. (1982). Flight encapsulation of particles. U.S. Patent. 4, 353, 962.

Hornbaker, A. L. and Hladik, W. B. (1970). Toxicant carrier granules from Kansas bentonite with volcanic ash additive. *Kans. State Geol. Sur. Bull.*, **199**(1): 11-13.

Hytte, J. M. and Pevedic, Sle (2002). Investigation of water dispersible tablets—technology for crop protection and environmental sciences. Book of Abstracts. Topics 1-4, Vol. I, 10th IUPAC International Congress on the Chemistry of Crop Protection, Basel, Switzerland. p. 398.

ISI (1980). Specification for blank granules. IS: 9666-1980. Indian Standards Institution, New Delhi.

ISI (1982). Methods of tests for pesticides and their formulations IS: 6940-1982, Indian Standards Institution, New Delhi.

ISI (1984). Specification for chlorfenvinphos granules. IS: 11009-1984. Indian Standards Institution, New Delhi.

ISI (1987). Specification for fenvalerate emulsifiable concentrate. IS: 11997-1987. Indian Standards Institution, New Delhi.

Iyer, V. and Parmar, B. S. (1987). Effect of pesticide formulation solvent on isomalathion formation in malathion solutions. *Curr. Sci.*, **56**(9): 416-417

Jatkenak, J. (1986). Preparation of silica fine powder as carrier for pharmaceuticals and pesticides. *Japan Patent,* **60**: 226, 826.

Johnson, V., O'Callaghan, M. and Hackson, T. (2002). Controlled release formulations for delivery of biopesticides. Book of Abstracts. Topics 1-4, Vol. I, 10th IUPAC International Congress on the Chemistry of Crop Protection, Basel, Switzerland. p. 394.

Joseph, W. N. and Edward, A. N. (Jr.) (1966). Pesticidal formulations containing a complex of calcium carbonate and silica. US Patent 3, 194, 730.

Karl, U. (1964). Surface acidity of mineral carrier substances and its influence on the formulation of plant protection and parasite control agents. *Chem. Tech.*, **16**(5): 263-66.

Kousaka, Y., Okyyama, K. and Payatakes, A. C. (1981). Physical meaning and evaluation of dynamic shape factor of aggregate particles. *J. Colloid and Interface Sci.*, **81**(1) 91-99.

Kumar, J., Chalapathi Rao, N. B. V., Singh, V. S. and Parmar, B. S. (2002b). Field appraisal of controlled release formulation of phorate against the rice leaf folder, *Cnaphalocrocis medinalis* (Guenee). *Ann. Pl. Protec. Sci.*, **11**(1): 129-133

Kumar, J. and Parmar, B. S. (1999). Stabilization of azadirachtan-A in neem formulation: effect of some solid carriers, neem oil and stabilizers. *J. Agri. Food Chem.*, **47**(4): 1735-1739.

Kumar, J., Singh, G., Dhandapani, A. and Parmar, B. S. (2002a). Release of butachlor from polymeric controlled release formulations in water. *Pesticide Res. J.*, **14**(1): 139-143.

Kumar, L. and Parmar, B. S. (2000). Effect of emulsion size and shelf life of azadirachtan-A on the bioefficacy of neem (*Azadirachta indica* A. Juss) emulsifiable concentrates. *J. Agri. Food Chem.*, **48**: 3666-3672.

Landham, R. R. (2002a). Microcapsule tapes. Book of Abstracts. Topics 1-4, Vol. I, 10th IUPAC International Congress on the Chemistry of Crop Protection, Basel, Switzerland. 401.

Landham, R. R. (2002b). Solid formulation with microbubbles for improved spray application. Book of Abstracts. Topics 1-4, Vol. I, 10th IUPAC International Congress on the Chemistry of Crop Protection, Basel, Switzerland. p. 402.

Lewis, D. H. and Cowsar, D. R. (1971). Principles of controlled release pesticides. In: *Controlled Release Pesticide,* (ed.) Herbert B. Scher. ACS Symp Series 53.

Malina, M. A. (1960). Flowability of dust formulations. *Agric. Chem.*, **15**(9): 49-95.

Marel, H. W. and Bentelspacher, H. (1976). Atlas of infrared spectroscopy of clay minerals and their admixtures. Elsevier Pub., N.Y., p. 387.

Marshal, C. E. (1975). *The Physical Chemistry and Mineralogy of Soils.* Kriegur Publication Co., N.Y., p. 366.

Matsumoto, S. (1958). Acidity of mineral carriers and the effects of hydrogen-ion on the decomposition of organophosphorus dust formulations. Studies on organophosphorus insecticide. V. Butyo-Kagaku **23**(2): 81-89.

Matsumoto, S., Okubo, T. and Honda, I. (1957). Stability of methyl parathion dust formulations and the effect of stabilizers on insecticides, IV. *Butyo-Kagaku* **22**(4): 74-81.

Matsumoto, S., Okubo, T., Yueda, I. and Torikai, Y. (1957). Effect of mineral carriers on deactivation of organo phosphorus dust formulations. Studies on organophosphorus insecticide, II. *Butyo-Kagaku*, **23**(2): 327-332.

Mortland, M. M. (1968). Pyridinium montmorillonite complexes with ethyl N, N-di-n-propyl thiocarbamate (EPTC). *J. Agric. Food Chem.*, **16**: 706-707.

Mortland, M. M. (1970). Clay organic complexes and interactions. *Adv. Agronom.*, **22**: 75-117.

Okuyama, K., Kousaka, Y. and Payatakes, A. C. (1981). Evaluation of affect of non-sphericity of fine aggregate particles in Brownian coagulation. *J. Colloid and Interface Sci.*, **81**(1): 21-31.

Oulton, T. D. (1967). *Encyclopaedia of Industrial Chemical Analysis.* Vol. 4, Wiley, New York, pp. 440-442.

Panwar, V. P. S., Kumar, J. and Parmar, B. S. (1999). Field appraisal of controlled release formulations of phorate against the shootfly infesting maize. *Ann. Pl. Protec. Sci,* **7**(1): 80-86.

Park, S. S., Lee, J. Y., Chum, K. T. and Lee, D. S. (1967). Carriers for pesticides. *Nongso Sihom Yocigu Pago*, **10**(3): 129-39.

Parmar, B. S. (1985). Improved DDT emulsifiable concentrates. *Intern. J. Trop. Agri.*, **3**(3): 138-145.

Parmar, B. S. and Dutta, Shiv (1990). Floating granule—a new pesticide formulation. *Pesticide Res. J.* **2**: 165-166.

Parmar, B. S., Kumar, J. and Panwar, V. P. S. (1998). A process for the preparation of a matrix containing phorate (O, O-dimethyl, S-(ethyl thiomethyl) phosphoro-dithioate). Indian Patant No. 188097 (2002).

Parmar, B. S. and Srivastava, K. P. (1986). Developement of some neem formulations for the control of *Spilosoma obliqua* in the laboratory and *Euchrysops cnejus* in the field. *In: Natural Pesticides from the Neem Tree (Azadirachta indica* A. Juss) *and other Tropical Plants.* (eds.) H. Schmutterer and K. R. S. Ascher. *Proc. 3rd Intern. Neem Conf.* Nairobi, Kenya. GTZ, Eschborn, West Germany, 1987.

Polon, J. A. (1973). Formulation of pesticidal dusts, wettable powder and granules. In: *Pesticide Formulations,* (ed.) Wade Van Valkenburg. Marcel Dekker Inc., New York, pp. 143-212.

Polon, J. A. and Sowyer, E. W. (1962). The use of stabilizing agents to decrease decomposition of malathion on high sorptive carriers. *J. Agric. Food Chem.*, **10**: 244.

Powar, S. L. and Parmar, B. S. (1995). Effect of granule carriers, binders and encapsulation materials on carbofuran release. *Pesticide Res. J.*, **7**: 119-124.

Rao, K. N. and Parmar, B. S. (1986). Fly ash as a carrier for phorate granules. *Intern. J. Trop. Agri.* **4**(4): 371-374.

Rao, K. N., Srivastava, K. P. and Parmar, B. S. (1989). Development of controlled release phorate formulations and their emulsion for pest control and grain yield on sorgham. *Pesticide Res. J.*, **1**(1): 7-11.

Rengsamy, S. and Parmar, B. S. (1988). Investigation of some factors influencing isomalathion formation in malathion products. *J. Agric. Food Chem.*, **36**: 1025-1030.

Rengsamy, S. and Parmar, B. S. (1989). Dissipation of isomalathion on solid pesticide carriers, container surfaces and leaves and some degradation products of isomalathion on carriers. *J. Agric. Food Chem.*, **37**: 430-433.

Rosenfield, C. and Valkenburg, W. (1965). Decomposition of Ronnel absorbed on bentonite and other clays. *J. Agric. Food Chem.*, **13**: 68.

Ross, H. W. (1983). Available granular carriers properties and general processing methods. ASTM Special Technical Publication, pp. 32-44.

Saltzman, S. and Yariv, S. (1976). IR and X-ray study of parathion-montmorillonite sorptive complexes. *J. Soil Sci. Soc.*, **40**: 34.

Sarup, P. (1967). Effect of formulation on the toxicity of insecticidal dusts to *Prodenia litura* Fabricius (Noctuidae: Lepidoptera). Ph.D. Thesis. Post Graduate School, Indian Agricultural Research Institute, New Delhi.

Sarup, P. (1970). Effect of formulation on the toxicity of insecticidal dusts to *Prodenia litura* Fabricius choice of diluents. *Indian J. Ent.*, **32**(4): 356-375.

Sawaguchi, K. and Taninani, H. (1977). Sintered granular base for pesticide formulations. *Japan Patent*. 76, 115, 928.

Scheiber, M. M., Shasha, B. S., Trimnell, D. and White, M. D. (1987). Methods of applying herbicides. In: *Controlled Release Herbicides*, WSSA Monograph 4, Weed Science, Society of America, p. 197.

Scholze, H. and Schmidt, H. (1982). Slow release formulation based on expanded perlite or similar porous materials. *U. S. Patent*: 6, 393, 011.

Schulze, D. G. (1989). An introduction to soil minerology. In: *Minerals in Soil Environment*, (eds.) J. B. Dixon and S. B. Seed, Soil Science Society of America, Madison.

Schwint, I. A. (1966). Stabilized pesticide compositions containing attapulgite clay. *United States Patent*: 2, 232, 831.

Searle, A. B. and Grimshaw, R. W. (1959). The Chemistry and Physics of Clays and other Ceramic Materials. Earnest Benn Limied, London.

Sharma, D. K. and Parmar, B. S. (1990). Effect of different solvents and their fractions on physico-chemical and biological performance of fenvalerate emulsifiable concentrates. *Pesticide Res. J.*, **2**(2): 69-82.

Shasha, B. S. (1998). Economics and techniques of controlled release formulations of pesticides. In: *Pesticide Formulation, Recent Developments and their Application in Developing Countries*, (eds.) Wade Van Valkenburg, B. Sugavanam and S. K. Khetan. UNIDO Publication, New Age International (P) Ltd. Publishers, New Delhi, India. pp. 146-168.

Shasha, B. S. Trimnell, D. and Otey, F. H. (1981). Encapsulation of pesticides in a starch-calcium adduct. *J. Polymer Science*, **19**: 1891.

Shasha, B. S., Doane, W. M. and Russell, C. R. (1976). Starch encapsulated pesticides for slow release. *J. Polymer Science*, **14**: 417.

Shasha, B. S., Trimnell, D. and Otey, F. H. (1981). Encapsulation of pesticides in a starch-calcium adduct. *J. Polymer Science*, **19**: 1891.

Sircar, P. (1975). Effect of formulation on the toxicity of insecticidal granules to some important crop pests. Ph.D. Thesis. Post Graduate School, Indian Agricultural Research Institute, New Delhi.

Stout, E. I., Shasha, B. S. and Doane, W. M. (1979). Pilot plant process for starch xanthide encapsulated pesticides. *J. Applied Polymer Science*, **24**: 153.

Strain N. H. (1942). Chromatographic adsorption analysis, Interscience Publishers Inc. New York, Heidelberg.

Synnatschke, G. and Guckel, W. (1972). The influence of pKa values of carriers on storage stability of active ingredients in plant protection. Proc. 2nd Int. IUPAC Cong. Pestic. Chem. V. A. S. Tahori (ed.). Gordon & Beach Publishers Inc., Israel.

Trademan, L., Malina, M. A., Wilks, W. and Wilks, L. P. (1960). Insecticide formulations and methods of making the same. US Patent 2, 927, 882.

Trimnell, D., Shasha, B. S., Wing, R. E. and Otey, F. H. (1982). Pesticide encapsulation using a starch borate complex as well as material. *J. Applied Polymer Soc.*, **27**: 3919.

Uejima, Toshiharu (1983). On new formulations of pesticides in Japan. *J. Pesticide Sci.*, **8**(1): 125-130.

Ullrich, K. (1964). *Chem. Tech.*, **16**: 263 cited from Synnatschke, G. and Giickel W. (1972) The influence of pka values of carriers on storage stability of active ingredients in plant protection. *Proc. 2nd. Int. IUPAC Cong. Pesti. Chem. (V.)* A. S. Tahori *(ed.)* Gordon & Breach Publications Inc., Israel.

Vogel, A. I. (1964). *A Text Book of Quantitative Inorganic Analysis*. The English Language, Book Society & Longmans, Green & Co., London.

Walling, C. (1950). The acid strengths of surfaces. *J. Am. Chem. Soc.*, **72**(3): 1164-1168.

Whitehead, D. (1976). The formulation and manufacture of granular pesticides. In: *Granular Pesticides, Proc. Symp. Granular Pesticide*, Br. Crop Prot. Council Monogr. No. 18.

Wing, R. E. (1998). Cited from B. S. Shasha. Economies and technologies of controlled release formulation of pesticides. In: *Pesticide Formulation: Recent Developments and their Applications in Developing Countries*, (eds.) Wade Van Valkenburg, B. Sugavanam and Sushil K. Khaitan. UNIDO. New Age International (P) Limited, Publishers. New Delhi pp. 146-168.

Wohleben G. *In*: Handbuch der Lebensmittel Chemie Vol. II, Part I Springer Verlag, Berlin.

World Health Organization (1985). Specifications for pesticides used in public health, Insecticides-Molluscicides-Repellents: Methods, 6th edn. World Health Organization.

CHAPTER 3

Adjuvants

The active ingredient(s), surfactant(s) and carrier or diluent(s) are the primary constituents of pesticide formulations. However, there are several situations where the formulations may employ additionally or exclusively various semiochemicals such as attractants, repellents, antifeedants etc. Besides, several other formulants namely binders, stickers, encapsulants antioxidants, sensitizers, stabilizers, deactivators, antifreeze, anti-transpirants, penetration aids, anti-cakes, lubricants, anti-dust, anti-foam, etc. find need based use as adjuvant in pesticide formulation. For a better clarity to the readers, the solid and liquid carriers and diluents have been discussed along with their respective formulations in Chapter 2. The other important adjuvants are briefly discussed in this chapter.

SURFACTANTS

A formulant which reduces the interfacial tension of two boundary surfaces, thereby increasing the emulsifying, spreading, dispersibility and/or wetting properties of liquids or solids

—(FAO, 1999)

Surfactants or surface-active agents (tensides or amphiphiles) are chemicals which possess capacity to alter the surface energy (surface tension) of solid or liquid. These are formulants incorporated to reduce the interfacial tension of two boundary surfaces or improving emulsifying, spreading, dispersibility or wetting properties of liquids or solids. In other words, surfactants are a type of adjuvant designed to improve absorption, dispersion, emulsification, penetration, spreading, sticking or other property related to surface behaviour of liquid or solid formulation.

Although present in small amounts, a surfactant exerts a marked effect on surface behaviour of a system. These are essentially responsible for producing changes in surface energy of liquid or solid surfaces and their ability to cause these changes is associated with their tendency to migrate to the interface between two phases. Consequently, they are of potential interest, wherever there are interactions between solid-solid, solid-liquid, solid-gas, liquid-liquid or liquid-gas interfaces in a system.

Surfactants consist of two parts i.e. (*i*) hydrophilic (water loving) or lipophobic (oil hating), and (*ii*) hydrophobic (water hating) or lipophilic (oil loving). Such additive components in any formulation, on account of their polar and apolar nature, work as wetting agents, spreaders, penetrants etc., which facilitate behaviour of a chemical in a particular phase and translocation of pesticide in different parts of plant or insect cuticle. The surfactant solutions exhibit functional properties like detergency, emulsification, dispersion, foaming, wetting etc. These are characterised by features such as solubility in at least one of the phases of the liquid system; amphipathic structure (combination of groups with opposing characteristics i.e. hydrophobic hydrocarbon chain in combination with a hydrophilic or water soluble ionic group); adsorption and orientation at phase interface (resulting in higher concentration at phase interface than in solution and subsequently in lowering surface tension and formation of oriented monolayers); and micelle formation (formation of aggregates of molecules or ions). Critical micelle concentration (cmc), a fundamental property of each solute-solvent system, is a limit at and above which micelle formation in any colloidal solution occurs and beyond which the micelles increase the solubility of many insoluble substances in water.

If any one of the above attributes is wanting, it may not be appropriate to call the substance a surfactant. Bentonite, for example, exhibits some of the above traits but may not be termed as surfactant because of insolubility in commonly used solvents.

The presence of two structurally dissimilar groups, both polar and apolar, within a single molecule, their make up, the relative size, location and the solubility characteristics influence the surface activity of the surfactant molecule. The surface activity is exhibited in all combinations of aqueous and non- aqueous phases known to occur. Because water is present as the major solvent phase in most of the commercial sprays and formulation systems, its presence is often assumed. Therefore, the water soluble amphipathic groups are often referred to as "solubilizing" groups.

Surfactants have an important role to play in penetration through cuticle. In case of polymeric ethylene glycols, it has been observed that both the hydrophilic constitution (hydroxy and ethylene oxide functions) and hydrophobic (hydrocarbon) portion of the molecule influence the bio-activity of a particular formulation.

Nomenclature of Amphipathic Groups

The terms hydrophillic, lipophobic, oleophobic, polar are usually used to describe water loving and oil hating groups. The oil loving (water hating) groups are described as hydrophobic, lipophilic, oleophilic, apolar etc. Generally, the hydrophobic groups consist of a hydrocarbon chain containing approximately 10-20 carbon atoms. It may also contain oxygen atoms, and groups like amide, ester and other functional groups with double bonds and a benzene ring. A propylene oxide hydrophobe can be considered as a hydrocarbon chain with methylene oxide group (CH_2O) at every third carbon. In some cases there may be substituents like halogens. Siloxane chains have also served as hydrophobe in some new surfactants. Carboxylates, sulphonates, sulphates and phosphates are examples of hydrophilic solubilizing groups of anionic surfactants. The cationics are solubilized by amino and ammonium groups. Ethylene oxide chains and hydroxyl groups are the solubilizing groups in nonionic surfactants. Combinations of anionic and cationic moieties form solubilizing groups in the amphoteric surfactants.

The molecular weights of surfactants range from approximately 200 to a few thousands. The optimum chain length in relation to surfactant activity will be governed by nature of polar group(s) but generally a C_{12} straight chain hydrophobe and a solubilizing group (ionic moiety) are very effective. In application and use of surfactants, their physico-chemical properties, and not their chemical homogeneity are of primary concern. Generally, the surfactants constitute polydisperse mixtures containing all molecules of the same molecular species and differing only in chain length or in some other structural detail e. g. natural fats as precursors of carboxylate surfactants. Coconut oil contains glycerol esters of fatty acids containing C_6 to C_{18} atoms, and this range varies according to the distillate treatment of the oil. Non-ionic surfactants of the alcohol ethoxylate are polydisperse type not only with respect to the hydrophobe but also to length of ethylene oxide chain. In case of alkylbenzenes, another surfactant raw material of commercial significance, the alkyl substituents not only contain different chain lengths but also carry benzene ring at all but the terminal positions along the hydrocarbon chain.

The surfactants have been found to affect activity of spray materials on cuticle surface, and within the cuticle layers. There may be some other sites of activity also namely the surfaces of living cells in the cuticle or within the cuticle surface. The physico-chemical and biological properties of soil are influenced by application of pesticides and the additives.

Classification

The commrcial surfactants, being complicated mixtures, the nomenclatures such as that of IUPAC and Chemical Abstracts, which are designed to express pure molecular species, are of little significance. American Oil Chemists Society recognises the value of trivial names such as lauryl and myristyl in designating a mixture of chain lengths with a distribution around C_{12} and C_{14} respectively. The terms coco or tallow, describing a mixture of acids derived from coconut oil and tallow respectively are even less systematic but still useful. In analogy with the chemically precise acyl groups, the acid radicals from coconut and tallow fatty acids are sometimes referred to as cocoyl and talloyl, respectively.

Surfactants are classified primarily on the basis of charge on the surface-active moiety forming the larger part of the molecule and sub-classified on the basis of the skeleton of the alkyl, aryl or heterocyclic structure and the main functional group. Four major classes of surfactants are:

(*i*) **Anionic Surfactants:** The major part contains net negative charge as in soaps e.g. $C_{17}H_{35}COO^-Na^+$.

(*ii*) **Cationic Surfactants:** The major part contains net positive charge generally on a quaternary ammonium moiety e.g. $(C_{18}H_{37})_2N^+(CH_3)_2Cl^-$

(*iii*) **Non-ionic Surfactants:** These are not ionic, hence carry the solubilizing contribution made by a chain of ethylene oxide or ester groups as in polyethylene glycols and sorbitans e.g. $C_{15}H_{33}(OC_2H_4)_7OH$.

(iv) **Amphoteric Surfactants:** These are basically long chained dipolar molecules containing both positive and negative moieties. They possess both the anionic and cationic properties depending on pH as in $C_{12}H_{25}N^+(CH_3)_2CH_2COO^-$ (Schick, 1966).

Anionic Surfactants

In anionic surfactants, the hydrophilic moiety is a polar group that is negatively charged in aqueous solutions or dispersions. These are mostly alkali metal salts of organic acids e.g. carboxylates, sulphonates, sulphates or phosphates and are accordingly sub-classified. The solubilizing power of sodium salts of the above four groups is almost equal in alkaline medium but that of carboxylates is much less in neutral and/or acidic media.

The ionic environment influences the properties of anionic surfactants in solution. Sodium or potassium salts are generally more soluble in water and less in hydrocarbons. The Ca, Ba and Mg salts are more compatible with hydrocarbon solvents and less so with water. Ammonium and amine salts like those of trimethanolamine, improve the compatibility of anionics with water and hydrocarbons and are widely used in emulsification and detergent applications. The total ionic strengths are usually associated with lower solubilities of anionic surfactants. This effect is offset by lowering the molecular weight of the hydrophobe in products to be used in electrolyte solutions of higher concentrations. Miceller solubilization by anionics is markedly affected by total ionic strength and also by the identity of associated cations.

These surfactants have been sub classified into various groups as under:

Carboxylates: Soaps and aminocarboxylates are the only commercial, man made carboxylate surfactants. Chemically, these belong to the type $(RCOO)^-(M)^+$ where, R is an alkyl group usually in C_{9-21} range and M is metal ion or ammonium ion. The soaps exhibit excellent surfactant properties in soft water but in hard water, the presence of polyvalent metal ions like Mg^{2+}, Ca^{2+}, Ba^{2+}, Fe^{3+}, Al^{3+} etc., causes formation of 'curds' or 'lime soap' precipitates. It lowers the effective concentration of the surfactants and their performance. The 'curd' formation can be prevented by addition of sequestering agents like ethylenediamine tetracetic acid (EDTA) and nitrilotriacetic acid, which tie up the heavy metal ions in non-ionizing complexes. However, it results in increase of cost. Soaps are favoured because of their easy biodegradability, (in waste water treatment plants) which is even faster than alkylbenzene sulphonates. Also, the polyphosphate residue, often associated with alkylbenzene sulphonates in household detergents (alkylbenzene sulphonates plus alkanolamides; polyphosphates and carboxymethyl cellulose) contaminates water due to excessive algal growth.

The sodium and potassium fatty acid soaps of coco are superior as lathering and cleansing agents in soft to medium-hard water. The soap containing sodium salts of C_{14}-C_{18} fatty acids are effective laundry and industrial detergents in soft to medium-hard water. The amine salts are excellent emulsifiers, dispersants and solubilizing agents. In contact with skin, the soaps have an emollient action and leave a soft feel on textile fabrics. Soaps also find several uses such as in ore floatation, wetting of asphalt, cleaning and thickening of woolens and emulsifying agents for polymerization.

Sulphonates: These are metal salts of alkyl or alkyl aryl sulphonic acids. The sulphonates have the general formula RSO_3M; where, R is an alkyl or alkyl aryl hydrocarbon group and M a metal ion like sodium or calcium. Unlike soaps, the surface activity of SO_3^- group is not over sensitive to pH variation or heavy metal ions and C-S-linkage is not susceptible to hydrolysis or oxidation under normal conditions of use.

Sulphonates can be sub grouped into various classes depending on the organic moiety:

(i) *Linear Alkylbenzenesulphonates (LAS):* These are metal salts of phenyl sulphonic acid containing a linear alkyl substituent in the ring. Dodecylbenzenesulphonates rank next to soaps in total usage. Linear alkyl group is more easily biodegradable than the branched chain isomers used earlier. These are light coloured flakes, granules or beads at 90 + percent assay and viscous liquids, pastes or slurries at lower concentrations.

    Performance of alkylbenzenesulphonates as compared with that of aliphatic sulphonates reveals that the effect of the benzene ring is approximately equivalent to 3 carbon atoms of an aliphatic chain. Being strong organic acids, alkylbenzenesulphonic acids form neutral alkali metal salts having good solubility in aqueous solutions for use at various concentrations over the entire pH range. These acids are not sensitive to precipitation by the natural hardness of waters but their alkaline earth metal salts are less water soluble than the alkali metal or amine salts. The calcium salts are sufficiently soluble in hydrocarbons. Alkylbenzenesulphonates are among chemically the most stable types of surfactants, which are stable to acidic or alkaline hydrolysis and strong oxidizing agents.

    The independent use of LAS stems from their simple preparation, low cost, reproducible quality, adequate supply, light colour, low odour and excellent response in formulations.

(ii) *Lignosulphonates:* These compounds neither form an oriented layer in phase interfaces, nor micelles, and they do not even reduce surface tensions at low concentrations to levels attained by other surfactants. However, these products do perform surfactant properties.

(iii) *Petroleum Sulphonates:* Synthesised by direct sulphonation of appropriate petroleum fraction, the surfactants belonging to this class are predominantly used in non-aqueous systems. These are produced as co-products in the refining of certain petroleum fractions. There are two broad classes namely the water-soluble types (Green soaps) and the oil soluble types (Mahogany soaps). The former are rarely used but the latter are more useful in solubilization, detergency, dispersion, emulsification and corrosion inhibition.

(iv) *Dialkyl Sulphosuccinates:* These are metal salts of symmetrical diesters, prepared by the esterification of maleic anhydride using conventional technology followed by addition of sodium bisulphite across the olefinic linkage.

    Sodium dialkyl sulfosuccinates are highly surface active but ester linkage is susceptible to acidic or alkaline hydrolysis, which limits their use. They are usually good wetting agents.

(v) **Naphthalenesulphonates:** These are metal salts of naphthalene sulphonic acids or their derivatives. Four series exhibiting specially surfactant properties are the salts of naphthylsulphonates, alkylnaphthylsulphonates, tetrahydronaphthyl sulphonates and sulphonates of formaldehyde-naphthalene condensates. These are mostly odourless, light grey solids, highly soluble in water; possess limited surface activity in soft water, stable to acid/alkali hydrolysis and are sensitive to oxidation. They are generally used as wetting and dispersing agents, and stabilizers.

(vi) **N-acyl N-alkyl Taurates:** These are derivatives of 2-aminoethanesulphonic acid (taurine) and have the general formula $RCO(R')NCH_2CH_2SO_3M$; where RCO is an acyl group and R' an alkyl moiety. These find use as surfactants on a limited scale only owing to the high raw material cost. These compounds are stable to acidic and alkaline hydrolysis. There is no loss of performance when used in hard water. These possess soap like biodegradability but their residue persists on the fabrics. Their molecular structure is capable of yielding strong wetting or detergent configurations e.g. in structure $RCON(R') CH_2CH_2SO_3Na$, if the acyl chain has more than ten carbons (R = 11-17, R' = CH_3 or C_2H_5), then it behaves as a strong detergent, if both acyl and alkyl chains have less than ten carbons (R = R' = C_6 to C_{-9}) then the molecule is a strong wetting agent.

(vii) **2-Sulphoethyl Esters of Fatty Acids:** Commercially known as β-sulphoesters, these are unaffected by hard water. Other properties are governed by fatty acid chain. Ester linkage being sensitive to hydrolysis, these are not used widely.

(viii) **Olefin Sulphonates or Alpha-olefene Sulphonates (AOS):** These are sulphonates of olefenes and are, produced by reaction of sulphur trioxide and air with a olefene followed by neutralization of the intermediate with an alkali. Low cost availability of linear 1-olefins in C_{14-18} range is of main interest. However, the sulphonation occurs with difficulty and the products so obtained are always mixtures. These sulphonation mixtures are called a-olefin sulphonate or AOS and possess detergency and foaming abilities similar to LAS. These are superior in performance to similar products made from linear straight chain olefins. These are less toxic, cause lesser skin irritation. Biodegradation of AOS is better than LAS.

Sulphates and Sulphated Products: Chemically these compounds can be considered to be S-alkoxy sulphonates. In these molecules the hydrophilic group, SO_3^-, of the half sulphonate ester surfactants is attached through an oxygen atom to a carbon atom of the hydrophobic moiety to form the sulphate (OSO_3) hydrophille. The additional oxygen makes sulphate a stronger solubilizing group than the sulphonate but the C-O-S linkage of the molecule is more easily hydrolysed than the C-S linkage of sulphonates, which limits their use.

(i) **Alkyl Sulphates (Sulphated Alcohols):** Sulphates obtained from normal primary alcohols are similar in performance to soaps of corresponding molecular weights. The branched chain alkyl sulphates have strong wetting properties. Alkyl sulphates are light coloured pastes or almost colourless solutions. They are stable to hard water but sensitive to hot acidic or alkaline medium. A common example is lauryl sulphates e.g. sodium lauryl sulphate.

(ii) Sulphated Natural Fats and Oils: The sulphates of most animal, vegetable and fish oils are useful surfactants. Ricinoleic acid (12-hydroxy-9, 10-octadecenoic acid), which contains a hydroxyl group and a double bond, is a desirable constituent of oils for sulphation. Oleic acid is also satisfactory. Esters of these acids can usually be sulphated with a minimum of hydrolysis. Polyunsaturated fatty acid moieties are undesirable components of glycerides for sulphation since the resulting surfactants are usually dark in colour and sensitive to acidifcation.

Sulphated natural fats and oils find use as emulsifying agents, wetting agents, detergents, penetrants, dispersants, textile softners and textile lubricants. Sulphated oleic acid (disodium salt) is used in large volume in textile industry as dispersing agent for lime soaps.

(iii) Sulphated Alkanolamides: These are good foaming agents and detergents. A typical product is sodium salt of the half sulphate ester of lauric ethanolamide $(C_{11}H_{23}CONHCH_2CH_2OSO_3Na)$. Several variations have been tried.

(iv) Sulphated Esters: These are sulphonates of naturally occurring fatty acids prepared by addition of sodium bisulphate to unsaturated fatty acids. Oleic and ricinoleic acids are esterified with alcohols of low molecular weight and then sulphated, to yield useful surfactants (e.g. $CH_3(CH_2)_7CH_2CH(OSO_3Na)CH_2(CH_2)_5CH_2COOR$; where, R = C_2H_5, C_3H_7, C_4H_9, C_5H_{11}). These products are marketed as yellow viscous liquids and possess good foaming, detergent and emulsifying properties.

(v) Etyoxylated and Sulphated Alkylphenols: These compounds have general formula $[R{-}C_6H_4{-}(OCH_2CH_2)_n{-}OSO_3M$; where, R = alkyl, M = NH_4, Na etc.]. Sodium salts of sulphate esters of nonylphenoxytri(ethyleneoxy)ethanols find wide application as detergents, wetting and foaming agents and emulsifiers.

(vi) Alcohols, Ethoxylated and Sulphated: Sodium or ammonium sulphates of ethoxylated alcohols have general formula $[R(OCH_2CH_2){-}OSO_3M]$; where, R = lauryl, myristyl etc. and M = NH_4, Na etc. The advantages of plain sulphated alcohols include the improved foaming (particularly in hard water), decreased irritation of eyes and skin, increased solubility in water and low raw material cost.

Phosphate Esters: Alkyl orthophosphates such as di(2-ethylhexyl) phosphate and alkyl polyphosphates such as $(2\text{-ethylhexyl})_5Na_5(P_3O_{10})_2$ are well known surfactants. These products are used in electroplating and as additives in dry-cleaning detergents, hard surface detergents, gelling agents, detergents in waterless hand cleaners and cosmetic products, dedusting agents for powdered alkaline detergent formulations and polymerization emulsifiers etc.

Non-Ionic Surfactants

Non-ionic surfactants bear no charge in solution. The hydrophilic tendency is due to oxygen in the molecule which hydrates by hydrogen bonding to water molecules. The strongest hydrophilic moieties in non-ionics are the oxo or ether linkages and hydroxyl groups. In addition, ester and amide linkages are also present in many non-ionics. Contribution of oxo group or the ether linkage in solubilization being weak necessitates the presence of multiple oxygen atoms in the molecule. The non-ionics are compatible with the ionic and amphoteric surfactants. These can be grouped into following classes:

(i) **Polyoxyethylene Surfactants:** The recurring ether linkages govern the solubility of these surfactants. The hydrophilic strength of one ethylene oxide unit ($-CH_2-CH_2-O-$) is approximately equal to that of one ether linkage between two methylene groups. Water solubility decreases as temperature increases. In a formulation, a minor proportion of anionics raises the cloud point (temperature at which a second phase appears). The performance of these surfactants is maximum just below the cloud point and is unaffected by hard water. These are moderate foamers; the polypropylene oxide derivatives being still low foaming. These are mostly liquids and may be safely incorporated in dry free flowing powders. Any desired hydrophilic-hydrophobic character can smoothly be incorporated in polyoxyethylene surfactants by proper selection of the hydrophobe.

$$Hydrophobe - O^- + H_2C \overset{O}{\overset{\diagup \diagdown}{-}} CH_2 \xrightarrow{\text{Slow}} Hydrophobe - O - CH_2CH_2OH$$

$$Hydrophobe - OCH_2CH_2OH + H_2C \overset{O}{\overset{\diagup \diagdown}{-}} CH_2 \xrightarrow{\text{Fast}}$$

$$Hydrophobe - OCH_2CH_2OCH_2CH_2OH$$

(ii) **Ethoxylated Alkylphenols:** The alkylphenols containing:

(a) 20-40%, polyoxyethylene are used as defoamers in surfactant solutions, and/or dispersing agents in petroleum oils, co-emulsifiers, intermediates for sulphonation.

(b) 40-60% polyoxyethylene are used for oil soluble detergents, dispersants and emulsifiers in pesticide emulsions and as intermediates for sulphonation.

(c) 60-70% polyoxyethylene are used as textile detergents and processing auxiliaries, pitch control in manufacture of paper pulp, rewetting agents in paper towels, emulsifiers for pesticide emulsions, detergents, wetting agents etc.

(d) 70-80% polyoxyethylene are used as detergents and wetting agents at high temperature and/or in electrolytes during plating, emulsifiers for fats, oils, waxes, stabilizers, wetters, penetrants etc.

(e) 80-95% polyoxyethylene used as stabilizers for synthetic latex, as emulsifiers for emulsion polymerizations, dyeing and leveling assistants and lime-soap dispersants.

(iii) **Ethoxylated Aliphatic Alcohols:** Their hydrophobes are generally mixtures of straight chain alcohols (C_{12}-C_{18}) and the mole ratio of combined ethylene oxide to hydrophobe vary from 1 to 50. The viscosity increases with polyoxyethylene content. The alcohols usually employed are, oleyl, lauryl, cetyl, stearyl, tridecyl, myristyl etc. Their functional properties and uses are similar to polyoxyethylene alkylphenols.

Carboxylic Esters: Carboxylic esters can be sub-grouped into various classes depending on the esterification of organic moiety.

(i) **Glycerol Esters:** Mostly mixtures of mono- and di-glycerides, these differ in respect of position of hydroxyl groups that are esterified. Usually their name is based on the most abundant isomer. The esters of saturated fatty acids are light coloured solids with melting points between 25-85°C. 1-Mono-glycerides are higher melting than the corresponding 2- monoglycerides. These find use as emulsifiers, suspending and dispersing agents, solubilizers, lubricants, food and pharmaceutical additives etc. The products intended for ingestion are prepared from edible fats. Alcoholysis of fats with glycerol is the most important industrial method for preparation of partial fatty acid esters of glycerol. Common esters are monolaurate, monoricinoleate, monostearate and mono- and/or dicarboxylates of fatty acids derived from coconut, cotton, safflower and many other edible or non-edible seed oils.

(ii) **Polyethylene Glycol Esters:** This series of surfactants comprises of esters of polyethylene glycol and fatty acids including alicyclic carboxylic acids related to abietic acid. The former are mixtures containing varying proportions of the mono esters $RCOO(OCH_2CH_2)_nOH$ and diesters $RCOO(OCH_2CH_2)_nOCOR'$. By regulating reaction conditions, the reaction can be moulded towards mono- or di-esters.

These may be mostly free flowing liquids or slurries depending on polyethylene content. The ester linkage is slightly hydrophilic. Emulsification is their key property. For good emulsification, about 60% by weight of polyoxyethylene is required to stabilize the saturated fatty acids in water. These are also low foaming. However, they are susceptible to hot acidic or alkaline media.

Polyethylene glycol esters are synthesized by alkali catalyzed reaction of fatty acids with ethylene oxide or by esterification with a polyethylene glycol in presence of an acid catalyst. Compounds prepared using lauric, oleic, stearic, (coconut, tallow oil), ricinoleic acids etc. are generally used. Abietic and related acids are major constituents of pine rosin. Tallow oil is a mixture of unsaturated fatty acids with alicyclic acids of the abietic family. The polyoxyethylene derivatives of the rosin acids are generally similar to the corresponding polyoxyethylene fatty acids. The surfactants can be obtained from these products by earlier described processes. These are similar in surface modifying properties but are more stable towards hydrolysis than simple fatty acid esters.

(iii) **Anhydrosorbitol Esters:** These form the second largest class of polyol surfactants. The important commercial groups are mono-, di- or triesters of sorbitan and fatty acids, where sorbitan is a mixture of anhydrosorbitols, 1, 4- sorbitan and isosorbide being the principal constituents.

1, 4-Sorbitan   Isosorbide   Spans

Sorbitan is not a strong hydrophilic group. Its derivatives are, therefore, not water-soluble and are lipophilic emulsifiers, solubilizers, softners and fibre lubricants. Sorbitan oleate and mono-laurate are pale yellow liquids and palmitates and stearates are light tan solids. The commercial products available under brand name 'Spans' are anhydrosobitol esters. These esters are prepared by the direct esterification of sorbitol with a fatty acid using an acidic catalyst at 225-250°C.

(iv) Ethoxylated Anhydrosorbitol Esters: Ethoxylation of sorbitan fatty acid esters leads to a series of more hydrophilic surfactants. The products find use as emulsifiers, antistats, softners, fibre lubricants and solubilizers. Often used as co-emulsifiers with the unethoxylated sorbitan fatty acid esters or the glycerol partial fatty acid esters. 'Tweens', the well-known commercial products belong to this group.

(v) Glycol Esters of Fatty Acids: Ethylene glycol, propylene glycol and 1, 2-propenediol esters of fatty acids are also used widely as surfactants.

(vi) Ethoxylated Natural Fats, Oils and Waxes: These are fatty acid esters of polyoxyethylene-derived alcohols; for example, ethoxylated castor oil (has high content of ricinoleic acid) containing polyoxyethylene 60-70% by weight. It is a water-soluble complex surfactant mixture. Depending on polyoxyethylene content to oil ratio (20-200), a number of products are available commercially.

Carboxylic Amides: The carboxylic amides are the amide derivatives of various organic moieties and their bases are sub-grouped into various classes as mentioned below.

(i) Diethanolamine Condensates: These are divided into regular (2/1, amine/acid) and super (1/1 amides). The common fatty acids are coco, lauric, oleic, stearic. The products are used as foam stabilizers and detergents.

(ii) Monoalkanolamine Condensates: Amides obtained from coco, lauric, oleic and stearic acids and monoethanol amine and/or mono-isopropanolamine, with amine ratios in the range from 2/1 to 1/2 are principal surfactants of this group. These are usually water insoluble solids.

(iii) Polyoxyethylene Fatty Acid Amides: These are mono- and di- adducts obtained by ethoxylation of a fatty acid amide (obtained by reacting ethylene oxide with alkyl amides) that are equivalent in gross composition to the mono- and diethanolamide condensates of the corresponding fatty acids. The condensates, obtained from reaction of alkyl epoxides with alkyl amides, are more widely used than above adducts because the latter have better properties for most uses. Ethoxylates of fatty acid amides are predominantly secondary amides (the second hydrogen on the amide being less reactive than the first towards ethylene oxide/epoxide, the tertiary products are seldom reported).

(*iv*) *Polyalkylene Oxide Block Copolymers:* These are reaction products of propylene oxide, butylene oxide, styrene oxide or cyclohexene oxide with compounds containing active hydrogen in a manner analogous to the reaction of ethylene oxide with alcohol or a phenol (active hydrogen). An aspect, which needs to be kept in mind to prepare a desired surfactant, is that in contrast to the effect of the added ethylene oxide units (or epoxide units), the higher alkene oxide units are more hydrophobic.

(*v*) *Polyoxypropylene-Polyoxyethylene Nonionics:* A series of polyoxypropylene hydrophobes (with molecular weights of 950-3250) are formed by reaction of propylene oxide with propylene glycol and ethoxylation to polyoxyethylene contents that range from 20-90% of the total weight. In spite of low activity, this group exhibits useful surfactant properties and is used commercially in detergents and pest control formulations.

Cationic Surfactants

The hydrophilic moieties in catonic surfactants are amino or quaternary nitrogen atoms. These form cations (ions bearing a positive charge) when dissolved in aqueous medium. Normally, single amino nitrogen has hydrophilic tendency capable of solubilizing a lipophilic group in the surfactant molecular weight range, in dilute acids. Introduction of additional primary, secondary or tertiary amino groups tends to increase water solubility. It is also achieved by quaternizing the amino group with an alkyl group of low molecular weight e.g. methyl, ethyl or hydroxyethyl. Most quaternary ammmonium surfactants are soluble even in aqueous alkaline solutions. Polyoxyethylenated cationics behave like non-ionics in alkaline solutions and like cationics in acidic solutions. Cationic surfactants find use as dispersants, emulsifiers, wetting agents, sanitizers, dye fixing agents, textile lubricants, textile softeners, foam stabilizers and corrosion inhibitors. This pattern of use is similar to that of anionics in neutral and alkaline solutions. Cationics being antiseptic in nature, are preferred for use in skin care and cure products. These are rather expensive to use in pesticide formulations. The cationic surfactants belong to following main classes.

(*i*) *Amines Containing no Oxygen:* Oxygen free aliphatic mono-, di- and poly-amines, and rosin-derived amines are the important surfactant compounds of this class. Usual product forms are acetates, naphthenates and oleates. Their principal uses are as floatation agents, corrosion inhibitors and dispersing agents. Some typical products, other than mono- and di-alkylamino derivatives of the type: $RNHCH_2CH_2CH_2NH_2$, obtained from N-alkyl ethylene or propylene diamine or N-alkyltrimethylenediamine or imidazoline (where, R is coco alkyl, talloyl, soya-alkyl or 9-octadecenyl) are 2-alkyl-2-imidazoline (a) and (2-aminoethyl) -2-alkyl-2-imidazoline (b); where, R = 8-heptadecenyl, heptadecyl, nonyl or mixed alkyl.

(a)   (b)

(ii) **Amines (Except Amides) Containing Oxygen:** Amine oxides, polyoxyethylene alkyl-amines, 1-(2-hydroxyethyl)-2-alkyl-2-imidazolines and N, N, N', N'-tetrakis— substituted ethylenediamine derivatives are of economic significance.

(a) **Amine Oxides:** These N-oxides of tertiary alkylamines are polar with maximum electron density at the oxygen atom. These exhibit strong hydrogen-bonding tendency and are hygroscopic. Most products are liquids or pastes.

$$R - \underset{\underset{CH_3}{|}}{\overset{\overset{CH_3}{|}}{N}} \rightarrow O \quad \text{or} \quad R - \underset{\underset{CH_2CH_2OH}{|}}{\overset{\overset{CH_2CH_2OH}{|}}{N}} \rightarrow O$$

The fatty amine oxides act as cationics in acidic solution and non-ionics in neutral or alkaline solution.

$$R(CH_3)_2N \rightarrow O + H^+ \longrightarrow R(CH_3)_2NOH^+$$

Amine oxide   Hydroxyl ammonium ion

where, R = aliphatic hydrocaxbon residue similar to cetyl, lauryl, myristyl, stearyl, coco, hexadecyl, octadecyl, decyl, oleyl etc.

The products are used as foam builders, effective shampoo detergents, wetting agents and emulsifiers.

(b) **Polyoxyethylene Alkyl and Alicyclic Amines:** These are derivatives of amines possessing aliphatic residue like that in coco, soybean and tallow etc. fatty acids. Three series of products namely polyoxyethylene linear alkyl amines, polyoxyethylene aliphatic tert-alkylamines and polyoxyethylene dehydroabietylamines have similar properties, with specific gravity of 0.9-1.15 at room temperature. The N-C linkage is resistant to hydrolysis. The products are used as emulsifiers, dispersants and wetting agents.

(c) **2-Alkyl-1-Hydroxyethyl-2-Imidazolines:** These surfactants have a general structure as given below: -

$$R - C \underset{\underset{\underset{CH_2CH_2OH}{|}}{N - CH_2}}{\overset{N - CH_2}{<}}$$

where, R = 8 heptadecenyl, heptadecyl, heptadecadienyl, undecyl etc.

The products find use as wetting agents, emulsifiers, dispersants, corrosion inhibitors and detergents. These can be ethoxylated to yield more hydrophilic products. These can also be reacted with dimethyl sulphate or benzyl chloride to form quaternary amines and then oxidized with H_2O_2 to yield amine oxides (both involve substitution on 1-position in the imidazoline ring).

N, N, N', N'-Tetrakis-substituted Ethylenediamines: These are compounds of the type shown below.

$$H(CH_2CH_2O)_y \, (OH_6C_3)_x$$
$$H(CH_2CH_2O)_y \, (OH_6C_3)_x \Big\rangle NCH_2CH_2N \Big\langle \begin{array}{l} (C_3H_6O)_x(CH_2CH_2O)_yH \\ (C_3H_6O)_x(CH_2CH_2O)_yH \end{array}$$

Hydrophobic   Hydrophilic

These are weakly cationic products. The uses include wetting, emulsification, dispersion, corrosion inhibition, detergency etc.

(iii) ***Amines having Amide Bridge:*** These are relatively cheap products which include compounds of the type $RCONHCH_2CH_2NHCH_2CH_2NH_2$ or $RCON(CH_2CH_2NH_2)_2$; where, RCO is acyl moiety derived from coconut, oleic, stearic and tallow oil acids.

(iv) ***Quaternary Ammonium Salts:*** These are quaternary ammonium compounds of the type RNR'_2X that are derived from tertiary amines. Quaternary ammonium ion is a much stronger hydrophilic moiety than a primary, secondary or tertiary amino group. The strong basicity or cationic activity makes them effective even in alkaline medium. These also possess some biocidal activity and textile softening properties.

Amphoterics

These molecules contain both an acidic and a basic hydrophilic moiety in the structure, which may behave as cationic or anionic depending on the environment of the system. Many amphoterics also contain oxygen atoms of ether or hydroxyl moiety that strengthen their hydrophilic tendency. These are marketed under trade names, usually with ambiguous details. Functional properties are in the order of decreasing importance e.g. detergency, emulsification, wetting, hair conditioning, textile softening, antiseptic protection and foaming. Some common examples are acids and sodium salts of 2-alkyl-1-carboxymethyl-1-hydroxyethyl-2-imidazolinium hydroxide and N-alkyl-3- aminopropionic acids, where, alkyl residues are similar to alkyl chain in fatty acids i.e. undecyl, heptadecyl etc. (Grayson, 1983).

where, R = nonyl, undecyl, heptadecyl, M = H or Na

$$R - NH\,CH_2\,CH_2\,COONa$$

$$R - N \begin{cases} CH_2CH_2COONa \\ CH_2CH_2COONa \end{cases}$$

where, R = Fatty acyl, talloyl, lauryl etc.

Effect on Properties of Solutions

The amphipathic structure of surfactants results in (i) adsorption at interfaces in solutions, (ii) orientation of the adsorbed surfactant ions or molecules, (iii) micelle formation in the bulk of the solution and (iv) orientation of the surfactant ions or molecules in the micelles. Due to this, the effect of surfactants is far in excess of the concentration used. An efficient surfactant is relatively insoluble as individual ion or molecule in the bulk of a solution, e.g. 10^{-2}-10^{-4} mole l^{-1}.

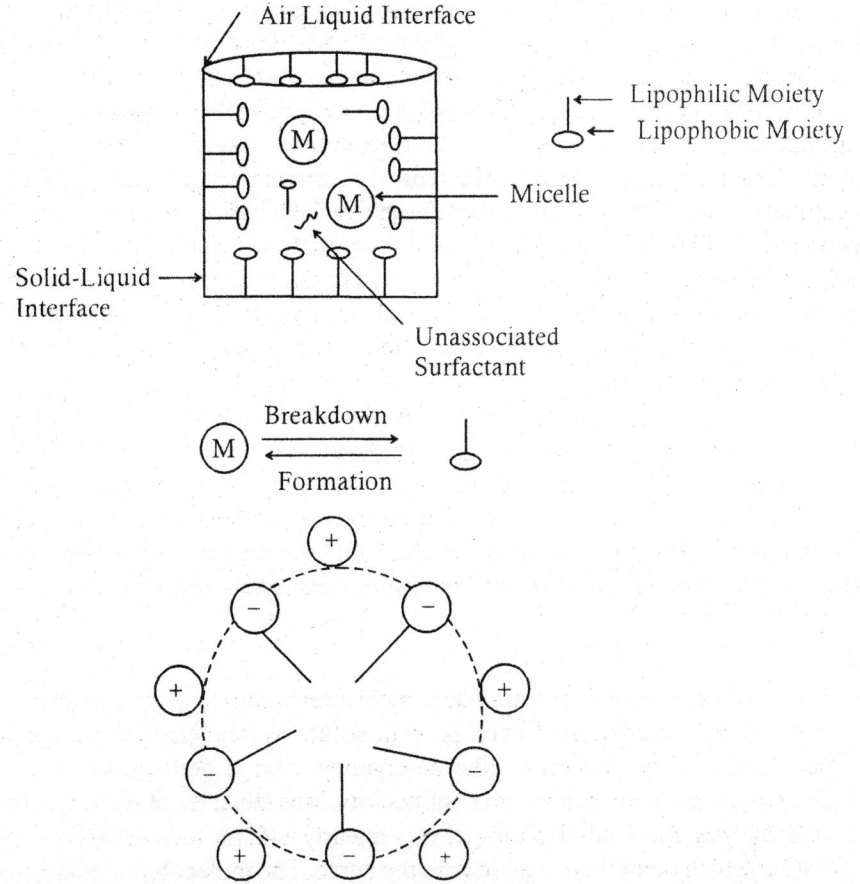

Spherical Micelle of an Anionic Surfactant

Fig. 3.1: Interface, Micelle Formation and a Typical Micelle of an Anionic Surfactant with its Counter Ions

There is a relatively larger number of adsorbed molecules at the air-liquid and solid-liquid interfaces than in the solution (Fig. 3.1) and the adsorbed molecules are so oriented that their lyophilic (solvent liking) moieties are in the solvent and the lyophobic (solvent hating) moieties in the liquid-air interface and the solid-liquid interface. The surfactant molecules aggregated in the micelles are oriented with their lyophilic moieties exposed to the solvent and shielding the lyophobic moieties in the centre of the micelles. The adsorbed molecules are in equilibrium with the micelles through relatively small concentration of individual solute molecules in the bulk of the solution. The liquid-liquid interfaces and the liquid-solid interfaces are encountered in emulsified droplets and suspended solids respectively.

The adsorption at the liquid-air, liquid-liquid or solid-liquid interface may be positive or negative. Low attractive force between the molecules or ions of a surfactant as compared to forces between the molecules or ions of a solvent are responsible for positive adsorption. The other situation leads to a negative adsorption and is associated with surface tension slightly higher than pure water as in concentrated aqueous solutions of inorganic compounds like sodium hydroxide. The adsorbed layer has closely packed molecules forming multilayers.

The force holding an adsorbate on the adsorbent may be physical, ionic or chemical. The physical adsorption is weak attraction primarily owing to van der Waals forces. The ionic adsorption is between charged sites on the substrate and oppositely charged surfactant ions. The linkage due to covalent bonds or forces of comparable strength causes chemisorption and it is not of much significance in case of surfactants. The smoothness, cleanness, particle size, packing, capillaries etc. of the solid adsorbent effect the adsorption characteristics. Typical substrates on which effects of adsorption are important include metals, glass, plastics, textile fibres, sand, crushed minerals, plant foliage, paper etc. The shape of adsorption isotherms helps in studies on adsorption mechanisms. However, their exact mode of action has not been established for the numerous absorbent-absorbate combinations and their interactions in multicomponent surfactant systems.

The physical and ionic adsorption can be both mono- and multi-layer. Capillaries of 1-2 molecular diameter exhibit hysteresis effect on desorption. The extent of surfactant adsorption on metals, glass, textile, plastics, paper and many mineral surfaces generally follow the order: cationic, anionic, nonionic. The adsorption of ionic and amphoteric surfactants is affected by pH; the latter being least soluble at their isoelectric point, the degree of adsorption and speed being increased by the presence of dissolved inorganic salts in surfactant solutions.

Micelles

In a defined physical environment, the surfactant molecules or ions aggregate to form micelles at concentrations characteristics of each solvent-solute system (called critical micelle concentration, CMC). The properties like detergency, charge density, surface tension, conductivity, osmotic pressure, equivalent conductivity, interfacial tension, refractive index, light scattering, dialysis, dye solubilization etc. vary linearly with the increasing concentration up to the CMC at which point there is a break in the curve. The molecular orientation will be reversed in a non-aqueous system.

Micelle size is expressed as the micellar molecular weight or the aggregation number (the number of monomers making up the micelle). The aggregation numbers generally lie between

20 and 100 for single chain anionic and cationic, and over 1000 for the non-ionic surfactants (near the cloud point).

In a surfactant series with a constant hydrophilic group, increasing size of the hydrophobic group decrease CMC values for both ionic and non-ionic surfactants. At constant hydrophobe size, CMC values decrease with the decreasing ethylene oxide content of non-ionic surfactants. Increasing the electrolyte concentration lowers the CMC for both anionic and non-ionic surfactants, especially for the former. The CMC of anionic surfactants increases with temperature and that of non-ionic surfactants decreases. Under ideal mixing conditions and plotting a phase separation model, prediction for surface tensions and the CMC of mixtures can be made.

Small micelles in dilute solutions close to the CMC are generally spherical. Under other conditions, these may be oblate and prolate spheroids, vesicles (double layers), rods and lamellae.

Hydrophile-Lipophile Balance (HLB)

Hydrophilic-lipophilic nature of surface active agents is responsible for many of their characteristic properties in solutions. It is determined by assigning a HLB number to each surfactant, which is related by a scale (Table 3.1) to suitable applications.

TABLE 3.1. HLB *VIS A VIS* POSSIBLE APPLICATION OF THE EMULSIFIER

HLB	Application
3-6	W/o emulsifier
7-9	Wetting agent
8-18	O/w emulsifier
13-15	Detergent
15-18	Solubilizer

The scale is devised so that the more hydrophilic surfactants have the higher HLB numbers. For most polyhydric alcohols, fatty acid esters, approximate values can be obtained from the formula:

$HLB = 20(1 - S/A)$, where, S is saponification number of the ester and A, acid number of the acid.

For many fatty acid esters it is difficult to get good saponification numbers. For these, the equation $HLB = (E + P)/5$ is used, where, E is % weight of oxyethylene content and P is % weight of the polyhydric alcohol.

When only ethylene oxide is used to form the hydrophilic portion of fatty alcohol-ethylene oxide condensation products, the above equation becomes, $HLB = E/5$. A range approximation of HLB can be obtained by observing the water solubility of the compound, e.g. 1-4, no dispersibility in water, 4-6, poor dispersion, 6-8, milky dispersion after vigorous agitation, 8-10, stable milky dispersion; 10-13, from translucent to clear and above 13, for clear solution.

While using mixtures of surfactants of known HLB, fairly accurate products of desired HLB can be prepared by the equation:

$$HLB = (HLB_1 N_1 + HLB_2 N_2)/100.$$

where, HLB_1 and HLB_2 are the HLB values and N_1 and N_2 are weight percentage of the two surfactants.

For example if Span 60 having HLB of 4.7 and Tween 60 with HLB of 14.9 are to be used to form a product of HLB 10, then

$$\frac{4.7 X N_1 + 14.9 \times N_2}{100} = 10$$

The percentages of the two surfactants in the mixture are 48.04 and 51.96% respectively. Now computer software have been developed for the purpose.

Role in Pesticide Formulation

Surfactants play a very important role in pesticide formulation. Their choice may often make or mar the practical pest control potential of a product. The role, however, differs from product to product. For example, in wettable powders and water dispersible granules, they are wetting, dispersing and suspending agents. In emulsifiable concentrates, they act as emulsifiers intended to yield instant emulsion of the product on dilution with water. The suspension concentrates need surfactants to remain in suspension, and so on.

Wettable Powders

Surfactants facilitate suspension of particles in water and prevent their flocculation for a long time. The flocculation is the result of van der Waals forces between particles in close proximity. The surfactants counteract these forces by providing electrostatic or steric repulsive forces (Fig. 3.2). The ionic surfactants account for the electrostatic repulsion by forming electrical double layer.

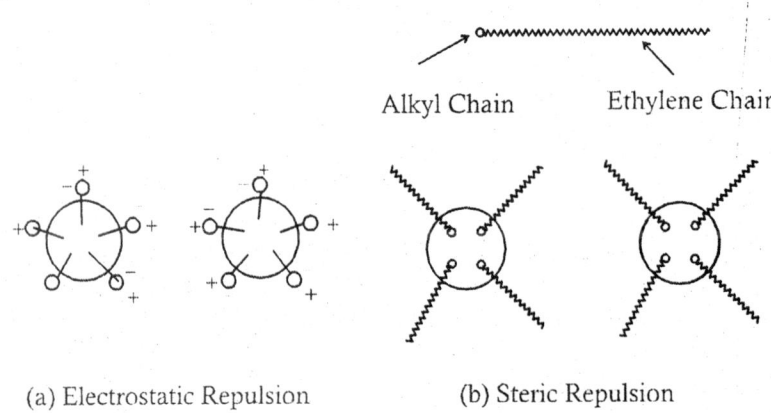

Alkyl Chain   Ethylene Chain

(a) Electrostatic Repulsion   (b) Steric Repulsion

Fig. 3.2: (a) Electrostatic Repulsion by Ionic, (b) Steric Repulsion by Non Ionic Surfactants

The non-ionic surfactants account for steric repulsion by adsorbed layer of surfactants or macromolecules. In general, anionic or a combination of anionics are reported to give a satisfactory performance with respect to dispersing and suspending of the WPs in water.

Emulsifiable Concentrates

These are intended to emulsify the toxicant in water before application and to keep the emulsion stable by preventing creaming, sedimentation, coalescence etc. Creaming or sedimentation can be controlled if KT is more than $4/3a^3\Delta\rho gL$ where, K = Boltzmann constant, T = temperature, a = droplet radius, $\Delta\rho = \rho - \rho_w$ (difference between densities of the particle and water), g = acceleration due to gravity, L = length of the container.

Coalescence is the result of thinning and disruption of liquid films between emulsion droplets. This stretches the film leading to increased surface area (Fig. 3.3). The gaps created on film by stretching are occupied by the surfactant after a finite time by which the fluctuations are dampened (Gibbs-Maragonni effect). The prevention of coalescence can be achieved in practice by using mixture of non-ionic surfactants like Spans and Tweens. They provide films, which provide steric interactions that prevent thinning and disruption.

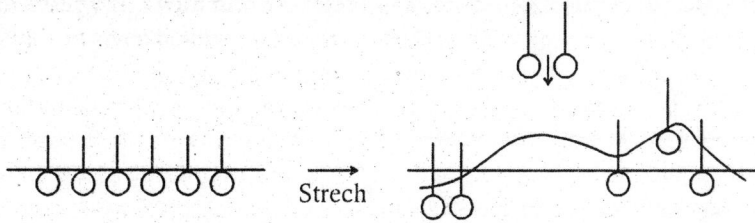

Fig. 3.3: Thinning and Disruption of Emulsion Droplets Leading to Film Stretching

Ostwald ripening is another course of emulsion instability. It is based on the fact that no two liquids (inner and outer phases of emulsion) are completely immiscible. Hence, transfer of molecules between droplets through the continuous phase is possible from small to large droplets. The driving force is the difference in solubility of small and large droplets. The presence of adsorbed surfactants helps in reducing this driving fore.

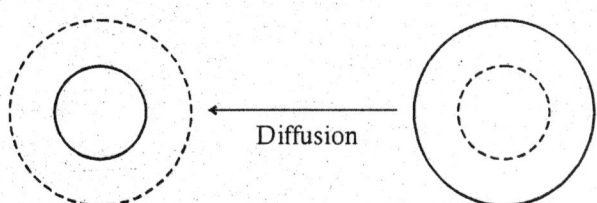

Fig. 3.4: Ostwald Ripening in Emulsions

Suspension Concentrates

These employ surfactants, which adsorb on the particles rendering them stable in the colloidal system. The chains of surfactants are essentially so strongly adsorbed on the particle surface as not to allow the movement or the displacement when the particles approach each other particularly in a concentrated suspension. Further, the adsorption per unit area should be sufficiently high to ensure complete coverage of the particles and provision of sufficiently thick barrier to prevent flocculation on storage. This is the basis of a novel range of surfactants called 'hypermers and polymeric surfactants'. For an effective wetting of solid actives, low foaming ethoxylated alkylaryl phenols (0.5-3.00%) are used. Foam should be avoided in SC production. The wetting agents have a cloud point above 60°C, a temperature that may be reached during milling. If the cloud point is lower, desorption of the surfactant from the particles may occur at the cloud point. The surfactants also need to overcome van der Waals forces to prevent flocculation.

The role of surfactants in SC includes wetting, air displacement, preparation of pumpable premix (before milling), quick wetting of newly formed particles (during milling); and physical stability of suspension by creating electrostatic or steric repulsion between the particles (after milling).

General Considerations

Surfactants used as dispersing agents get adsorbed on to the surface of particles and help to prevent the particles from re-aggregating during preparation and storage by forming a charged or steric barrier around each particle. Sometimes, a mixture of anionic and non-ionic surfactants is used to modify the thickness of the adsorbed layer and increase physical stability. There is an optimum partition coefficient for maximum performance of a pesticide against a given weed or insect. The effect of surfactant is to increase the apparent hydrophobicity of the pesticides. Thus if the compound were slightly too hydrophilic to possess optimum partition coefficient, the surfactant would decrease the observed biological activity. If, however, the candidate pesticide were slightly too lipophilic, the decrease in apparent coefficient as affected by surfactant would cause an improvement in biological response. Thus, a given surfactant can either increase or decrease apparent biological activity, depending upon the relationship between the partition coefficient of a given pesticide and an idealized optimum partition coefficient.

Two important parameters for dispersing power of a surfactant are the molecular weight and the nature of the hydrophobic group. Generally, the higher molecular weight hydrophobes are slower to diffuse in the particle surface but are strongly adsorbed into it because they can provide many more anchoring points on the surface. They are less likely to be displaced from the surface and, therefore, can impart better long-term stability.

This effect of increasing molecular weight is being witnessed from the shift in use of conventional surfactants. Alkyl sulphates (e.g. sodium dodecyl sulphate) and alkylaryl ethylene oxide condensates (e.g. nonyl phenol ethoxylates), which have hydrophobic groups with an average molecular weight around 200-300 units are being replaced by polyelectrolytes of higher molecular weight such as naphthalene formaldehyde condensates (2200 units), or lingosulphonates (upto 20,000 units). These poly-electrolytes are now frequently used in preparation of suspension concentrate formulations. These are also very useful for preparation of water dispersible granules where their good water solubility makes them ideal both as dispersing agent and binder. In particular, molecular weight and the degree of sulphonation in the molecule can modify the hydrophobicity and the water solubility of lignosulphonates.

High molecular weight polymeric surfactants have been developed as dispersing agents to improve storage stability and in some cases, for preparation of highly concentrated suspension concentrate. These polymeric surfactants have a very long hydrophobic "backbone" to form the teeth of these so-called "comb" or "rake" surfactants. Polymeric surfactants, therefore, have good water solubility and at the same time are adsorbed very strongly on to particle surfaces because of the many anchoring points provided by the high-molecular weight hydrophobic group. In some cases the amount of polymeric surfactant absorbed per square meter of surface may be up to ten times higher than with conventional surfactants. Polymeric surfactants are generally non-ionic and on adsorption form more compact layers than anionic surfactants. This effect enables preparation of suspension concentrates, suspoemulsions, and oil/water emulsions of higher concentrations.

Three major types of polymeric surfactants, random, ordinate and comb, based on their chemical structure are gaining prominence in new formulations. Random structured surfactants are derived from polyalkene glycols, polyols, aliphatic carboxylic acids, aliphatic and/or aromatic polycarboxylic acids or anhydrides. These products have three-dimensional networks. An example of this type is Atlox 4914 with HLB value of 6. They form diffused layers in which the concentration of the polymer is not significantly greater than that in the solution.

Comb or graft copolymers are structured polymers that comprise a continuous absorbing polymer backbone into which water-soluble segments are combined like teeth in the comb. They are highly effective in keeping the particles in suspension. Adsorption of a continuous hydrophobic backbone on to a particle surface is stronger. They give more concentrated layers than the random type surfactants and the difference between the equilibrium concentrations of the layer and solution is significantly higher. Atlox 4913 is a comb type surfactant.

In silicone surfactants the composition of hydrophobic polymeric backbone ensures strong adsorption of a special type of graft polymers in which the polymeric carbon chain backbone is replaced with a silicone chain. They impart extremely low surface tension (<25 mN^{-1}m), interfacial tension and contact angles ($<20°$) and are extremely good wetting agents. Alkylated (graft) vinyl pyrrolidone copolymers, such as Agrimore range products are other examples of surface-active comb polymers. Alkylpolysaccharides are special class of new generation surfactants with qualities of anionic and cationic surfactants that get degraded rapidly in environment.

Some surfactants possess unfavourable environmental properties e.g. tallow amine ethoxylates have greater fish toxicity than many active ingredients. Alkylphenol ethoxylates possess estrogenic side effects and some of the cationic surfactants cause skin and eye irritation. Therefore there is a gradual shift towards newer biodegradable products that are non-toxic to mammals and fish. The high molecular weight surfactants such as lignosulphonates, some sugar based ethers and alkyl polysaccharide surfactants are being tried by various formulators.

SYNERGISTS

A substance, which while formally inactive or weakly active, can significantly enhance the activity of the active ingredient in a formulation —IUPAC (1996)

A pesticide synergist may be defined as a chemical which may or may not be toxic in itself but when added to a toxicant, increases its potency in such a way that the total biological activity attained is much more than that could be accounted from a simple arithmetic summation of the activities of the individual components. Attempt to supplement the powerful but transient paralytic action of pyrethrum with a more permanent toxic effect by adding rotenone to pyrethrum sprays seems to be the beginning of research on pesticide synergists. The report that N-isobutyl-undecyleneamide (IN-930) considerably increased the toxicity of pyrethrum sprays to which it was added appears to be the first step in search of a true synergist (Weed, 1938).

Unlike rotenone, this substance is itself, non-toxic. It is desirable for general acceptance that the synergist should be comparatively non-toxic, cheap and easily available and have the least detrimental effect on pesticide formulation.

The era of pyrethrum synergists actually started with the discovery that sesame oil, a non-toxic chemical, when used alone or diluted with kerosene, at 5 per cent by volume, in a kerosene based pyrethrum spray, increased the mortality of the houseflies from 8 to 88 per cent (Eagleson, 1942). These findings along with the discovery of other chemical constituents of sesame oil (sesamin, **SMN**, sesamolin, **SMO** etc.) seem to have given birth to the idea of synergism in the field of pesticides.

The main advantages/reasons of using a synergist are to increase efficiency and safety of pesticidal products and to reduce cost (economy).

Classification

The synergists can be classified both on the basis of the toxicant they synergise (i.e. pyrethrum synergists, DDT synergists etc.) and on the chemical detoxifying enzymes they inhibit (*viz.* mixed function oxidase inhibitors, hydrolase inhibitors etc). However, chemical classification based on their chemical nature or structure of the synergist molecule is the most preferred one. On the basis of their structure, various synergistic compounds have been grouped as follows:

1, 2, 3-Benzthiadiazole Synergists

Substituted 1, 2, 3-benzthiadiazole derivatives inhibit phenolase enzymes and act as synergists for pyrethrins. 5, 6-Dichloro-1, 2, 3-benzthiadiazole **DCBD** and 5-chloro-6-methyl-1, 2, 3-benzthiadiazole **CMBD** are effective pyrethrum synergists.

Analogue Synergists

These are compounds that resemble in structure to a particular insecticide and synergise its activity e.g. 4, 4-dichlorophenyl methyl carbinol (**DMC**) and p-chlorophenyl N, N-dibutyl sulphonamide (**WARF-Antiresistant**) are synergists for DDT; salithion, triphenyl phosphate and O, O, O-tri(2-hydroxyphenyl) phosphate are in general synergists for organo-phosphorus compounds, and so on.

Benzo-1, 3-dioxole synergists

Methylenedioxyphenyl or MDP pyrethrum synergists are derivatives of methylenedioxybenzene e.g. piperonyl butoxide (**PBO**), sulfoxide (**SFO**), dillapiole (**DA**)etc. This group of compounds inhibits mixed function oxidase and synergizes, besides pyrethrins, organophosphorus, organocarbamate and organochlorine insecticides also.

Imidazole Synergists

Imidazole derivatives, substituted at 1, 2 or 4 positions, inhibit microsomal oxidase and synergise insecticides liable to quick detoxification by oxidation in the living organism. For example 1-(2, 3-dimethyl phenyl) imidazole (**DPI**); 1-(1-naphthyl) imidazole (**NI**) etc. are synergists for pyrethrins.

Aryl 2-propynyl ether synergists

Aryl 2-propynyl ethers of substituted phenols are inhibitors of oxidases and some other insecticide metabolising enzymes. They synergizse N-methyl carbamate insecticides. The most active compound of the series is 3-(2-nitro-4-chlorophenoxyl) prop-1-yne **(NPE)**.

Miscellaneous

All those compounds, which are not a part of a big group or are under experimentation, can be put in miscellaneous sub-group e.g. MGK- 264, numerous alpha, beta-unsaturated carbonyl compounds, acylated 1, 3-indandiones etc. Each of the above groups may be further sub-classified depending upon the major skeleton or the important functional group present in the molecule *viz*. ethers, esters etc.

Structures of some synergists are given in Fig. 3.5. Some important synergists for insecticides are listed in Table 3.2.

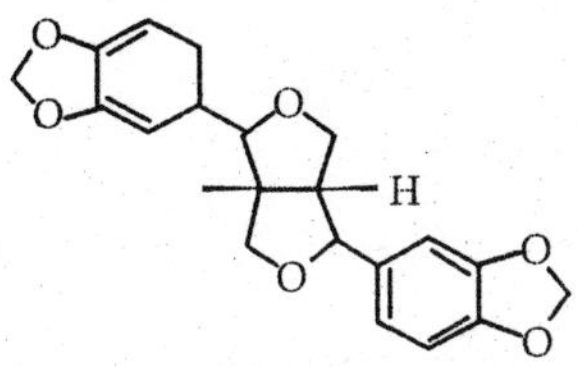

SMN

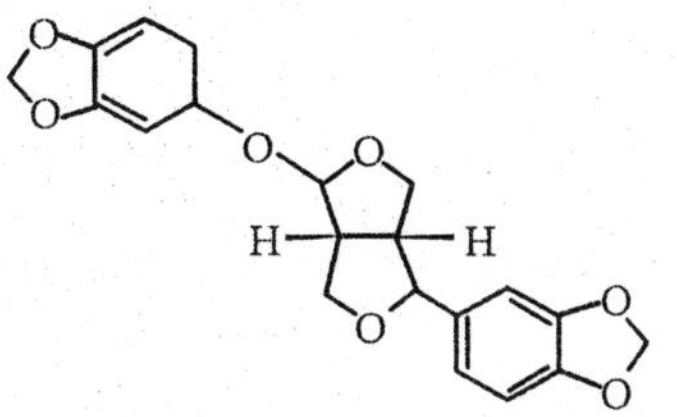

SMO

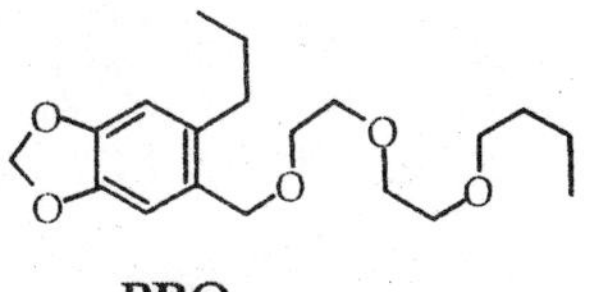

PBO

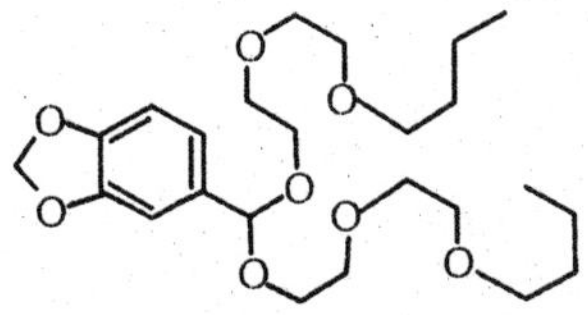

TPI

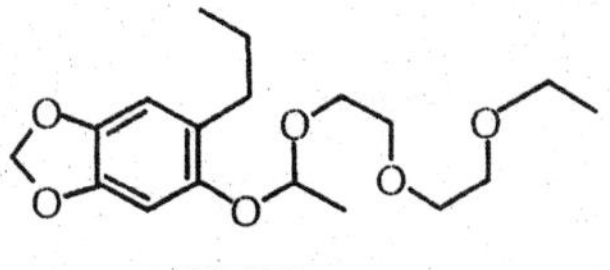

SME

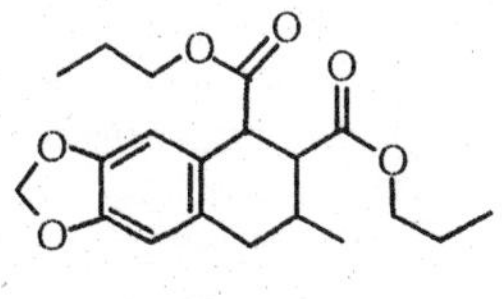

PRI

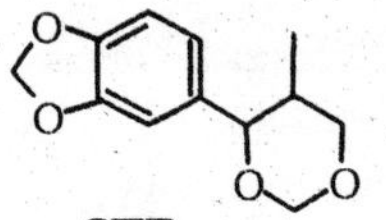

SFR

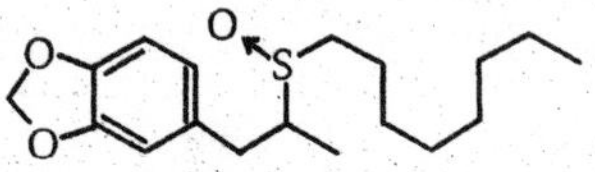

SFO

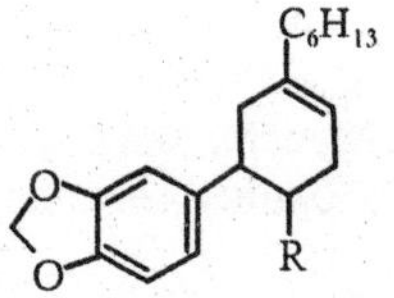

$R = H$ and $COOC_2H_5$

PCN

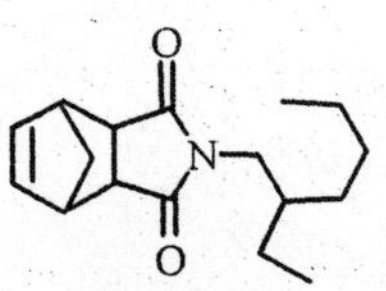

MGK

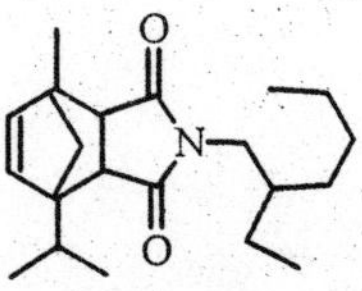

SPR

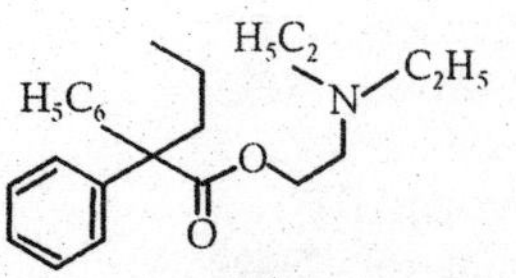

SKF

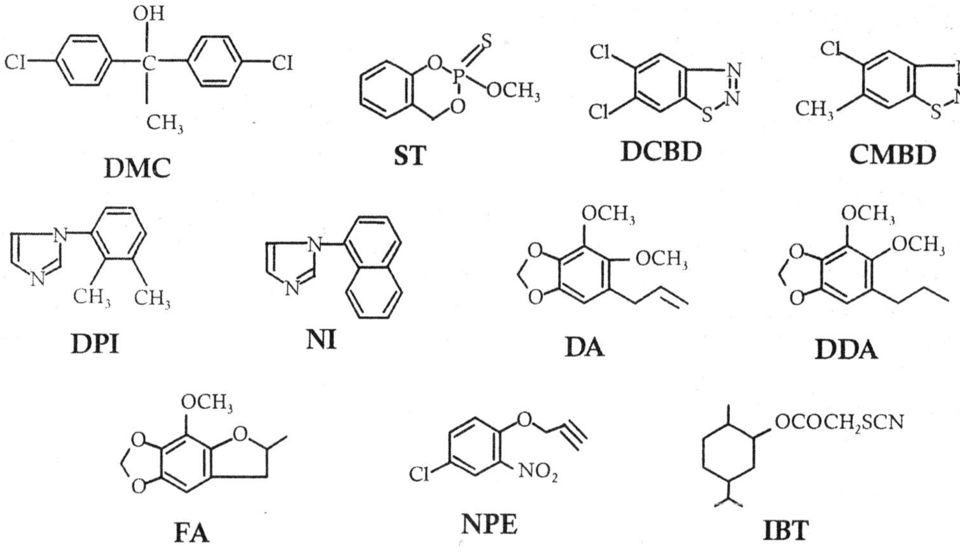

Fig. 3.5: Structures of Some Synergists

TABLE 3.2. SOME IMPORTANT SYNERGISTS FOR INSECTICIDES

Common name/Code	Insecticide Synergized
Chloromethylbenzthiadiazole (**CMBD**)	Pyrethrins
Dichlorobenzthiadiazole (**DCBD**)	Pyrethrins
Dichlorophenyl methylcarbinol (**DMC**)	DDT
Dillapiole (**DA**)	Pyrethrins
Dihydrodillapiole (**DDA**)	Pyrethrins
Dimethyl phenylimidazole (**DPI**)	Pyrethrins
Isobornyl thiocyanoacetate (**IBT**)	Pyrethrins, Carbaryl
Furapiole (**FA**)	Pyrethrins
MGK - 264 (**MGK**)	Pyrethrins, Organochlorine
Naphthylimidazole (**NI**)	Pyrethrins
Nitrochlorophenyl propynyl ether (**NPE**)	Carbamate
Piperonyl butoxide (**PBO**)	Pyrethrins, Organochlorines, Organophosphates
Piperonyl cyclonene (**PCN**)	Pyrethrins
n-Propyl isome (**PRI**)	Pyrethrins
Safroxan (**SFR**)	Pyrethrins
Salithion (**ST**)	Organophosphates
Sesamex (**SME**)	Pyrethrins
Synepirin (**SPR**)	Pyrethrins
Sulfoxide (**SFO**)	Pyrethrins
SKF 5254 (**SKF**)	Pyrethrins
Tropital (**TPT**)	Pyrethrins
Sesamin (**SMN**)	Pyrethrins
Sesamolin (**SMO**)	Pyrethrins

Structure Activity Relationship

The synergistic ability of a chemical towards a toxicant depends largely on its molecular structure. All synergists need to reach the site of action at the same time or a little before the toxicant so as to engage or exhaust the enzyme(s) responsible for detoxification of the toxicants and act as alternate substrate for the detoxifying enzymes. Most of the findings on structure-activity relationship among synergists are based on studies on benzo-1, 3-dioxoles as pyrethrum synergists and aryl-2-propynyl ethers as synergists for N-methyl carbamates.

Benzo-1,3-dioxoles: Among the benzodioxole or MDP synergist molecules, optimum synergism requires only one methylenedioxy function and a C_3 to C_5 carbon side chain alkyl substituent in the benzene nucleus. One to three ether linkages or alkoxy groups in the side chain may help in synergism by imparting appropriate hydrophile-lipophile character. An additional methylenedioxyphenyl moiety generally results in loss of synergism with pyrethrins (Winkinson, 1971; Parmar *et al.*, 1974; Tomar *et al.*, 1979; Parmar and Tomar, 1983). Substitution of oxygens in methylenedioxy moiety by $-CH_2$ group results in loss of activity though substitution of O by S results in lowering but not loss of activity. Substitution of methylenic hydrogens by CH_3 or substitution of methylene by ethylene groups also results in loss of activity. Intactness of methylenic hydrogens, and to some extent of oxygen in methylenedioxy moiety is underlined.

Aryl 2-Propynyl Ethers: This series of synergists reveals maximum activity when alkynyloxy side chain is attached to naphthalene nucleus or a substituted phenyl nucleus. In the side chain, the tripple bond (acetylenic bond) is a necessary requirement. This bond should not be more than three carbon away from the ether oxygen. The activity is in the triple bond *per se* and not the acetylenic hydrogens. Replacement of hydrogen by iodo- or methyl group results in higher activity (Barnes and Felling, 1969).

Slight activity of 1-propynyl naphthalene indicates that the requirement for ethereal linkage is not absolute. Also, the replacement of ether oxygen atom by a sulphur atom in the molecule does not result in much loss of activity; however, this loss of activity is more in phenyl 2-propynyl thioethers than that in the corresponding naphthalene derivatives.

Analogous to 1, 3-benzodioxoles, the synergistic activity of the phenyl (but not naphthyl) propynyl ethers is dependent on aryl substitution, suggesting the influence of appropriate lipophilicity of the molecule on the synergistic efficacy.

Mode of Action

Though the mechanism of synergism in insecticides has been much investigated, many details remain to be worked out. Broadly speaking, in the absence of synergist, the insect is able to metabolize some or most of the dose of the insecticide to non-toxic compounds or compounds much less toxic than the insecticide. The synergist reduces the metabolic rate of detoxification by offering itself to the detoxifying enzymes as an alternative substrate. Thus, a greater portion of dose of the insecticide remains available for a toxic effect in the presence of synergist than in its absence, resulting in death/control of the insect. The sites of detoxification and of toxic action being separate, an understanding of the synergism does not depend merely upon the knowledge of the mode of action of the insecticide. Moreover, the same enzyme complex within the insect appears to be responsible for oxidative detoxification of insecticides of widely different chemical structures, so that a given synergist may synergise insecticides differing widely in their mode of action. In a broad sense, the synergistic action of benzodioxoles (mdp synergists) towards carbamates, pyrethroids, cyclodienes and certain organophosphorus compounds, can be accounted for by the fact that these depress the metabolism (oxidative detoxification) of the insecticides. Oxidative metabolism converts certain other organophosphorus insecticides into compounds of higher insecticidal potency than the original insecticides, whence benzodioxole compounds are prone to antagonise these. Thus, piperonyl butoxide may synergise pyrethrins but will antagonise malathion. DDT is metabolised by dehydrochlorination and/or oxidation to dicofol like substances, especially in resistant strains. Synergists such as DMC depress the dehydrochlorination. WARF- antiresistant appears to synergise DDT by depressing both dehydrochlorination and/or oxidation. One of the mechanisms for detoxification of organophosphorus insecticides (such as parathion, malathion etc.) is by carboxyesterase hydrolysis, especially in resistant strains of insects. Correspondingly, such insecticides are synergised by triphenyl phosphate, O, O, O-tri (2-hydroxyphenyl) phosphate, and a number of other related compounds like salithion.

Metcalf (1967) and Casida (1970) have made substantial contribution relating to mode of action of benzodioxoles pertaining to depression of oxidative detoxification of many insecticides *in vivo*. They have shown that insect microsomal systems metabolise such derivatives by reactions involving the active methylene group (i.e. the $-CH_2-$ between the two oxygen atoms of the dioxole moiety) to yield catechols via the formation of unstable benzodioxonium ion which itself may form complex with certain essential metal ions like Fe^{2+} or Cu^{2+}, or through its unstable hydroxy intermediate. The catechols, in turn, disrupt the electron transport chain in the biological system by oxidation to quinones, which in presence of hydrogen and electrons, reform catechols; thus disruptining the metabolic chain.

Practical Utility

The purpose of adding synergist to a formulation will depend upon the toxicant being used and efficiency and economy of the final product. Most of the synergists have been used with pyrethrins and a fewer with DDT. The purpose for synergists in pyrethroid formulations is to keep down the cost and increase the efficiency while that for including them in DDT formulations was to control DDT-resistant strains of insects and to prevent the appearance of such strains in future.

The three major advantages of using synergists are economy, efficiency and safety in use of pesticides. These advantages can be achieved through proper selection and blending of the adjuvants with toxicant.

(i) *Economics:* The use of synergists is recommended for very costly pesticides causing short term effect. They are of particular importance where availability of the toxicant is scarce as in case of natural pyrethrins. The proportion of blending synergist to insecticide has to be worked out carefully for the economic use of insecticides as per isobologram of Hewlett (1968) in Fig. 3.6.

An elegant application of isobolograms occurs in determination of insecticide-synergist mixture of minimum cost. Instead of plotting the doses of insecticide and synergist in ordinary scientific units, these doses can be plotted in units of cost. Fig. 3.7 shows a cost isobologram.

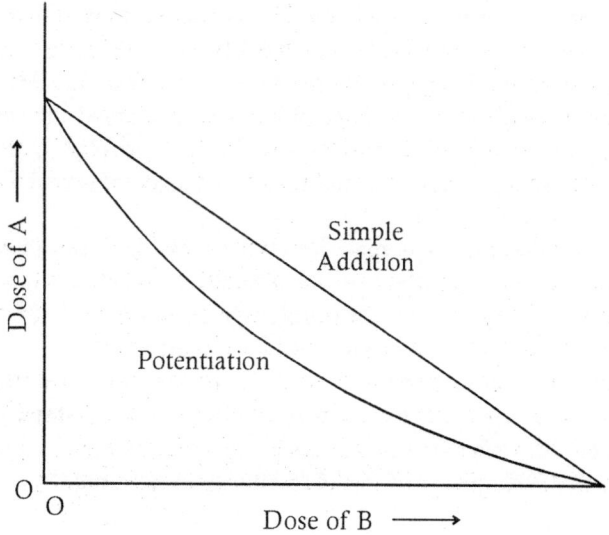

Fig. 3.6: Isoboles for Addition and Potentiation

Curves II, III and I are hypothetical isoboles for different synergists. The cost of a formulation containing both synergist and insecticide is $(x + y)$. If the use of a synergist is to be economic at all, the isobole must dip below the line $x + y = y_0$ (Curve I) where, x and y are costs of synergist and insecticide respectively.

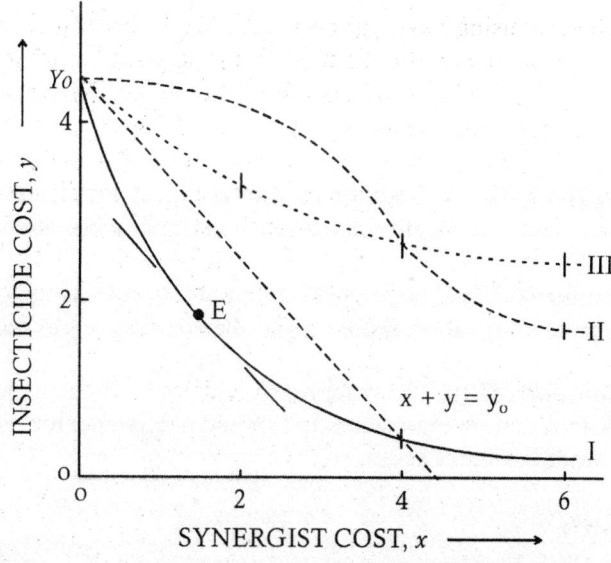

Fig. 3.7: Cost Isobologram for Insecticide Synergist Mixture

(ii) Efficiency: The unwanted qualities like development of resistance towards pesticides can be minimized by use of synergists, since their addition provides a new or altered mode of action for the toxicant. They not only make the toxicants more efficient but also extend their commercial lives. The synergists may also broaden the spectrum of activity of a pesticide e.g. DDT which is known to be less active against the body louse as well as the housefly reportedly controls them effectively if mixed with DMC. These also help in restoring the activity of a pesticide against the resistant strains of insect pests e.g. the use of WARF-antiresistant brings down the resistance to DDT in DDT-resistant houseflies. This also enables an extension of useful life span of pesticides.

(iii) Safety: The addition of synergists results in a reduction of amount of the active ingredient and thus smaller quantities of the pesticides are required for the same degree of effectiveness. This mitigates the problem of excessive pesticide residues and makes the use of pesticides safe for mankind and the environment.

Besides the above practical considerations, the fundamental investigation of synergism and synergists has led to a much better understanding of the mechanism of action and detoxification of insecticides in insects and basic biochemical processes involved in development of insect resistance to insecticides.

Requirements

Along with the biological potency, a synergist should meet the requirements of cost, mammalian toxicology and compatibility. The synergist, as far as possible, should improve the physical and chemical properties of insecticide and should have no deteriorating effect on shelf life of the formulation. Therefore, following basic considerations should be kept in mind before selecting a synergist.

(i) Cost: Main object of using a synergist is to reduce the cost of insecticide without any loss of potency. If the material to be used as a synergist is costly in itself, it may not serve the desired purpose e.g. synergists like DMC or WARF-antiresistant for DDT could not become practical owning to cost.

(ii) Mammalian Toxicity: The addition of synergist must not increase the mammalian toxicity of the product. Chronic toxicity should be paid particular attention in this regard.

(iii) Physical Compatibility: The physical compatibility such as solvency, stability etc., desirable for a given formulation, should not be altered negatively by the use of a synergist.

(iv) Chemical Compatibility: The synergist should not react chemically with either the toxicant or the formulation ingredients and should not hamper the movement of active ingredient towards the site of action.

Quantitative Aspects

Level of Synergist: For a fixed level of response (knockdown or kill, say 50%), plot a graph between the insecticide and synergist doses. A curve of the type shown in Fig. 3.8 is obtained.

The isobole leaves a point on the ordinate (insecticide) axis to become asymptotic to a line parallel to the abscissa (synergist) axis. The ratio of the dose of insecticide without synergist to its symptotic dose, as above, is the maximum synergistic ratio possible.

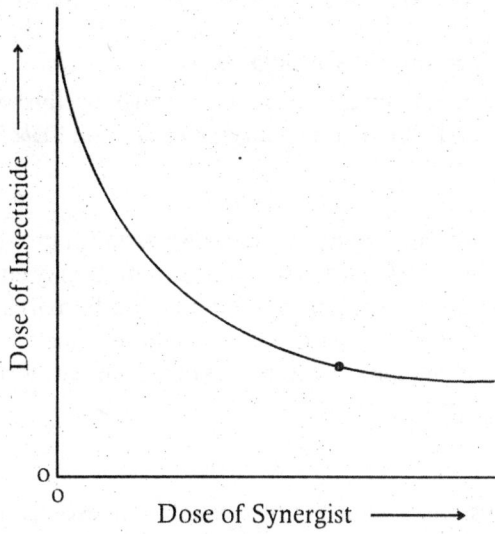

Fig. 3.8: Typical Isobole for Determining Optimum Levels of Insecticide and Synergist

Scale of Synergism: The degree of synergism is expressed in terms of factor of synergism (FS) or synergistic ratio (SR). This is defined as the ratio of ED_{50} of insecticide alone to that of the insecticide in the synergized formulation, where, ED_{50} is the effective dose which may cause 50% knockdown (KD_{50}), or 50% kill (oral, LD_{50} or lethal concentration, LC_{50} etc.) of the test insect.

Relative Potency of Synergists: The comparative efficacy of synergists for an insecticide is best expressed in terms of their relative potency. It is the ratio of ED_{50} for a standard synergist and toxicant combination to the ED_{50} of the experimental synergist plus the toxicant. The ratio provides an idea about the potency of new material in relation to a standard when used at the same level.

$$\text{Relatively potency} = \frac{ED_{50}\text{of insecticide } + \text{Reference synergist}}{ED_{50}\text{of insecticide } + \text{Experimental synergist}}$$

Synergistic Response: Various attempts have been made to fit a mathematical relationship to the results of synergistic efficacy changes with the varying ratio of pyrethrins to synergist concentration in a toxicant formulation. Using a fixed dose of toxicant, the toxicity can be varied by altering the dosage of the synergist.

$$y = a + b\log x_T + c\log x_S$$

where, a, b and c are constants and x_T and x_S are the concentrations of the toxicant and synergist respectively.

Alternatively, a different and empirical approach providing an approximate linear relationship between the factor of synergism and the ratio of synergist to pyrethrins can also be reported as given below:

Log (factor of synergism) $= a + b \log(x_S/x_T)$.

This equation has been manipulated from studies on houseflies with cinerin and pyrethrin I using piperonyl butoxide (PBO) on *Calandra granaria* L. and *Tribolium castaneum* Herbst.

Limitations

A synergist is generally effective for a particular insecticide or group of insecticides and hence it cannot synergise all the insecticides to the same extent. It may enhance the activity of some other groups of insecticides by engaging or inhibiting the detoxifying enzymes but this alone does not qualify the compound to be incorporated in the formulation. The study of joint toxicity and toxicology of the combination and economics of the ultimate formulation are the important guiding factors in this regard.

SEMIOCHEMICALS

These are chemical substances, which induce behavioral response or communicate messages amongst insects. The prefix *semio* is derived from *semeon* (German) meaning a mark or signal (Law and Regnier, 1971). These group semiochemicals includes all chemicals, which can induce any response, attraction, repulsion, fear, alarm etc. that can modify insect's behaviour whether within a group of particular species (intra species, *pheromones*) or in different species (inter species, *allelochemicals*). These are sub classified into several groups depending on whether the response of the receiver is adaptively favourable to the emitter (*allomones*), is favourable to the receiver and not to the emitter (*kairomones*) or is favourable to both emitter and the receiver (*synmones*). Some of the important groups of allelochemicals are described below:

For the sake of practical utility, within semiochemicals, it is useful to sub-classify them as arrestants, attractants, repellents, deterrents, stimulants or by some other descriptive term. These terms can indicate what behaviour is involved in the response such as feeding deterrent (antifeedant) or flight arrestant (anti-migrant).

Attractants

Chemicals that elicit or induce in insects a response to make oriented movements towards their source of origin are called insect attractants. These influence both the gustatory (taste) and the olfactory (smell) receptors. Attractants can be used in conjunction with simple, inexpensive sticky traps for monitoring pest population in relation to the economic threshold, for removal/trapping of pests, for luring them to toxic bait or for creating mating confusion. These can be put into two major groups, pheromones and plant kairomones.

Pheromones

A pheromone is a chemical secreted by an insect to induce or elicit a response in another insect of the same species. A specific reaction takes place between the insects of the same species towards a developmental process. The term pheromone, derived from two Greek words *pheroom* (to carry) and *horman* (to excite or stimulate), was introduced by Karlson and Butenandt (1959). The pheromones, may be further classified on basis of the interaction mediated, as follows:

Aggregation Pheromones: These lead the insects to increase their population density near the sources of their origin. The insects related to orders Coleoptera. Diclyoptera, Hemiptera, Homeoptera give out the aggregation pheromones, which induce the insects to aggregate or collect near the pheromone source. The chemicals secreted help in the selection of mates and also provide defense against predators. The aggregation pheromones increase beetle density near the source on the host trees where successful copulation occurs. 2-Methyl-6-methylene-7-octen-4-ol (ipsenol) is an important aggregation pheromone isolated from Coleopteran species

Territorial Pheromones: Males of many species secrete these, e.g. males of bumblebees mark mating sites that attract both males and females. A mixture of esters of geraniol, citronellol, and farnesol e.g. geranylacetate; 2, 3-dihydrofarnesyl acetate and (E)-farnesyl acetate are among the distinctive sesqui-and di-terpenoid perfumes, utilized by these bees.

Alarm Pheromones: The alarm pheromones are volatile compounds of lower molecular weight and are given out by insects of the order Diclyoptera and Hemiptera. Such compounds exhibit a quick reaction that promotes the escaping tendency in the insects. (E)-2-Hexanol, secreted by the insects of Hemiptera, is present in the dorsal abdominal glands of the larvae.

Epideictic Pheromones: These pheromones act as dispersive agents so that the distances between the insects increase resulting in less intraspecific competition. The insects belonging to the groups Coleoptera, Homeoptera, Orthoptera etc., generate these.

Sex Pheromones: These are mating inducing chemicals that are secreted by the males or females. These help in successful mating of the insects depending on the species which secrete these pheromones. Mating dissuption has been adopted as an approach to control insects; and the first meaningful trial was conducted on mating disruption in Lepidopterous insects (Shorey and Mckelvey, 1977).

Trail Pheromones: These are used to recruit other insects to a new food source so that the emigration of insects is accelerated. These chemicals are present in insects of the order Dictyoptera and Lepidoptera.

Oviposition Deterring Pheromones: The chemicals drive away the female at the time of egg laying. 9-Oxo-(E)-2-decenoic acid present in mandibular glands of Hymenoptera is one such pheromone. Also called ovarian inhibitors, these are applied on a crop to reduce egg laying by the pest species or redirect them to other parts of the plants. Some insects themselves deposit pheromones with their eggs to reduce the laying of eggs by the females of that species.

Plant Kairomones

These botanical chemicals are used by insects in inter species communication for host plant selection. The compounds that work over a long range are called attractants and the others, which act over a shorter distance, are termed as arrestants. Owing to their nature of luring the insects near the site of their application/origin, the latter are put in two major groups *viz. Natural food lures and Oviposition lures.*

Natural Food Lures: These chemicals are present in plant and animal hosts and attract or lure insects for feeding. They stimulate olfactory receptors, and may be chemically a floral scent in case of nectar feeding insects; essential oils for the phytophagus insects; some decomposition products for the scavengers; and carbon dioxide, lactic acid or water for the blood-sucking insects.

Oviposition Lures: These are (natural) chemicals that govern the selection of suitable sites for oviposition or egg laying by the adult female. p-Methyl acetophenone attracts rice stem borer to oviposit and extracts of corn do so for the earworm, *Heliothis armigera*. Surfaces impregnated with these substances attract the insects to lay eggs on them; no matter if they are only a piece of wood or twine. Some of the naturally occurring (botanical) and synthetic products effective as practical food lures are given in Table 3.3 and 3.4 respectively.

TABLE 3.3. SOME NATURAL INSECT FOOD LURES

Insect	Source	Lure
Pests of Cruciferae	Brassica seeds	Isothiocyanates
Onion fly	Onion	Propyl mercaptan
Carrot fly	Umbelliferae	Phenyl propanoids
Lepidoptera	*Araujia sericofera*	Phenyl acetaldehyde
Bark beetle	Bark	Terpenoids
House fly	Sugar cane	Sugar, molasses
Oxythyrea sp.	Floral scent	Cinnamyl alcohol
Southern pine beetle	Pine wood	α-Pinene
Corn ear worm	Corn silk	Silk extract

Note: In nature pure compounds do seldom occur alone and it can be assumed that the insects in general are attracted to their hosts by a mixture of odours (lures)

TABLE 3.4. SOME SYNTHETIC INSECT LURES

Insect	Lure
Oriental fruit fly	Methyl eugenol
Melon fruit fly (*Dacus spp.*)	Ceu-lure (p-Acetoxyphenethyl ketone)
Mediterranian fruit fly	Trimedlure
Japanese beetle	Geraniol
Rhinoceros beetle	Ethyl-3-isobutyl-,2,2- dimethylcyclopropane carboxylate
Japanese beetle	Phenethyl propionate
Japanese beetle	Eugenol
Vespula sp.	Heptyl butyrate

Attractants and pheromones can also be classified on the basis of their molecular structure i.e. alcohols, esters, ketones or aliphatic carbonyls, epoxides etc. However, these being very specific chemicals, their use based classification *viz.* house fly attractant, or response base classification as described above is more popular.

Practical Utility in Pest Control: The population of insect pests is reduced by mass trapping with the help of insect's own pheromone as a lure. The crop may be sprayed with the feeding or oviposition deterrent chemicals for preventing the insects from laying eggs. The sex pheromones of Lepidoptera are used both for monitoring and for communication disruption.

The use of pheromone type compounds was examined in the 1940s in the USA when crude abdominal tips of female Gypsy moths were used in monitoring traps. However, it was not until around mid sixties that the possibility of using pheromones in crop protection was accepted, and the first meaningful field trial on mating disruption of a Lepidopterous pest was reported.

Since Butenandt *et al.* (1959) identified the first pheromone; many compounds have been reported for hundreds of species of insects. Early studies concentrated on pheromones used by social insects, but subsequently attention has turned to non-social, economically important species, especially members of the Coleoptera and Lepidoptera. These include the boll weevil *Anthonomus grandis* (Boheman); the cotton bollworm, *Heliothis zea* (Boddie); the tobacco budworm, *Heliothis virescens* (F.); the bollworm, *Heliothis armigera* (Hübner), the Egyptian cotton leafworm, *Spodoptera littoralis* (Boisd.); the Eastern spruce budworm, *Choristoneura fumigerna* (Clem) and the stripped rice stem borer, *Chilo suppressalis* (Wlk.). Slow release formulations like granules, plastic capsules, tapes and tubes releasing 2-3 μg of lure per hour are generally effective for 6-12 weeks. In other cases the dose is so calculated as to release the pheromones at a constant rate during the trapping period that generally lasts for few months. Mass trapping of insects with certain behavioral pattern is limited to those with a highly developed ability to respond to attractants. An equimolecular mixture of Z, Z and Z, E isomers of 9, 11-tetradecadienyl acetate acts as aggregation pheromone for both male and female boll weevil (*Anthonomus grandis*), a serious pest of cotton.

The value of pheromones in controlling these and other major insects has been examined during the last decade and successful products did emerge. But it is worth noting that the research findings on commercial applications of pheromones in monitoring, mass trapping and mating disruption have been published for less than 5% of species of which pheromones have been identified. Nonetheless, it should be emphasized that all commercially important insect pests appear to use pheromonal communication in some way, often in more than one stage in the life cycle.

Advantages and Disadvantage of Pheromones: The use of pheromones has a number of potential advantages. The compounds are naturally occurring, generally non-toxic and non-polluting to the environment. Pheromones are insect-specific and their safety to beneficial insects makes them ideal components of integrated pest management system. These are devoid of the problem of resistance development.

TABLE 3.5. SEX PHEROMONES OF SOME AGRICULTURAL INSECT PESTS

Common name	Pheromone component	Purpose
Gram pod borer	(Z) 11-Hexadecenal and (Z)-9-Hexadecenal (97 : 3)	Mating
Tobacco caterpillar	(Z, E) 9, 11-Tetradecadienyl acetate and (Z, E) 9, 12-Tetradecadienyl acetate (10 : 1)	Mating
Greasy cutworm	(Z) 7-Dodecenyl acetate and (Z) 9- Tetradecadienyl acetate	Mating
Indigo caterpillar	(Z, E) 9, 12-tetradecadienyl acetate	Mating
Spotted bollworm	(E, E) 10, 12-Hexadecadienal	Mating
Sorghum stem borer	(Z) 11-Hexadecenal and (Z) 11-Hexadecenal (50:50)	Mating
White stripped borer	(Z) 11-Hexadecenal and 13-Octadecenal	Mating
Cabbage looper	(Z) 7-Dodecenyl acetate	Mating
Oriental fruit moth	(E, Z)8-Dodecenyl acetate (7 : 43)	Mating
Oriental fruit fly	4-Allyl-1, 2-dimethoxybenzene	Mating
Melon fly	4-(p-Hydroxy phenyl)-2-butanoyl acetate	Mating
Codling moth	(E, E)8,10-Dodecadienol	Mating
Gypsy moth	(Z) 7, 8-Epoxy-2-methyl octadecane (7R, 8S)-epoxy-2-methyl octadecane	Mating
Stalk borer	(Z)-7-Dodecenyl acetate, (Z)-8-Tridecenyl acetate, (Z)-9-Tridecenyl acetate, (Z)-10-Tridecenyl acetate(1 : 1 : 1 : 1)	Mating
Corn earworm	(Z)-11-Hexadecenal	Mating
Tobacco budworm	(Z)-9-Hexadecenal	Mating
Tobacco budworm	(Z)-7-Hexadecenal	Mating
Navel orangeworm	(Z, Z)-11, 13-Hexadecenal	Mating
Codling moth	(E, E)8-10-Dodecalenol	Mating
Cabbage borer	(Z)-7-Dodecenyl acetate	Mating
Oriental fruit moth	(Z)-8-Dodecenyl acetate	Mating
Grape berry moth	(Z)-9-Dodecenyl acetate	Mating
Beet armyworm	(Z)-9-Tetradecenyl acetate	Mating
European corn borer	(E)-11-Tetradecenyl acetate	Mating
Almond moth	(Z, I)9, 12-Tetradecenyl acetate	Mating
Pink boll worm	(Z, Z)-7, 11-Hexadecadienyl acetate	Mating
Peach tree borer	(Z, Z)-3, 13-Octadecadienyl acetate	Mating
Lesser peach tree borer	(3E, 13Z)-Octadecadienyl acetate	Mating
Gypsy moth	(Z)-7, 8-Epoxy-2-methyloctadecane	Mating

Note: Chemicals can be used for mating disruption or mass trapping.

Some Important Pheromones

In recent years sex pheromones have been identified and synthesized, their functions understood and practical uses earmarked. The largest number of pheromones, known currently belongs to Lepidoptera.

The pheromones of scale insects, Hemiptera spp. are terpenoids or their derivatives containing 10-18 carbon atoms. The beetles and weevils belonging to Coleoptera emit both the

TABLE 3.6. SOME IMPORTANT ALARM PHEROMONES OF COMMON INSECTS

Insect	Chemical class	Chemical
Dolichoderine ants	2-Alkanones	2-Heptanone, 6-Methyl-5-heptan-2-one, 4-Mehtyl, 2-hexanone
Myrmicine ants	Ethyl-ketones (3-alkanones)	3-Octanone, 3-Nonanone, 3-Decanone
	Methyl-branched ketones	4-Methyl-3-hexanone, 4-Methyl-3-heptanone. 6-Methyl-3-octanone, 4, 6-Dimethyl-4-octen-3-one
Formicid ants Odontomachus spp.	Alkyl pyrazines	2, 5-Dimethyl-3-isopentyl pyrazine 2, 6-Dimethyl-3-n-pentyl pyrazine
Dolichoderine ants (genus Azteca)	Cyclic ketones and acetates	2-Methyl-cyclopentanone cis-1-Acetyl-2-methyl cyclopentene 2-Acetyl-3-methyl cyclopentene
Bees	Aldehyde 2-Alkanone Ester	Citral 2-Heptanone Isopentyl acetate
Aphids	Hydrocarbons	E-β-farnescene

Sarode et al. (1993).

TABLE 3 7. TRAIL PHEROMONES OF SOME INSECTS

Insect	Class	Chemical
Trigona spinipes	Alkanols	2-Heptanol, 2-Undecanol, 2-Tridecanol
T. subterranea	Aldehydes	Citral
Lasius fulglenosus	Alkanoic acids	Hexanoic, Heptanoic, Octanoic, Nonanoic, Decanoic, Dodecanoic acids
Zootermopsis nevadensis		Caproic acid
Atta texana	Heterocyclic esters	Methyl-4-methyl-pyrrole-2-carboxylate
Monomorium dharaonis	Indolizine alkaloids	3-Butyl-5-methylocta-hydroiondolizine
Pristomyrex pugens (Hymenoptera)	Fatty acids	C_{14}-C_{20} Unsaturated fatty acids
Reticulitemes flavipes (a termite)	Alkenols	(Z, Z, E)-3, 6, 8-Dodecatrien-1-ol
Nasuti-termes spp.	Terpenoids	Nonacambrene

Sarode et al. (1993).

sex and aggregation pheromones. Japanese beetle emits (Z)-5-(1-decenyl) dihydro-2-(3H) furanone as the sex pheromone. Most of the sex pheromones of Lepidoptran insects are unsaturated aliphatic straight-chain alcohols, acetates, aldehydes or ketone derivatives having 10 to 21carbon atoms. Some important sex pheromones of agricultural insects are given in Tables 3.5. Table 3.6 lists the alarm pheromones and 3.7 shows common trail pheromones of some insects.

Formulation of Pheromones: The following key requirements are envisaged in the formulation of pheromones:

(*i*) To release a constant amount per unit time
(*ii*) To provide different release rates to comply with the needs of different temperatures, humidity, light etc.
(*iii*) Protect the pheromone from degradation
(*iv*) Release all the pheromone
(*v*) Relatively easy to apply, and suitable for aerial application.
(*vi*) Non-toxic and biodegradable
(*vii*) Economic (cheap auxiliaries)

The research and development efforts have been concentrated mainly on four controlled-release formulations of pheromones *viz.* the hollow fibre, the twist-tie rope, the plastic laminate flake and microcapsules. Even though the controlled release formulation principles have been discussed in Chapter 2, the subject is briefly discussed with special reference to pheromone-based formulations.

(*i*) ***Hollow Fibre Controlled-released System:*** Developed by Albany international in mid 1974 as part of their effort in the area of reverse osmosis technology, the hollow fibres release volatile materials such as insect pheromones at a controlled rate with virtual zero-order kinetics. First governmental registration for the formulation was accorded in the USA in 1978.

Theoretical Aspects: Ideally, a controlled release device discharges all of its active material in a given time period, at a constant rate (zero-order kinetics). It is unlikely that the ideal controlled release can be achieved. However, the capillaries, after an initial burst, release it at a relatively constant rate (pseudo zero-order kinetics) and appear to be superior to impregnated plastic and rubber materials, plastic laminates, microcapsules etc., release for all of which follow the first order kinetics.

The mechanism of vapour release from capillaries, if trans-wall permeation is excluded, has been reported to be a simple three-stage process: (*a*) evaporation at the liquid-vapour interface, (*b*) diffusion through the vapour-air column to the end of the capillary, and (*c*) convection, away from the ends of the capillaries.

Release Devices: The hollow-fibre formulations to meet the monitoring and the mass trapping needs, require: (*a*) The complete pheromone blend, (*b*) The controlled-release device designed so as to release the blend at the correct release rate, component ratio, and have the desired field longevity, (*c*) an insect trap of optimized design, and (*d*) additives such as solvents for diluting the active ingredients to regulate longevity, antioxidants and ultra-violet stabilizers to protect the active materials from degradation, and in some cases a toxicant as killing agent in the trap.

To meet the need of an area-wide control system, these formulations require (*a*) the requisite pheromone as a major component, (*b*) a controlled-release device (*c*) an application device, (*d*) additives such as those described above with the addition of an adhesive to stick the devices to the plants, and (*e*) a monitoring system incorporating the first category.

There are basically two types of pheromonal hollow fiber products (*a*) tape dispensers and (*b*) chopped-fiber dispensers. The former consist of a parallel array of fibers (2-100) affixed to an adhesive tape and are primarily used as trap lures in monitoring and quarantine programmes, and for the timing of insecticide and/or pheromone application. They may also be used in mating disruption strategies with labour-intensive, high value crops.

The chopped fiber dispensers are used in area-wide broadcast control in classical mating disruption strategies. These interfere with the female to male sexual communication or provide attract and kill technique.

Manufacture: The hollow fibre controlled-release devices for area wide control (No Mate System) and insect monitoring (Scentry System) differ significantly in their methods of manufacture. In the former, the fibre primarily used is the highly crystalline celanese polycetyl copolymer celcon. Outlined in Fig. 3.9 is a flow diagram of the overall manufacturing process for No Mate System. The manufacturing process for Scentry System is illustrated in Fig. 3.10.

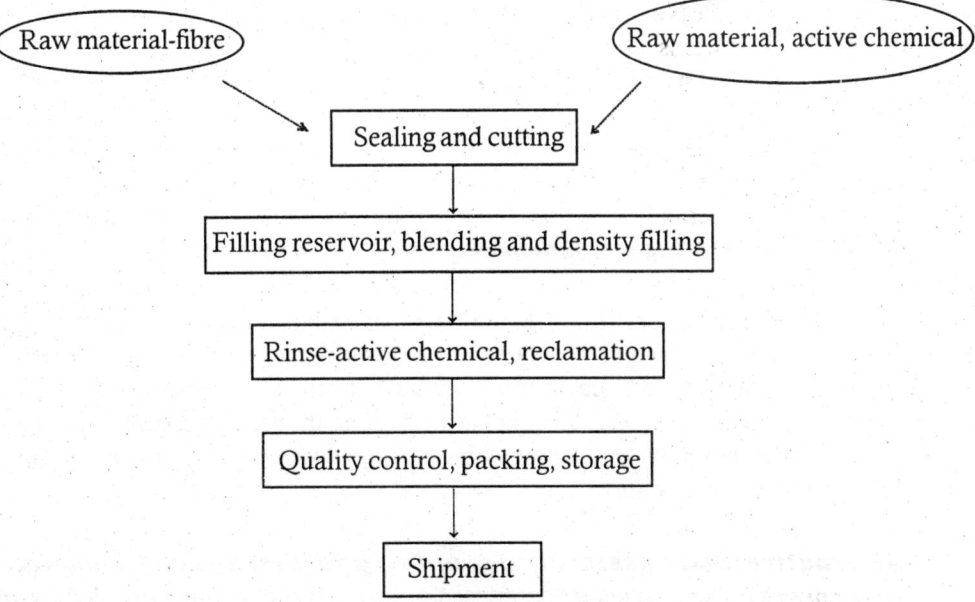

Fig. 3.9: Manufacturing Flow Chart for No Mate System

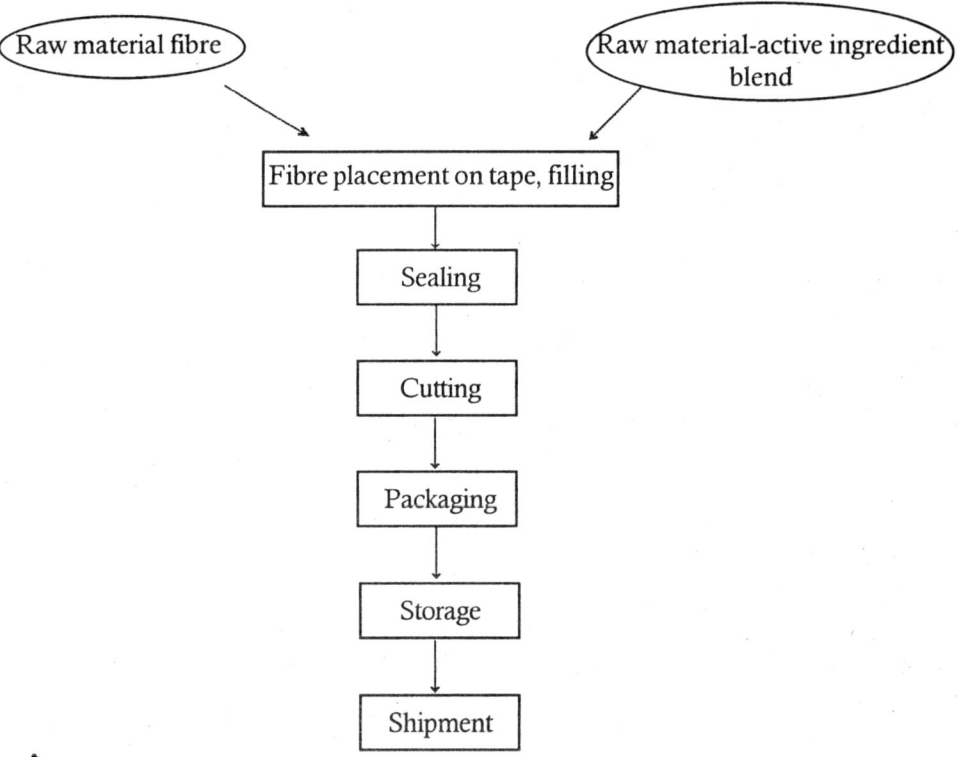

Fig. 3.10: Manufacturing Flow Chart for Scentry System

In contrast to the No Mate System where only one type of fibre is utilized, the Scentry Monitoring Systems employ two fibre types: Celcon, polyester terephthalate manufactured and supplied in continuous lengths. A parallel array of the chosen fibre type containing the desired number of individual fibres is fixed to an adhesive substrate. Normally, a 33 m length of this parallel array is placed into a length of stainless-steel tubing and epoxy-sealed, making certain that sufficient length of fibre is available to be subsequently placed into the semiochemical/solvent solution. The entire length of the fibre tape is filled using a pressure vessel, sealed, cut and packed.

(ii) Twist Tie Rope: The Shin-Etsu Chemical Company, Japan produces a variety of plastic fibres, sealed at both ends and containing a hollow channel with pheromone and a wire spine. The rope is about 15 cm long and must be attached/placed at a suitable place in the crop. In comparison with the other formulations, the rope places a very large initial reservoir of pheromone at one point, and the high dose per rope provides a relatively long persistence of release.

Alternative Dispensers for Trapping and Disruption (Commercial Dispensers): The systems are composed of various types of thin polyethylene tubes capable of loading 3-300 mg per tube of active material, with fixed release periods.

Tubes containing (Z)-11-tetradecenyl acetate (90 mg per tube), so prepared, are registered in Japan as a pesticide against both the tea tortrix (*Homona magnasima* Niet.) and the smaller tea tortrix (*Adoxophyes* spp.).

Rubber Septa: Of all the pheromone release devices, rubber septa (sleeve stoppers) are the most widely used as trapping baits by researchers, since they can be obtained from several supply houses, are easily loaded with pheromone and are convenient to handle. The septa are prepared by dissolving the lure in a solvent (e.g. dichloromethane) and placing the solution into the septum.

Polyvinylchloride Rods: A solid polyvinyl chloride slow release pheromone formulation, usually in the form of cylindrical pellets or rods has been used as trap bait for monitoring several insects, including *Trichoplusiani* Hübner, *Rhyancionia buoliona* and *Choristoneura fumiferana* Demens.

(iii) *Plastic Laminate Dispensers:* These consist of active ingredient implanted and protectively sealed between outer and inner plastic layer barriers. The specially formulated inner layer serves as reservoir of insect pheromones, which then migrate continuously, owing to imbalance of chemical potential through one or more initially inert, permeable barrier layers. Once at exposed surface, the active ingredient becomes available for biological action and is eventually removed by volatilization, ultraviolet light degradation, oxidation, acid or alkaline hydrolysis, or mechanical removal by wind, rain, insects, humans or other agents.

The Hecon Controlled-release Dispensing System: To make the laminated pheromone dispenser, a sheet of polymeric film is coated on one side with an attractant-impregnated polymeric mixture and then over layered with a second sheet of polymeric film. This three-layered structure is then allowed to set under suitable conditions of heat and pressure until an integral, firmly bonded product is obtained. The laminated sheet is then processed into its final form e.g. by slitting it into strips, ribbons or squares, or diecut into smaller flakes or confetti. More recently, multilayered dispensers have been produced with the pheromone and insecticide within the same reservoir for control of pink bollworm, *Pectinophora gossypiella* (Saunders), and American cockroach *Periplaneta* species. The encapsulated material is isolated to a greater or lesser extent from the surrounding environment, but it can be made available at a later stage by rupture of the capsule wall or passage through the wall.

(iv) *Microcapsules:* When technique of using synthetic pheromones to disrupt intraspecific communication of insect pests became a feasible proposition in early 1970s, the technical problems were seen to be those of applying expensive, unstable, water-insoluble, volatile, synthetic pheromones over large areas and to protect these materials from environmental factors for periods of at least several weeks, during which time they were released slowly into the atmosphere. Micro encapsulation seemed to be the only exceedingly appropriate process. For this technique to be put into practice, an aqueous or suitable non-aqueous suspension of microcapsules can be applied over large areas with the conventional equipment used for spraying insecticides. The micro capsule wall would provide some protection from environmental factors for the contents. A stabilizing agent needs to be mixed in with the contents or incorporated in the microcapsule walls. The physical and chemical properties of the microcapsules and their walls can be modified to provide appropriate release rate for the contents.

Methods of Micro Encapsulation: Microencapsulation processes can be categorised operationally into two general types, (*a*) physico-chemical and (*b*) mechanical.

(a) Physico-chemical Methods

(*i*) ***Wall Formation in the Presence of Polymeric Wall Material:*** In this process, the material to be encapsulated is dispersed in a continuous phase containing the wall-forming polymer and the latter is then deposited at the interface by coacervation, change in pH, chemical reaction, change in solvent or change in temperature.

Simple coacervation of a sol can be induced by addition of salts or other precipitants, by dilution or by change in pH. Wall materials include gelatin, gum arabic, polyvinyl alcohol, cellulose esters and ethers, styrene-maleic anhydride copolymers, polethylene oxides, polychlorotrifluoroethylene, vinylidene chloride-vinyl chloride copolymers and other polymers, which are soluble in solvents of appropriate volatility.

(*ii*) ***Wall Formation from Monomeric/Oligmeric Starting Materials:*** Processes pioneered by the Pennwalt Corporation (U.S.A.) have been extensively used for micro encapsulation of both insecticides and pheromones. The basic principle involves emulsification of the core material and a monomer in an aqueous solution of another monomer. Condensation occurs at the core/water interface. In a typical process, the core material mixed with adipoyl chloride is emulsified in water containing polyvinyl alcohol as emulsifying agent. An aqueous solution of hexamethylenediamine and sodium carbonate is then added so that a capsule wall of polyamide is formed at the interface, sodium carbonate neutralizing the acid liberated during the process.

(b) Mechanical Methods

(*i*) ***Spray Drying/Congealing:*** In spray drying, the core material is dissolved in a solution of a film-forming polymer. The dispersion is atomized into a stream of hot, inert gas which removes the solvent so that the polymer is deposited on the core droplets.

(*ii*) ***Multi-orifice Technique:*** In a well-established technique of centrifugal microencapsulation, the core material is fed into a rotating disc which breaks the stream into small droplets. These are propelled through orifices in the periphery of the centrifugal bowl, which are continuously supplied with a film of wall-forming material. Wall materials can be selected from a range of soluble or liquefiable natural or synthetic polymers including gelatin, waxes, cellulose, styrene/butadiene copolymers and acrylic polymers. The process is potentially suitable for micro encapsulation of pheromones.

(iii) *Electrostatic Processes:* Liquids as well as small particles can be encapsulated by atomizing core materials and wall material separately and giving them opposite electric charges with high-voltage so that they combine. The coating substance can be molten e.g. a phthalate ester or microcrystalline wax or a solution of a synthetic resin e.g. polyamide, polyvinyl esters, ethers, polystyrene or acrylates. Alternatively, the core and coating materials can contain appropriate monomers so that interfacial polycondensation occurs when they come in contact with each other.

Miscellaneous Devices: The use of pheromone traps for quarantine, monitoring or mass-trapping purposes has employed such substrates as polyethylene vials and vial caps, glass capillary tubes, dental wicks, cigarette filters and tubing made of polyethylene and rubber for dispensing.

Limitations of Using Pheromones: A number of problems are associated with the use of pheromones. The effectiveness of such compounds may be limited by the ratio of major and minor pheromone components used, timing of application and insect population density. These are highly specific and affect only the susceptible species. Most of the pheromones are expensive to manufacture. Still others are unstable, decomposing within minutes in the presence of light or oxygen The use of controlled release formulations is desirable for application of unstable volatile compounds. Such formulations permit control of duration of the chemical effect, deliver pheromone at target levels, maximise effectiveness with the minimum quantity of chemical and protect the active ingredient from light, oxidation and hydrolysis.

The release of a synthetic formulation of pheromones depends on the same factors, which control the degradation of other organic compounds *viz.*

(i) *Temperature:* Higher the temperature, more is the rate of release of the components.

(ii) *Vapour pressure:* The rise in temperature raises the vapour pressure of the volatile components, which volatilize away and are lost quickly.

(iii) *Velocity of wind:* Velocity of wind or air in the area increases the rate of release of the synthetic compounds because the surface above it containing the vapours is carried away so that fresh air moves in to take away more of the chemical.

(iv) *Humidity:* Humidity of air also influences release of the organic compound. Generally, lower the humidity higher is the rate of release. By controlling the release of pheromones, the chemical is given out very slowly and its effect carries on for a longer period.

There are a number of advantages from the application of pheromones or use of attractants such as monitoring, mass trapping, disruption of sexual mating or luring and killing. The advocacy of pheromones is based upon the social and also pristine benefits. The use has the limitation of insect pest specificity due to which the crop cannot be saved from damage by other insects/pests and, therefore, pesticides have to be sprayed for a total and effective control. High cost of pheromones is another disadvantage. Since a pheromone is specific for insect pests, the predators and parasites are not harmed. The use of pheromones for crop protection requires a high level of financial and capital support, which makes the returns very small. The various advantages of pheromones are also negated by obviate lengthy registration processes. It has not been possible to get a cent percent kill by a pheromone due to which the pesticide use may have to be relied upon, thus persisting with the problems of pollution etc.

Rate of application of pheromones is so low that it is not a simple job to assess their concentration and degradation products in soil, water and air. Very little progress has been made in the practical applications of pheromones in the field of insect control. Commercial applications of pheromones in monitoring, mass trapping have been identified only in few species. A proper strategy for their development and large scale adoption by the farmers needs to be formulated to derive maximum benefit through their use.

Food Lures

Poison Baits: Poison baits in form of pellets, granules or pastes consist of a mixture of food lures and insecticides. The effort is made to make the bait more attractive to insects than their natural food. A smaller quantity of attractant is required to attract a large number of insects. Baits are used when for some reason spraying or dusting of insecticides is not practicable e.g. for insects hidden under the soil, inside the fruits and vegetables or for household insects like ants, cockroaches and houseflies. Some bait formulations for different kinds of insects are given in Table 3.8.

TABLE 3.8. SOME BAIT FORMULATIONS FOR INSECTS

Insect	Poison baits
Biting insects	Moistened bran base + molasses + insecticide
Sucking insects	Sugar solution + insecticide
Fruit flies	Trimedlure + cuelure + methyl eugenol
Cockroach	White/yellow P + sweet syrup; Baygon + sweet syrup; Lindane + sugar
Sweet loving ants	Thallous sulphate + sugar (honey or glycerine) + water
Meet loving ants	Thallous sulphate + peanut butter
Subterranean insects	Vegetable pieces or germinating seeds + insecticide
Garden snails/slugs	2-3% Metaldehyde + wheat flour
Western corn root	Carbaryl + cucurbitacins + (1, 2, 4-trimethoxy benzene + indole + *trans* cinnamaldehyde, 1 : 1 : 1)
Rose chafer	Octyl butyrate, nonyl butyrate,
Macro-dactylus	1-nonanol, valeric + hexanoic acid
Oriental fruit fly	Methyl eugenol + sugar

Repellents

Repellents are chemicals that prevent insect damage to plants or animals by rendering them unattractive, unpalatable or offensive. These are non-poisonous or mildly poisonous chemicals that induce avoiding (oriented) movements in insects, away from the source or site of their application. For example, citronella oil is a powerful repellent for mosquitoes.

Classification: Repellents are primarily classified on the basis of their mode of action and subsequently each group is further sub classified on the basis of chemical entities. Broadly, these are classified as physical and chemical repellents.

    Physical Repellents: They induce repellence by physical means and are of the following kinds:

    (*a*) ***Contact Stimuli Repellents:*** These are the substances (e.g. dust, granule, water, oil, leaf axis, spines and waxes) that influence the surface-texture of the substrate (plants) to produce a disagreeable effect on the tactile sense of the insects.

    (*b*) ***Auditory Repellents:*** Devices like an electronic scare crow which employ sound to ward off birds/insects (by suitably amplified sound effect) have been found effective in repelling mosquitoes, pyralid moths and flies. Recorded vibrations of natural enemies, such as supersonic vibrations of bat, which predates upon cotton bollworm moths when relayed, scare these pests away from the cotton fields and thus act as an auditory repellent.

    (*c*) ***Barrier Repellents:*** Substances such as tar oil, painted as bands on trees and hedges, dusts on roadside trees, mosquito nets, wire-meshing on doors and windows or tightly enveloping plastic coverings over woolen clothes act as barrier repellent for insects.

    (*d*) ***Visual Repellents:*** While light normally attracts insects, yellow light is the least attractive and to some extent acts as a visual repellent to insects.

    Chemical Repellents: These are chemicals of natural or synthetic origin. Repellent action is primarily a property of the natural or synthetic compound and is dependent on the nature of the chemical. These are of three types:

    (*a*) ***Excitatory Repellents:*** Chemicals like pyrethrum, DDT, BHC etc. which excite the insects tarsi by stimulating the sensory nerves and force them to leave the treated surface are called excitatory repellents.

    (*b*) ***Feeding Repellents:*** Also called antifeedants or feeding arrestants, these chemicals inhibit feeding in insects. These chemicals inhibit feeding by insects or drive (repel) them away, when applied on the foliage (food), without impairing their appetite and gustatory receptors. They are also called gustatory repellents, feeding deterrents and rejectants. Their use may cause death of the insects slowly due to starvation e.g. neem products are feeding repellents for desert locust.

(c) **Olfactory Repellents:** These volatile chemicals, when introduced in the environment, above a particular concentration, act on the olfactory receptors and drive the insects away from the place of application, probably by inducing a sense of alarm. Most of the essential oils and synthetic esters fall in this category.

Chemical Classification: Each of the above groups of chemicals can be classified into two basic classes as mentioned below.

Botanicals or Repellents of Plant Origin: e.g. oil of citronella, camphor, oil of cedar wood etc. Pyrethrum repels blood-sucking insects.

Synthetic Repellents: e.g. synthetic compounds like dimethyl phthalate and benzyl benzoate are used to repel mosquito and fleas.

Either of the above groups can be further sub-classified in various sub groups depending upon their major structural skeleton.

Characteristics of an Ideal Repellent: A chemical repellent, besides being cheap and economic, should possess one or more of the following characteristics:

(i) It should be non-toxic to mammals. If applied as cream, it should not possess any dermal toxicity.
(ii) Be effective for a long time.
(iii) Be least affected by weathering.
(iv) Should be non-irritating to man and animals.
(v) Acceptability i.e. should have an acceptable odour, taste and touch.
(vi) Be harmless to clothes and non-corrosive to container.
(vii) Should be cheap.

Antifeedants: There are several groups of antifeedants. Depending on the importance in pest control, the major antifeedants are briefly described below.

Botanicals: Insects recognise their hosts both by visual and chemical means. The antifeedant compounds present in non-host plants actually do not impair the insect's ability to maintain their host specificity. A large number of plant products are antifeedant in nature e.g. pyrethrum to *Glossina* sp., neem products to desert locust, *Scistocerca gregaria*, apple factor phlorizin to *Myzus persicae*, *solanum* alkaloids (leptine, tomatine, solanine) to potato leafhopper (*Empoasca desvastans*) etc.

Synthetics: Several synthetic compounds are known to act as feeding deterrents. The major groups are described below:

(i) **Triazines:** The antifeedant activity of several triazines is known. Among them, the most active compound 4-(dimethyl triazino) acetanilide has antifeedant activity against chewing insects, caterpillars, roaches, beetles etc.

(ii) Organo-tins: Organic compounds of tin, namely triphenyltin acetate and triphenyltin hydroxide show antifeedant activity to foliage feeding insects e.g. cotton leafworm, Colorado potato beetle, caterpillars of *Agrotis* and grasshoppers.

(iii) Carbamates: It has been found that a sub lethal application of some of the carbamates like thiocarbamates and phenyl carbamates exhibit antifeedant activity when applied on foliage against foliage feeding insects like Colorado potato beetle, caterpillars etc. Baygon, a commercial carbamate insecticide, also acts as a systemic antifeedant for the boll weevils.

(iv) Amines: The quaternary derivatives of some commercially available heterocyclic secondary amines *viz.* benzyl amine and derivatives of benzylidene amine have been found to possess antifeedant activity against *Leptinotarsa decemlineata*.

Though large number of natural products have shown feeding deterrent activity against a number of insect species, actually none of them has been commercially exploited for insect control or in integrated pest management (plant protection). Some of the neem (*Azadirachta indica* A. Juss) derived products like azadirachtin-A and its hydrogenated products dihydro- and tetrahydro-azadirachtin-A have received a lot of attention recently in this regard. These were initially reported to possess antifeedancy against the desert locust *Schistocerca gregaria* but are now known to prevent feeding in several other insect species also. Their potency varies from species to species. For example, another African locust species, *Locusta migratoria*, is far more willing to eat azadirachtin than *S. gregaria*, and grasshoppers in North America show no antifeedant response at all. Azadirachtin is non-volatile, so an insect must taste it, rather than smell it, in order to respond to it. Taste of azadirachtin stimulates at least one 'deterrent neurone' in insects, which convey antifeeding response to the brain.

Azadirachtin, Dihydroazadirachtin (reduced double bond A)
and Tetrahydroazadirachtin (reduced double bonds A and B)

Although the antifeedant effect of azadirachtin may be useful for short-term crop protection, insects, which are exposed repeatedly to azadirachtin become less sensitive to it. For example, desert locust *Schistocerca gregaria* is reported to adapt completely to azadirachtin after 7 or 8 days of exposure. This is an example of habituation, which is a common response of many animals to repeated stimuli.

Substances other than azadirachtin can influence the strength of the antifeedant effect. In some insects the stimulatory effect of a preferred food source can overcome the antifeedant effect of azadirachtin. One report showed that sensitivity of *L. migratoria* to azadirachtin treated food decreased with sugar content of food. Some evidence suggests that non-azadirachtin components of neem may slow the ability of insects to habituate to azadirachtin. For example, Asian armyworm larvae, which normally habituate to low levels of purified azadirachtin after two days of exposure, did not habituate to a neem extract containing the same absolute amount of azadirachtin after three days of daily exposure.

TABLE 3.9. SOME IMPORTANT GENERAL-PURPOSE REPELLENTS

Repellent	Insect / Animal
Anthraquinone	Biting flies, gnats
Dipropyl isocinchomeronate	Housefly
Dimethyl phthalate	Mosquito
N, N-diethyl-*m*-toluamide (DEET)	Mosquito, flies, fleas, ticks,
N-butylacetanilide	Fleas
Diacetyl phthalate	Cattle fleas
Benzyl benzoate	Mites
Oil of tarweed	Mosquito (oviposition)
Thymol	Mosquito
Carvacrol	Mosquito
α-Cymene	Mosquito
Trans-cinnamaldehyde	Mosquito
1, 8-Cineole (Eucalyptol)	Mosquito
Eugenol	Mosquito
p-Menthane-3, 8-diol esters	Mosquito
Volgarone B	Gastropod
Thiazole derivatives	Thrip
Benzylamine and benzylidene triamine derivatives	Insects (antifeedant)
Dipropyl pyridine -2, 5-dicarboxylate	Housefly
GTA (Guaitazatine acetate)	Bird repellent
*MGK Repellent 874 [2-(Octylthio) ethanol]	Crawling insects
Ziram [Zinc bis (dimethyldithiocarbamate)]	Birds, rodents
Indalone (Butopyronoxyl) [Butyl 3, 4-dihydro -2, 2-dimethyl-4-oxo-2H-pyran-6-carboxylate]	Insects
DBP (Dibutyl phathalate, [dibutyl 1, 2-benzene dicaboxylate]	Insects
Dibutyladipate	Ticks
Dibutyl hexanedioate	Crawling insect
Tebutrax (Dibutyl butenedioate)	Insects
Dicyclopentadiene	Animal (rodents)
MGK repellent 11	Repellent for cockroach

*McLaughlin Gormley King Co. (MGK).

There are chemicals of both botanical and synthetic origin, which have been successfully utilized commercially against household pests like flies and mosquitoes. Some important general-purpose repellents are listed in Table 3.9. Table 3.10 gives some of the repellents commonly used in household formulations (creams and aerosols).

TABLE 3.10. SOME REPELLENTS USED IN HOUSEHOLD FORMULATIONS

Repellent	Insect Pests
Benzyl benzoate	Chigger mite (*Trombicula* spp.)
Butoxypoly (propylene glycol)	Fly repellent for cattle
N-butylacetanilide	Fleas, ticks and biting flies
2, 3, 4, 5-Bis (butyl-2-ene) tetrahydrofurfural	Ticks and fleas
Butoxypyronoxyl	Mosquito, fly
Dibutyl adipate	Tick
Di-n-butyl succinate	Fly on cattle
N, N-diethyl-3-methylbenzamide (DEET)	Fly, mosquito
Dimethyl phathalate	Fly, mosquito
2-Butyl-2-ethyl-1, 3-propane diol	Mosquito
2-Ethyl-1, 3-hexanediol	Mosquito
Di-n-propyl pyridine-2, 5-dicarboxylate	Flies on cattle
Trichlorobenzenes	Soil termites
Sodium fluorosilicate	Cattle moths
1, 4-Dichloro benzene	Cloth moths
Diphenyl ethers (Eulane)	Moths on woolens
Chlorinated diphenyl ethers and ureas	Moths on woolens

Coal-fire soot scattered along the row of seedlings is a very effective deterrent to slugs. Its effect is probably mainly mechanical, providing a very unsuitable surface for crawling. Pepper, as a repellent for cats, is effective only when scraping disturbs it. Evil-tasting oily products smeared on tree-boles, in so far as they are effective in preventing damage by deer and rodents, are distasteful to tongue and lips and even this may be partly mechanical. These compounds are required in comparatively larger doses for being effective. Hence, are not as active as olfactory repellents and seldom used in pesticide formulations.

Practical Utility: The market for repellents for insects on livestock is potentially very large. The animals suffer from many ectoparasites, such as ticks, fleas and lice, which are carriers of diseases. Also, the nuisance insects often pester them in much larger number than they do humans. The fly *Musca autumnalis*, a nectar feeder in its native Europe, was introduced into North America where it adopted a habit of sucking facial liquids of cattle. The face fly has become the major nuisance and animals have been known to die through exhaustion under the intolerable discomfort. Mosquito repellent creams and coils are the most widely used in developing countries. 9,10-Anthraquinone is used as a seed dressing to deter birds. An interesting compound is 4-aminopyridine which does not kill birds at the concentrations used for repellence but causes those which eat bait containing it, to behave in a manner which alarms the whole flock and makes them leave and not return. In some way, not understood, this compound creates a disturbance in birds and produces in them fear of treated areas.

The fungicides, tetramethyl thiuram disulfide and zinc dimethyldithiocarbamate-cyclohexylamine complex are used to repel deer and rabbits. Methyl nonyl ketone is used to repel cats and dogs whereas tertiary-butylsulphenyl dimethyldithiocarbamate is incorporated into rubber and plastic insulations for electric cables to repel rodents from gnawing at them.

Formulations of Repellents: Repellents are formulated as oils, creams or gels for hand or body application, or put up in self-dispensing aerosol packs. Creams and aerosols incorporating a sun screen with the repellents are also marketed. Smoke generator coils are particularly effective against mosquitoes, flies etc. and can be made more lethal through incorporation of suitable active ingredients. Slow/controlled release products offer promise under specific situations such as within barns and cattle inhabiting dens etc.

Limitations: Even though, a large number of synthetic organic compounds and bio-botanicals have been reported to possess repellent activity towards one insect or the other, the activity is generally very short lived and species specific. Therefore, most of the reported chemicals have neither been used in the field nor in household premises. Repellents and attractants might seem to be compounds evoking directly opposite responses. In fact, neither response is simple. Repulsion is a less complex behaviour. It is much more frequently evoked by synthetic compounds and is much less specific than attraction. It is partly for this reason that repellents are already manufactured in large amounts and widely used with good effect, while attractants are still in the developmental stage.

A repellent need not be effective over a great distance from its site of application. It would be preferable to drive the mosquitoes out of range of sight and hearing but the only really necessary effect is to stop them from biting. Only a very small proportion of the population of mosquitoes within flight range may come near enough to be attracted to blood or repelled by repellent. An attractant, to be useful, must lure a high proportion of the population to its destruction, and therefore, must be effective over a long range. It has a more difficult task to perform.

In household application, there are, two good reasons for more advanced state of repellents. First the safety and then the ease of application. The most important thing here is that repulsion, if efficient enough, is itself a useful means of pest control. However, insect repellents have found little use in modern agriculture for protection of growing crops. The large areas involved would probably make their use uneconomical even if it could be made effective. Their use is also incompatible with monoculture in large fields, within which there can be no choice of food for the insect. Under these conditions, only a repellent which also completely inhibits feeding, could be effective. Chemical repellents for birds and small mammals which cause tremendous damage to growing crops are acceptable, as killing of birds and other "pleasant" animals provokes violent public reaction. However, the public sentiment views destruction of insects or even rodents, with equanimity. Therefore, repellents are not much sought after for their control.

ENCAPSULANTS

These are materials used for encapsulation or formation of macro/microcapsules. Polymers (both natural and synthetic), which are used as coating material or membrane include carboxymethyl-cellulose, ethyl cellulose, ethylene vinyl acetate, gelatin, nitrocellulose, polyvinyl alcohol, various waxes, polylactic acid, polyamides etc. Several other products, which can also be used, are listed under the controlled release formulations (refer Chapter 2). Encapsulated materials have been used for various purposes in different fields of science. The technique has been used in drug delivery systems, cosmetic preparations, dyes, flavours etc. The capsule walls can be made solvent resistant or solvent degradable so that release of the encapsulated material takes place at a convenient time, in the required dose and at the appropriate place. The polymeric membrane may release its contents through rupure, diffusion outwards or inwards, or solution. Variation of membrane material or geometry of the wall will alter permeability and hence the release rate of contents from the microcapsule. Generally, increase in density, crystallinity, orientation and cross linking and low plasticizer level in the polymeric wall material will decrease permeability. Use of a strong solvent in film preparation will also decrease its permeability. Similarly, lower permeability can be achieved by increasing size and wall thickness, a spherical conformity and using post cross linking and multiple coatings.

Practical Utility

Pencap™, a formulation of methyl parathion was the first commercial encapsulated insecticide used against mosquito larvae. In the control of cotton bollworm, the optimum degree of cross-linking in the microcapaule wall is 25% with a polyamide - polyurea wall composition. The capsule wall of Pencap-M is a cross-linked polyamide, and is prepared by reaction of sebacoyl chloride and polymethylenepolyphenyl isocyanate (PAPI) with an ethylene diamine/diethylenetriamine mixture at the interface of a microdroplet of the agent in water.

The controlled release formulations of diazinon and chlorpyrifos macrocapsules obtained from flowable slurries or dry-coated granular products have been reported. The starch-encapsulated diazinon invovling introduction of sulphur bonds in soluble starch by xanthating it, and then cross-linking the starch xanthate to encapsulate the pesticide, have also been emphasized.

Polyurethane coated fenitrothion has been reported to possess a superior residual activity than the EC formulation and equivalent activity to the WP formulation. Its advantage over the latter is that there is no dustiness or contamination on sprayed surface. The larger ratio of fenitrothion to the wall material gives better control of *Culex pipiens and Blatella germanica*. The microcapsules are prepared by dispersing an organic mixture (fenitrothion, polyisocyanate, cyclohexanone) into aqueous solution of gum Arabic (2%), which acts as a dispersing agent. Prescribed amount of ethylene glycol is added to the resulting dispersion with light agitation. The mixture is heated for two hours at 60°C to give wall formation. Total weight is adjusted according to the concentration required by adding water. The medial diameter of the capsules is 40-50 μm. The range of particle size is 5-110 μm.

The polyurea micro-encapsulation of herbicides that are immiscible in water has also been achieved by dispersing polymethylenepolyphenylene isocyanate (PAPI) dissolved in materials immiscible with water into aqueous ammonium, potassium, magnesium, calcium or sodium lignosulphate, followed by addition of polyamine which reacts with PAPI.

Diazinon and chlorpyrifos have been individually and together encapsulated in casein or casein derivatives. The product when suspended in a non-ionic surfactant and spray dried gives a controlled-release formulation that is readily suspensable in water and free from aggregation. Microcapsules so prepared are effective in 100% control of cockroaches for nearly 74 weeks.

A high degree of encapsulation (about 99%) is obtained by dispersing a mixture of a hydrophobic core material and an oil-soluble polyisocyanate, polycarboxylic acid chloride and/or polysulphonic acid chloride in water containing a soluble polymer prepared from substituted methacrylamide (containing up to C_8 alkyl substituent) and treating the dispersion with a water soluble polyamine and/or polyol to form the interface.

Pyrethroids have been successfully encapsulated by the interfacial condensation of complementary organic polycondensate-forming intermediates, which react to produce a polyamide, polyamide-polyurea, polysulfonamide, polyester, polycarbonate or polyurethane cover/capsule.

Micro-encapsulation of biological control agents can also be carried out. Preparation of hydrogel capsules containing an insecticidally effective amount of a nematode (capable of infecting an insect host) is an example. The capsule has sufficient hydration to maintain the viability of the organism. For example, a suspension of *Neocaplactana carpocapsae* (about 400 nematodes per ml.) is mixed with a solution of sodium alginate (2 g) in water (100 ml) and followed by treatment with 100 millimoles of calcium chloride to get the capsules. The capsules, even after nine months of storage, are suitable to control *Galleria mellonella* spp.

SENSITIZERS AND STABILIZERS

Sesnsitizers

These are chemicals that induce the degradation of pesticides. These find application while formulating hard pesticides like chlorinated hydrocarbons to induce degradation. Photo-sensitizers have received specific attention for the light induced degradation of such pesticides. Photo-degradation is induced normally by visible or ultraviolet light, involving both intra- and inter-molecular reactions. The absorption of light energy raises a molecule from its ground electronic state to the rather short lived excited one. Therefore, the photochemical reactions tend to be either intramolecular or restricted to reactions with adjacent molecules (solvent). The products of a photo-excited solid, whether as pure material or adsorbed on a surface to prevent diffusion before reaction, will be dependent on its immediate molecular environment. Photochemical reactions may also occur at the ground level through the interaction of sensitizer if the triplet excitation energy of the sensitizer is equal to the requisite energy for the transformation. If energy transfer from the sensitizer to a second molecular species occurs, the photochemical products characteristic of the energy acceptor species may result. In solutions, the solvent may also participate to influence the destiny of the product e.g. photo-excitation of an aromatic chlorinated compound may yield a phenyl radical which, in some solvents like methanol or cyclohexane, abstracts hydrogen to yield a dehalogenated product. Not many sensitizers have actually been incorporated in pesticide formulations, though a number of them have been used in degradation studies. However, practicality of their use in DDT formulations has been demonstrated (Parmar *et al*, 1976; Sarkar and Parmar, 1979) A list of common sensitizers (along with the triplet excitation energy, E_T) is given in Table 3.11.

TABLE 3.11. SOME SENSITIZERS AND THEIR TRIPLET EXCITATION ENERGIES (E_T)

Compound	E_T (Kilo cal mol^{-1})
Aldehydes	
Benzaldehyde	71.3
2-Hydroxybenzaldehyde	71.0
4-Chlorobenzaldehyde	70.8
3-Iodobenzaldehyde	70.8
Chlorobenzaldehyde	69.6
Acrylaldehydes	~ 69.0
2-Naphthaldehyde	59.5
1-Naphthaldehyde	56.3
Glyoxal	56.0
9-Anthraldehyde	40.0
Ketones	
Acetone	~ 78.0
Propiophenone	74.6
Xanthone	74.2(70.9)
Phenyl cyclopropyl ketone	74.0
Di-isopropyl ketone	74.0
Acetophenone	73.6
1, 3, 5-Triacetylbenzene	73.3
Isobutyrophenone	73.1
4-Methyl acetophenone	73.0
1, 3-Diphenyl-2-propanone	72.2
4-Chloracetophenone	72.0
4-Methoxyacetophenone	72.0
Anthrone	71.9
4-Bromoacetophenone	71.0
3, 5-Dimethylacetophenone	71.0
2-Acetyl pyridine	71.0
3-Acetylpyridine	71.0
Triphenyl methyl phenyl ketone	70.8
Sodium benzophenone-3-sulphonate	70.2
4-Acetylpyridine	70.0
3-Cyanoacetophenone	70.0
4, 4-Dimethoxybenzophenone	70.0
4, 4-Dimethylbenzophenone	69.0
4, 4-Diphenylcyclohexadienone	69.0
2-Benzoylbenzophenone	68.7
Benzophenone	68.5
4, 4'-Dichlorobenzophenone	68.0
4-Trifluoromethylbenzophenone	68.0
4-Hydroxybenzophenone	68.0
Pyruvic acid (CH_3COCO_2H)	68.0
4-Acetylacetophenone	67.6
4-Aminobenzophenone	67.0
9-Benzoylfluorene	66.8
4-Cyanobenzophenone	66.4

Thioxanthone	65.5
4-Aminoacetophenone	65.0
Ethyl pyruvate	65.0
Phenylglyoxylic acid ($C_6H_5COCO_2H$)	63.0
Phenylglyoxal	62.5
Anthraquinone	62.4
α-Naphthoflavone	62.2
Flavone	62.0
Ethyl phenylglyoxylate	61.9
Michler's ketone [4, 4'-bis-(dimethylamino) benzophenone]	61.0
4-Acetylbiphenyl	61.0
β-Naphthyl phenyl ketone	59.6
β-Acetonaphthone	59.3
α-Naphthyl phenyl ketone	57.5
Ethyl naphthylglyoxylate	57.0
α-Acetonaphthone	56.4
5, 12-Naphthacenequinone	55.8
Biacetyl	54.9
Acetylpropionyl ($CH_3COCOC_2H_5$)	54.7
Benzil	53.7
Fluorenone	53.3
Camphorquinone	50.0
Phenanthrenequinone	49.0
3-Acetylpyrene	46.0
Thiobenzophenone	40.0
Phenol	82.0
Benzamide	79.0
Benzoic acid	78.0
Benzonitrile	77.0
Sodium triphenylene-2-sulfonate	65.0
2-Naphthoic acid	60.0
Disodium naphthalene-2, 6-disulfonate	60.0
1-Naphthoic acid	58.0

Hydrocarbons, Heterocyclics and Derivatives

Benzene	85.0
Pyridine	82.0
Aniline	77.0
Carbazole	70.3
Diphenylene oxide	70.1
Hexachlorobenzene	70.0
Dibenzothiophene	69.7
Cyclopentadiene	58.3
Chlorobutadiene	57.4
cis-Stilbene	57.0
cis-I, 3-Pentadiene	56.9
1-Methoxybutadiene	56.6
1, 3-Cyclohexadiene	53.5
Thiophene	69.0
Fluorene	67.6

Triphenylene	66.6
Biphenyl	66.0
Phenanthrene	62.0
Quinoline	62.0
Naphthalene	61.2
1-Methylnaphthalene	60.0
Nitrobenzene	60.0
Acenaphthene	59.3
1-Chloronaphthalene	59.2
1-Bromonaphthalene	59.0
1-Iodonaphthalene	58.6
Chrysene	56.6
Coronene	55.0
1, 2, 5, 6-Dibenzanthracene	52.2
1, 2, 3, 4-Dibenzanthracene	50.8
Pyrene	48.7
Pentaphene	48.0
1, 2-Benzanthracene	47.0
11, 12-Trimethylenetetraphene	46.0
1, 12-Benzperylene	46.0
Acridine	45.3
Phenazine	44.0
3, 4-Benzpyrene	42.0
Anthracene	42.0
Azulene	31-39
Naphthacene	29.0
1, 4-Diphenyl-1, 3-butadiene	52.0
trans-4-Nitrostilbene	50.0
trans-Stilbene	49.0
trans, trans-1, 3, 5-Hexatriene	47.5
1, 3, 5, 7-Octatetraene (all trans)	39.0

Miscellaneous

Eosin	43.0
Crystal violet	39.0
Oxygen (O_2)	23.0
Diphenyl sulfide	74.0
Diphenyl selenium	72.0
Acridene yellow	58.0
Fluoroscein	51.0
Phenyl azide	78.0
Aniline	77.0
Cyanobenzene	77.0
Diphenylamine	72.0
Triphenylamine	70.0
Nitrobenzene	60.0

Gordon and Ford (1972).

Antioxidants and Stabilizers

Opposite to the sensitizers are the stabilizers which attempt to delay or arrest the degradation of an active ingredient. Like sensitizers, they are specific to the component or formulation and cannot be generalized into any chemical group. Pesticides, especially those of biological origin, get degraded by air (oxidation), light (photochemical) and by water (hydrolysis). Stabilizers and antioxidants are used in formulations to provide the active ingredient more stability towards above processes of degradation.

Antioxidants

These are chemicals that delay or retard aerial oxidation of pesticide molecules. These compounds get oxidized preferentially over the oxidation prone pesticide molecules. Many polyhydroxy alcohols and phenols are being used for this purpose. Antioxidants, commonly used in pesticide formulations, are listed in Table 3.12. Antioxidants used in tape releasing and fumigant formulations are given in Table 3.13.

TABLE 3.12. COMMON OXIDANTS USED IN PESTICIDE FORMULATIONS

Antioxidant	Pesticide
BHA [Butylated hydroxyanisole, (2-t-Butyl-4-methoxy phenol)]	Pyrethroids, pheromones
BHT (2, 6-Di-t-butyl-4-methyl phenol)	Pyrethroids, pheromones
TBH (2-t-Butyl-4-hydroxyphenol)	Pyrethrins and allethrin
Propyl gallate	Synthetic pesticides
Sodium thiosulphate	Paraquat
Tocopherol	Paraquat
Poly (oxyethylene)	Microbicidal
Dodecylamino ether	Microbicidal
Pyrogallol	Synthetic pesticides
Epichlorohydrin	Insecticides

TABLE 3.13. ANTIOXIDANTS IN TAPE RELEASING AND FUMIGANT FORMULATIONS

S. No.	Antioxidant
(i)	2, 2'-Methyene bis-(4-ethyl-6-*tert* butyl phenol)
(ii)	2, 2'-Methyene bis-(4-ethyl-6-*tert* butyl phenol)
(iii)	Stearyl-(3-5-di-*tert* butyl-4-hydroxyl phenol propionate)
(iv)	4, 4'-Thio bis-(3-methyl-6-*tert* butyl phenol)
(v)	1, 1'-Methylene bis-(2-methyl-6-*tert* butyl phenol)
(vi)	4, 4'-Butylene bis-(3-methyl-6-*tert* butyl phenol)
(vii)	1, 3, 5-Trimethyl -2,4,6-tris (3,5-di-*tert* butyl-4-hydroxybenzyl) benzene
(viii)	1, 1, 3-Tris-2(2-methyl-5-*tert* butyl phenol)
(ix)	4, 4'-Methylene bis (2,6-di-*tert* butyl phenol)
(x)	Tetrakis methylene (3, 4-di-*tert* butyl-4-hydroxy cinnamate) methane

Note: The above compounds are non-volatile, heat stable antioxidants. For slow release formulations of attractants, generally a tape (10 nm wide, 0.1 mm thick, 360 mm long) is impregnated with a hexane solution containing appropriate amount of the active ingredient, 80 mg pyramin and 10 mg of an oxidant, dried and cut into requisite lengths.

Photostabilizers

It has been reported that significant photo-stabilization of a molecule may be attained by use of additives like monocyclic aromatic compounds having extended *pie* electron system, a double bond in conjugation with a non bonding pair of electrons or any other electron donating group attached to the monocyclic aromatic ring. Nitro group attached to the benzene ring requires another electron donating group attached to *ortho* or *para* to the nitro group (electron donating group with a Hammet value of 6 and pie value of no more than -0.2 is desirable).

Wavelength of available sunlight, which affects the chemical bonds usually found in polymer backbones, is in the region 300-400 nm. The most widely used absorbers have been the salicylates (I), o-hydroxybenzophenones, (II) and benzotriazoles (III).

$$\text{I} \qquad \text{II} \qquad \text{III}$$

Salicylates are still available but little used. *O*-Hydroxybenzophenones and *O*-hydroxy phenyl benzotriazoles are extremely light stable and are most widely used stabilizers. The *ortho*-hydroxy isomers are much superior to the *para* isomer and this observation has led to postulation of a general stabilization mechanism involving intramolecular hydrogen bonding in a six membered ring. The following cyclic mechanism has been proposed and is supported by several authors.

Keto   Enol

R_1, R_2 = Alkyl, Halo, Nitro etc.

The enol form is assumed to be favoured energetically in the ground state but the keto form is preferred in the first excited singlet state. Charge transfer in the excited state permits the formation of a more stable keto form. The resulting decrease in energy difference between the ground and first excited state means that internal conversion becomes the favoured de-excitation mode and intersystem crossing to the chemically reactive triplet is markedly decreased.

An alternative explanation of the stability of the benzotriazole series involving steric inhibition of rotation of the *ortho*-hydroxyphenyl ring with respect to the triazole ring has been offered. It is qualitatively established that there are effects due to quenching as well as filtering in the benzophenone and benzotriazole stabilizers.

There are other heterocyclics, which are potentially capable of enol-keto tautomerism. They have defined photostability (inverse of the quantum yield of photochemical reaction) and have correlated this property with the change in resonance energy in going from enol to keto form in the first singlet state. Thus quinolines are more photostable than pyridines, and quinoxalines than pyrazines. Quinazoline and 5, 5'-bipyrimidine are two useful photostabilizers.

BINDERS AND STICKERS

Binders

Binders are resinous chemicals or other cementing materials used to hold the particles together. These provide mechanical strength and ensure uniform consistency and solidification on addition to a surface coating.

Many organic and inorganic compounds are used as binders in pesticide formulation. Some examples are given below.

Organic Binders Used in Sticking Method of Granular Formulation

Casein	a protein isolated from milk;
Blood albumin	a protein, isolated from slaughterhouse blood;
Starch and dextrin	obtained from plant sources;
Animal or fish glue	obtained from animal sources;
Natural gum	a complex polysaccharide (e.g. gum Arabic, gum Acacia, guar gum etc.);
Shellac	a product produced by lac insect, *Ceria lacca*;
Resins	polymeric synthetic organic compounds;
Beta-Lactoglobulin	a protein obtained from milk;
Lignosulphonates and polyvinyl alcohols	organic polymers (used in extruded granular formulations, aqueous solution is added to well-pulverized mixture of technical material and carrier, the kneaded mixture on extrusion granulation forms granules);
Polyethylene glycols	synthetic polymeric alcohols (PEGs, are liquid binders used as aqueous solution or solution in some organic solvent in spray formulations);
Machine oils	a mixture of petroleum products;
Coal tar	a product obtained from destructive distillation of coal;

Inorganic Binders Used in Sticking Method of Granular Pesticide Formulation

In sticking method of granular pesticide formulation the active toxicant is stuck to the outside of the carrier with the help of a binder. It is achieved by either spraying a suspension of powdered toxicant on a granulated carrier (carrier blank) and then spraying the binder material on it later as a post spray, or the granules are first sprayed with the binder solution and the sticky granules thus prepared are rolled with toxicant powder or the granules are sprayed with binder solution to make them sticky and the sticky granules are then rolled with toxicant powder and again post spraying of the material with the binder solution is carried out. Examples, asphalt, sodium silicate, etc.

Clay Binders

Clay colloids can be successfully used in extrusion method of granular pesticide formulation. Colloids of montmorillonite and attapulgite are largely used. But if dry pellet strength is not sufficient, a separate binder will have to be added. A common example of 30/60 mesh granule formulation consists of pesticide—20%, attapulgite carrier—60%, water—8% and attapulgite colloidal binder—12%.

Stickers

Stickers are adjuvants added to a formulation to improve its retention on the solid surface on which it is applied. The mechanism of sticking action can be explained based on the Furmidge equation i.e.

where, $$MIR = \theta_m \sqrt{\gamma(\cos\theta_r - \cos\theta_a)d}$$

where, MIR is the maximum initial retention of the material which occurs immediately after its application on a surface, θ_m = mean contact angle, θ_r = receding contact angle, θ_a = advancing contact angle, γ = surface tension and d = density of the formulation.

This equation states that as θ_r decreases, $\cos\theta_r$ increases, and hence MIR increases. So, stickers actually decrease the receding contact angle or increase the advancing contact angle and thus increase the retention of the formulation on a surface. However, chemically, there is little difference between stickers and binders.

Some of the materials used as stickers are mentioned below.

Flours: Pastes of flours like wheat and soybean flour are common stickers. The idea of gumming the formulation to the foliage forms the basis of using such materials. Flour paste controls red spider mite (*Tetranychus bimaculatus*) not because of direct toxicity but because it sticks the mite to the leaves. However, use of such adhesives never became popular for pest control.

Peps: It is chemically polyethylene polysulphide– a popular adhesive and sticker.

Powdered Skimmed Milk and Casein: These milk products are used as sticker in some baits.

Fish and Vegetable Oils: Can be used as stickers in various formulations.

Polyvinylacetate: A synthetic resin of latex type, used as sticker in the formulation of fungicides.

Polybutenes: These are available as viscous, water insoluble liquids, which are also known for acaricidal activity. They improve the tenacity of pesticides leaving a solid residue on plant surface.

Gums: The natural gums like gum Arabic, gum Tragacanth, gum Karanya, guar gum, etc. are used as stickers in emulsions. These complex polysaccharides increase the viscosity of emulsions and improve the retention of the formulation on plant surface.

Protective Colloids: These are high molecular weight polymeric materials, used as stickers in liquid formulations e.g. suspensions or aqueous dilutions of WP formulations. They are soluble or dispersible in the continuous phase of suspension and inhibit the agglomeration and sedimentation of dispersed particles. They increase the viscosity and surface adsorption whereby the solid particles become surrounded by a liquid shell of density similar to the surrounding liquid. Protective colloids may be organic or inorganic. Some important organic colloids are methylcellulose, sodium carboxymethyicellulose, polyvinylpyrrolidone, blood albumin, collagen etc. Some inorganic protective colloids are water-swelling bentonites, calcium carbonate and silica.

Spreader-Stickers: These are usually a combination of compounds, one of which serves to cause the droplets to spread over the surface of a leaf and the other helps in retaining the pesticide on the leaf during rainfall. These are marketed as liquids (HLB value 7 to 9) and used in water at concentration of 0.24 to 0.481 380 l^{-1}. These are mainly used in WP formulations of herbicides. Some such sticker products are available under commercial trade names like Du Pont spreader sticker, Miller Nu-Film, Ortho spray sticker, X-77 spreader, Spreader-sticker, Spreader-activator (Chipman and DePaster), Triton B-1956, Plyar spreader sticker, Sherwin Williams Spread-Rite, Multifilm X-77, De-Pester sticker etc.

CROP SAFENERS

"Safener" or more specifically "crop safener" (also called herbicide antidote) generally refers to a chemical compound which has limited phyto-toxicity of its own and selectively protects the crop plants against herbicide injury without protecting weeds. The selectivity arises from either its selective placement such as seed dressing or some very specific crop-herbicide-safener interactions (Hatzios and Hoagland 1989; Parker, 1983). There are objections against the term antidote because an antidote reverses intoxication while the safener protects the plant before the herbicides can cause injury to it. It cannot reverse the effect of herbicide (toxicant) as such. As a result, the term "safener" or crop safener has been widely accepted.

Much progress has been made over the past few decades in the development of selective herbicides. The selectivity depends on complex interactions between crops, weeds and their environment with the applied herbicide. Many of the currently available herbicides are either non-selective or have marginal selectivity, hence, the importance of safeners. Perhaps the greatest challenge in weed control lies in the control of the species related botanically to the crop plant, e.g., wild oat (*Avena fatua*) in cultivated oat (*Avena sativa*). Another problem of weed control lies in case of crop rotation systems wherein the preceding crop in rotation gives rise to "volunteer weeds" in the next crop. For instance, "volunteer" wheat as a weed in barley crop of a wheat-barley rotation. In both these cases, safeners can play a crucial role in enhancing herbicide selectivity or in introducing a selectivity factor, not attainable through conventional approaches.

Advantages

Some of the advantages underlying use of safeners are as follows:

(*i*) A selective chemical control of weeds in botanically related crops.
(*ii*) Selectivity in use of non-selective herbicides.
(*iii*) Counteraction or reduction of residual activity of soil-applied persistent herbicides in crop rotation system.
(*iv*) Widening of spectrum of herbicidal activity for weed control in "minor" crops.
(*v*) Enhanced use and marketability of conventional herbicides.
(*vi*) Reduction in cost of chemical weed control.

Development of herbicide safeners has proved to be quite an expensive process, justified only by the offered economic returns. Industrial establishments have played a leading role in their development, usually with the objective of increasing the marketability of the herbicides they produce.

The empirical, imitative and rational methods have been utilised to study the crop safening activity of candidate chemicals. Candidate chemicals are first screened for their effectiveness in protecting one or more important crops against herbicide(s) at the laboratory or greenhouse level. This primary safener screen not only characterises its activity but also provides details of the structure-activity relationship of the safener. The secondary screen consists of field trials of the successful chemicals.

Key parameters of a Potential Safener

Critical characteristics of potential safeners are: market need/opportunities for new safeners or herbicide/safener package, degree of crop safening (efficacy), chemical and botanical specificity of candidate safeners, ratio of safener to herbicide dose, reliability of safening effect under field conditions, suitability of active ingredient for proper formulation, potential phytotoxicity of safener, toxicology of the safener, environmental fate and residual effects of the safener, interactions of candidate safener with other biological systems, and the registration requirements.

Classification

Crop safeners for herbicides can be classified based on the structural framework of the molecule and further sub classified on the basis of functional groups present in them. Currently available crop safeners can be grouped into following classes:

1. Naphthapyranone derivatives (e.g. naphthalic anhydride, **NA**)
2. Chloroacetamides (e.g. dichlormid, **DC**).
3. Dichloromethyldioxolane (e.g. **MG-191**).
4. Oxime ether derivatives (e.g. cyometrinil, **CONCEP I**)
5. Derivatives of 2, 4-disubstituted 5-thiazolecarboxylates (e.g. flurazole, **FA**).
6. Substituted phenyl pyrimidines (e.g. fenclorim, **CGA 12407**).
7. N-Phyenylmaleamic acids and their progenitors [e.g. N-(4-chlorophenyl) maleimide, **CPMI**].

N-phenylmaleimides and N-phenyl-isomaleimides hydrolyse rapidly to their corresponding N-phenylmaleamic acids. Therefore, N-phenylmaleimides and N-phenyl isomaleimides are pro-safeners whereas, N- phenylmaleamic acids are actual crop safeners for protecting sorghum from alachlor injury (Rubin *et al.*, 1985). Fig. 3.11 depicts structures of a few common crop safeners.

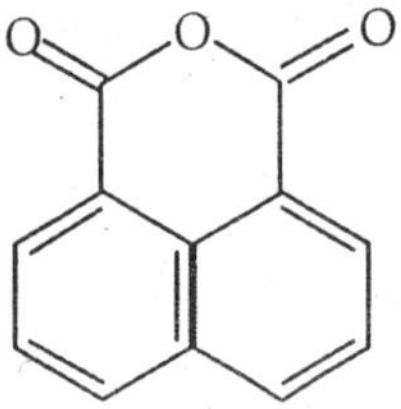

NA

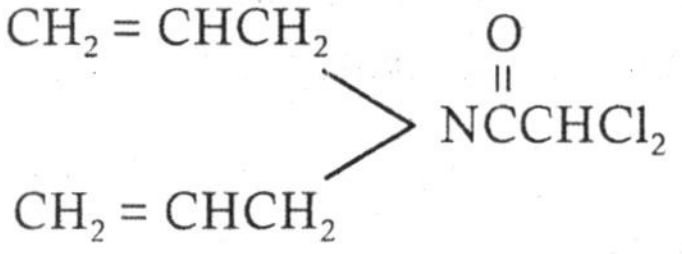

DC

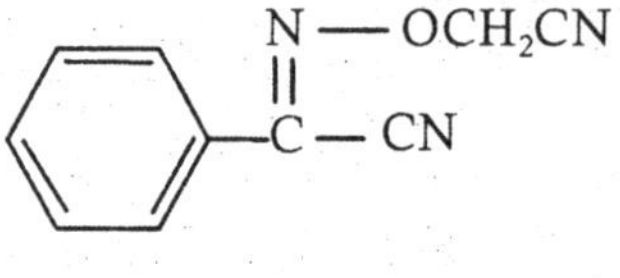

CONCEP–I

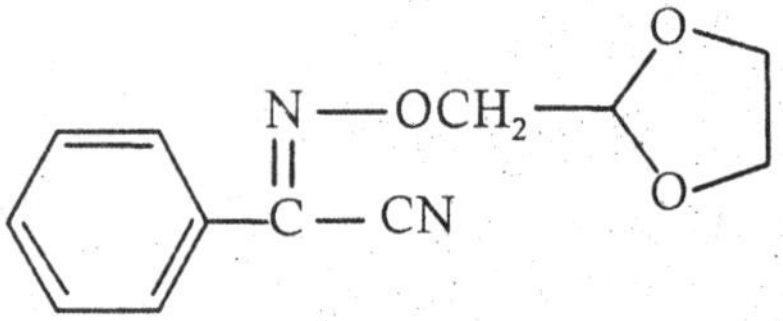

CONCEP–II

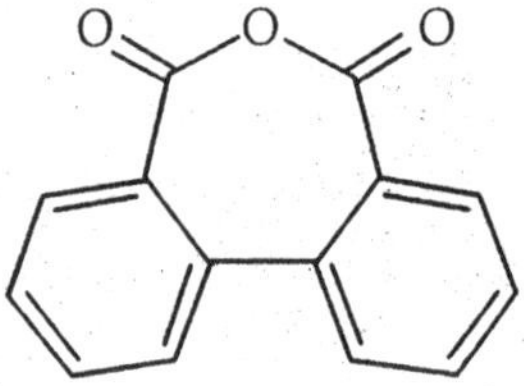

DA

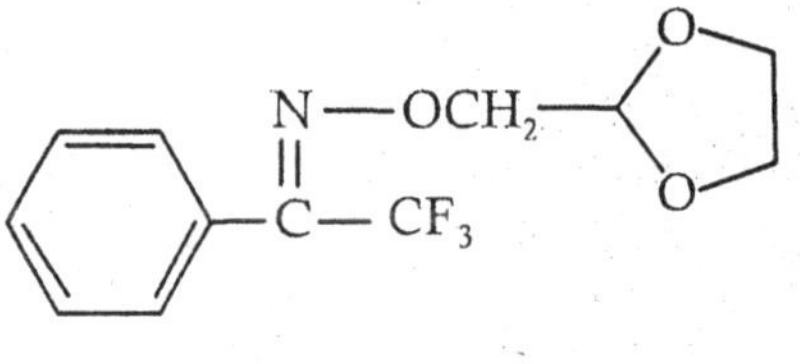

CONCEP-III

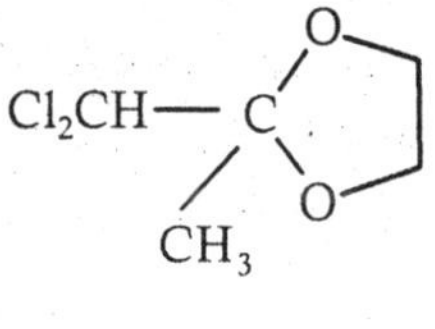

MG-191

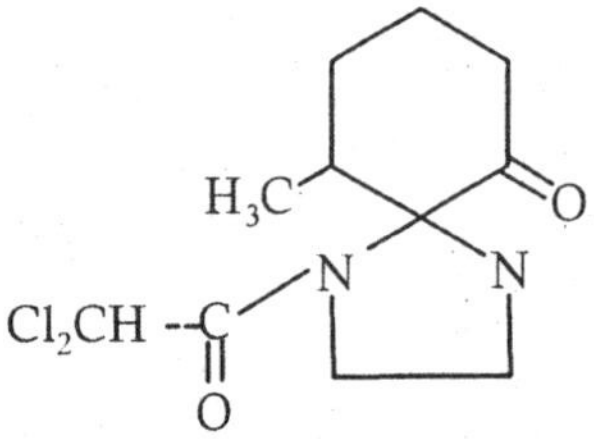

LAB-1

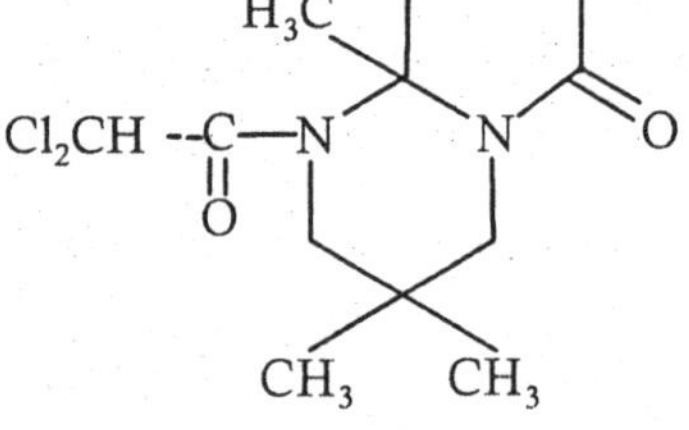

LAB-2

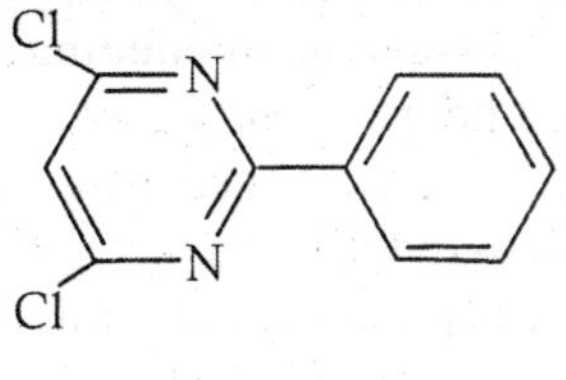

CGA-12407

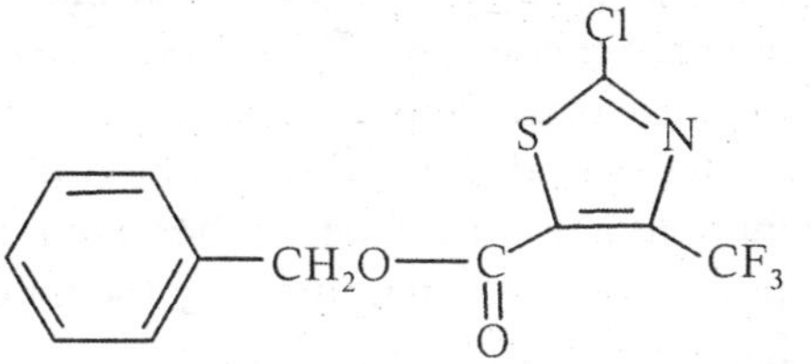

FA

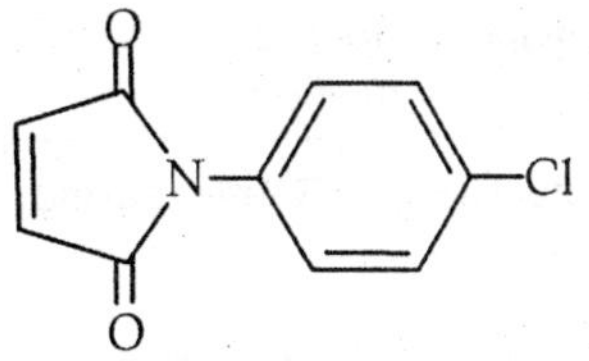

CPMI

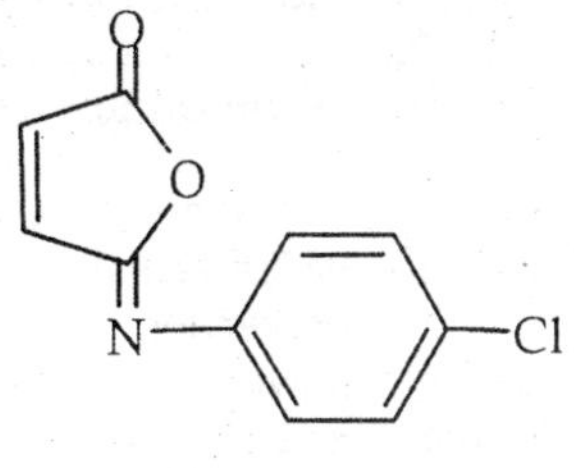

CPIM

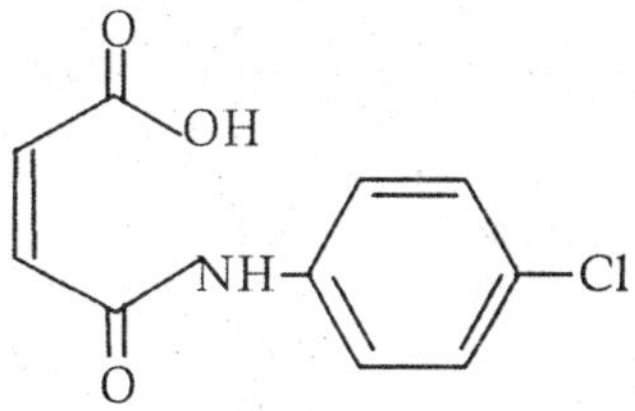

CPMA

Fig. 3.11: Structures of some Crop Safeners

Practical Application

The major agricultural uses, formulations and methods of application of some common safeners including those undergoing development for commercialization are given in Table 3.14.

Under practical conditions, the application of herbicide safeners in the field does not involve any extra operation, because the respective companies that manufacture herbicide safeners market them as either pre-packaged tank mixtures with the herbicide or seed treatments (seldom supply them as treated seeds). A basic reason for the above methods of application is the specificity of the crop, herbicide and safener. However, the other dimensions such as suitability of the active ingredient for formulation, storage stability, etc. must be studied before hand in detail.

Some applications of safeners on specific crops are discussed below.

Corn

Initially, NA was used to protect corn against thiocarbamate herbicides. Its versatility, however, extended to other unrelated groups as well e.g. phenylcarbamates, dithiocarbamates, chloroacetanilides, sulfonylureas, imidazolinones, cyclohexanones, and arylophenoxyalkanoic acids. In addition, NA is the only safener capable of providing protection to grass crops such as corn and oats against post-emergence application of selective herbicides like chlorsulfuron, diclofop-methyl etc.

Thiocarbamate protection is now given to corn by use of dichlormid, which is remarkable in its chemical and botanical specificity. It is primarily used as pre-packaged tank mixture with the thiocarbamate herbicide. Some such commercial products are Eradicane (EPTC +

dichlormid), Sutan (butylate + dichlormid) and Vernam (vernolate + dichlormid). More recently, an experimental chemical LAB-145138 has shown excellent activity in safening corn against injury by chloroacetanilide herbicide metazachlor and is under commercialization by BASF. This is a selective safener and it will be formulated as a pre-packaged tank mixture with metazachlor in a 1 : 6 safener to herbicide ratio. CGA-154281, a corn safener against metolachlor, is under development by the Ciba-Geigy Corporation as a tank mixture in a 1 : 30 safener to herbicide ratio.

TABLE 3.14. AGRICULTURAL USES, FORMULATIONS AND APPLICATION METHODS OF SOME COMMON HERBICIDE SAFENERS

Safener	Protected Crop	Herbicide counteracted	Formulation	Trademark	Application method
Napthalic anhydride (**NA**)	Corn	EPTC, butylate vernolate	95% Seed protectant powder	Protect	Seed dressing
Diphenic anhydride (**DA**)	Corn	EPTC, butylate vernolate	Seed protectant powder	—	Seed dressing
Dichlormid (**DC**)	Corn	EPTC, butylate vernolate	Tank mixture with herbicide	Eradicane, Sutan, Vernam	Pre-plant incorporation
LAB-147886 (**LAB-1**)	Corn	Triallate	20% Seed protectant powder	—	Seed dressing
LAB-145138	Corn	Metazachlor	Tank mixture with herbicide	—	Pre-emergence
MG-191	Corn	EPTC	Tank mixture with herbicide	—	Pre-emergence
Cyometrinil (**CONCEP I**)	Grain sorghum	Metolachlor	2.09 S	Concep I	Seed dressing
Oxabetrinil (**CONCEPII**)	Grain	Metolachlor	70 WP or 50 SD	Concep II	Seed dressing
CGA-133205 (**CONCEPIII**)	Grain sorghum	Metolachlor	70 WP	Concep III	Seed dressing
Flurazole (**FA**)	Grain sorghum	Alachlor	80 WP	Screen	Seed dressing
Fenclorim (**CGA12407**)	Rice	Pretilachlor	Tank mixture with herbicide	Sofit	Pre-emergence
CPIM	Sorghum	Alachlor	Tank mixture with herbicide	—	Pre-emergence
CPMI	Sorghum	Alachlor	Tank mixture with herbicide	—	Pre-emergence
CPMA	Sorghum	Alachlor	Tank mixture with herbicide	—	Pre-emergence

Sorghum

While NA is capable of safening sorghum against injury from several herbicides, it has not been commercialized.

Oxime ether safeners have been used for safe use of chloroacetanilide herbicides in sorghum. Ciba-Geigy has released three safeners, CONCEP I, CONCEP II and CONCEP III. in the same series. CONCEP I or cyometrinil proved to be inconsistent in that often adverse effects were observed on seed germination. CONCEP II, its analogue, was thus developed which showed no phytotoxicity. However, undesirable associations with downy mildew of sorghum were observed. Concep III (CGA-133205) is expected to replace Concep II and does not possess the undesirable properties of its predecessors.

Flurazole (FA) was developed by Monsanto under the trade name 'Screen' to safeguard sorghum against alachlor and acetochlor. All sorghum safeners provide protection to weeds as well. As a result, they are used either as seed dressings or marketed as treated sorghum seeds.

Rice

Fenclorim (CGA 12407) is the first rice safener to be marketed. It protects rice from chloroacetanilide, pretilachlor.

Pre-packaged Safener-Herbicide Mixtures

A pre-packaged tank mixture of herbicide and safener should possess the following characteristics:

(i) Safener should remain in the protection zone for as long as the crop is susceptible to the herbicide.
(ii) It should not be more water-soluble than the herbicide.
(iii) It should not be more biodegradable than the herbicide.
(iv) The safener should not interfere with effectiveness of the herbicide on target weeds.
(v) It should be compatible with other pesticides that may be added to the spray tank or applied to the soil when the herbicide/safener mixture is present.

These mixture formulations have a distinct advantage over seed safeners because these combine technical and marketing concepts into one package. The industry controls all aspects of the formulation. It is also attractive to user as the desired attributes are available from a single package without the headache of choosing and matching herbicides and safeners.

Seed Safeners

The important requirements for a good seed safener formulation/technical product are:

(i) Safener should be presented in such form that it can be applied or distributed uniformly on all seeds.
(ii) It should be stable during processing and subsequent storage.
(iii) Safener should be held strongly enough to the surface of the treated seed so that it is not removed during handling and storage.
(iv) Safener and the process of seed coating should not affect the subsequent performance of the seed.

(*v*) Size and particle flow properties of the coated seed must be consistent so that they do not interfere with mechanical sowing.

Limitations

Currently used safeners act primarily by enhancing the rate of metabolic detoxification of selected herbicides. The botanical and chemical specificities of herbicide safeners have been often cited as factors increasing the difficulty for discovery of new safeners. Most of the currently marketed safeners are particularly effective against soil applied, shoot-absorbed herbicides such as the thiocarbamates and the chloroacetanilides. Success has been very limited in developing chemical safeners for the protection of any crop from photosynthesis inhibiting herbicides or broad-spectrum weed killers such as glyphosate and paraquat (Hatzios, 1989).

REFERENCES

Barnes, J. R. and Felling, J. (1969). Synergism of carbmate insecticides by phenyl-2-propynyl ethers. *J. Econ. Ent.*, **62**: 86-89.

Butenandt, A., Beckmann, R., Stemm, D. and Hecker, E. (1959). Uber der sexual Lockstoff des Seiden spinners *Bombyx mori*. Reindarstellung und Konstitution. *Z. Naturforsch.*, **2**(1): 7-9.

Casida, J. E. (1970). Mixed function oxidase involvement in the biochemistry of insecticide synergists. *J. Agric. Food Chem.* **18**: 75.

Eagleson, C. (1942). Sesame in insecticides. *Soap Sanit. Chem.*, **18**: 125.

FAO (1999). FAO Specification for Plant Protection Products. AGP: CP/30- 306.

Gordon, A. J. and Ford, R. A. (1972). *The Chemists Companion. A Handbook of Practical Data, Techniques and References*. John Wiley and Sons, New York.

Grayson, M. (ex. ed.) (1983). Kirk Othmer, *Encyclopedia of Chemical Technology*. 3rd ed. Vol. 22, John Wiley and Sons, New York, pp. .332-432.

Hatzios, K. K. (1989). Herbicide safeners: Progress and prospects. *In*: '*Crop Safeners for Herbicides, Development, Uses and Mechanism of Action*'. (eds. K. K. Hatzios and R. E. Hooglonk). Academic Press Inc., San Diego, California pp. 355-363.

Hatzios, K. K. and Hoagland, R. E. (1989). *Safeners for Herbicides, Development, Uses and Mechanisms of Action*. Academic Press Inc. San Diego, California.

Hewlett, P. S. (1968). Synersism and potentiation in insecticides. *Chem. Ind.*: 701-706.

IUPAC (1996) Glossary of terms relating to pesticides *Pure. Appl. Chem.* **68**: 1167-1193.

Karlson. P. and Butenandt, A. (1959) . Pheromones (Ectohormones) *Ann. Review Entomol.* **4**: 39-58.

Law, J. H. and Regnier, F. E. (1971). Pheromones. *Ann. Review Biochem,* **40**: 533-548.

Metcalf, R. L. (1967). Mode of action of insecticide synergists. *Ann. Rev. Ent.*, **12** : 229-238.

Parker (1983). Herbicides antidotes—A Review. *Pestic. Sci.* **14**: 40-48.

Parmar, B. S., Pandey, S. Y. and Mukerjee, S. K. (1976). Photodegradable formulations of DDT. *Experientia* **32**: 279.

Parrmar, B. S. and Tomar, S. S. (1983). Review of research on insecticide synergists in India— retrospect and prospect. *Intern. J. Trop. Agri.*, **I**: 7-17.

Parmar, B. S., Swamy, R. V., Gupta, S. C., Attri, B. S., Singh, R. P. and Mukerjee, S. K. (1974). Improvement in or relating to the preparation of substituted 3, 4-methylenedioxycinnamoyl derivatives as synergists for pyrethrins. *Indian Patent* No. 137926.

Rubin, B., Kirino, O. and Casida, J. E. (1985). Chemistry and action of N-phenylmaleamic acids and their progenitors as selective herbicide antidotes. *J. Agric. Food Chem.*, **33**: 489-494.

Sarkar, D.K. and Parmar, B.S. (1979) Nicotine sensitized DDT formulations. *Agric. Biol. Chem,* **43(11)**: 2291-2296.

Sarode, S. V., Supare, N. R. and Saxena, D. B. (1993). Semiochemicals. In: *Botanical and Biopesticides.* (eds. B. S. Parmar and C. Devakumar). SPS Publication No.4, Society of Pesticide Science, India, Westvill Publishing House, New Delhi., pp. 111-130.

Schick, M. J. (ed.) (1966). *Surfectant Science* Vol. 1, Marcel Dekker Inc., New York.

Shorey H. H. and Mckelvey, J. J. (eds.) (1977). *Chemical Control of Insect Behaviour: Theory and Application.* Wiley Inter-Science Publications, New York, p. 414.

Tomar, S. S., Maheshwari, M. L. and Mukerjee, S. K. (1979). Synthesis and synergistic activity of some pyrethrum synergists from dillapiole. *Agric. Biol. Chem.,* **43**: 1479-1483.

Weed, A. (1938). A new insecticide compound. *Soap Sanit. Chem.,* **14**: 133.

Wilkinson, C. F. (1971). Insecticide synergists and their mode of action. *Proc. 2ʳᵈ Intern. IUPAC Congress. Pestic. Chem.* Vol. II (ed. A. S. Tahori). Gordon and Breach Science Publishers, New York, London. p. 117.

CHAPTER 4

Pesticide Mixtures

Background

In the wake of the realization that development of newer pesticidal molecules may hence forth be scanty, a judicious use of the existing products has assumed significant importance. Furthermore, the problem of pest resistance to pesticide(s) has necessitated a rethinking on the manner of their formulation and use. Pesticide mixture is one promising option that has come to the fore. It has potential to increase the commercial lives of pesticides through their use in combinations, by complementing the bioactivity of the individual products and simultaneously lowering their use pressure on the one hand and broadening the spectrum of activity and overcoming pest resistance to individual pesticide, on the other. In view of the current interest on the subject, it has been briefly discussed in this chapter.

Pesticide mixtures may be 'tank mixtures', the ingredients for which are chosen for mixing by the users in the spray tank at the time of application, or 'pre-packed mixtures' that combine multiple pesticide ingredients or formulations in a single pack before marketing. In former, the mixed components may often not be compatible and lead to separation, flocculation, agglomeration, coagulation etc. of the ingredients, resulting in application or performance problems. The latter, on the other hand, are scientifically developed and tested products and are based on compatibility studies. The final product is a 'ready to use' material, which fulfills the various physico-chemical, shelf life, bioefficacy, phytocompatibility, toxicology and other related requirements, prescribed for pesticide formulations.

Advantages

Several advantages that are attributed to the use of pesticide mixtures include:

Broad Spectrum of Activity: More than one pest or pest species may come in the gambit of control than those with the individual components of the mixtures.

Synergistic Joint Action: As the choice of components of the mixture is based on well-planned scientific studies, synergistic joint action is often anticipated by the use of multi-ingredients of the mixture.

Economic Pest Control: Economy can be exercised by carefully choosing the components of a mixture. The use of auxiliaries such as synergists (particularly mildly pesticidal chemicals) considerably lowers the quantity as well as the cost of several rare and expensive active ingredients such as pyrethrins. The application cost would also stand reduced due to single application of the mixture over the repeated applications of the individual components.

Overcoming/delaying Pest Resistance to Pesticides: The use of mixtures minimizes the selection pressure of the individual chemicals and thus delays the development of resistance. The development of resistance is delayed or overcome by the multiple mode of action offered by the ingredients of the mixture. By acting on different sites in an organism, the problem of pest resistance including cross-resistance generally stands mitigated.

Preparation

Compatibility of all components is a pre-requisite for developing ready to use mixtures. The parameters governing physical compatibility include formation of agglomerates or crystals, phase separation of emulsions or suspensions, thickening of spray liquids, appearance of any type of extraneous matter in the combination products etc. The chemical compatibility is characterized by photochemical, thermal and hydrolytic stability of the ingredients of mixture. The physical properties of the constituents such as melting and boiling points, volatility, solubility etc. influence the overall physicochemical properties and performance of the mixtures. The biocompatibility can be judged by the nature of joint action (synergistic/antagonistic) of the formulation. The phytocompatibility is characterized by a non-phytotoxic action on the plants to be sprayed. The mixture should not adversely influence the phytocompatibility of individual components. Likewise, toxicological compatibility is important and is indexed by lack of undesirable effect(s) on the useful non-target organisms, mammals and the environment. However, the choice of ingredients of the mixture is often governed by the pest complex to be managed at a particular location and season.

The data requirement for registration of a combination of pesticides under section 9(3) of Indian Insecticides Act (1968) is given in Annexure I.

Examples

The pre-packed multi-pesticide mixtures may exist in a single phase or as multiphase systems. Multi-pesticide formulations containing up to six active ingredients are reportedly available in the European market (Anonymous, 1989). The major products include: synergistic combinations of organophosphorus with a carbamate or synthetic pyrethroid insecticide e.g. malathion and diazinon with a carbamate such as metolcarb, fenobucarb, xylylcarb and isoprocarb; chlorpyrifos methyl and dimethoate, chlorpyrifos ethyl and cypermethrin and malathion; fenitrothion and cypermethrin etc. The fungicidal mixtures include triadimefon and carbendazim, metalaxyl and mancozeb, benlate and captan, dicarboximides and folpet, thiram, carboxin etc. The insecto-fungicidal combinations may be exemplified by thiram with dichlofenthion and others. The herbicidal mixtures include 2, 4-D or 2, 4-DB and atrazine or dicamba, chorpham and picloram, chlorsulfuron and metsulfuron methyl etc.

The examples of multipesticide single phase formulations registered in India include: Carbaryl 4% + Lindane 4% (GR), Deltamethrin 1% + Triazophos 36% (EC), Profenfos 40% + Cypermethrin 4% (EC), Chlorpyriphos 50% + Cypermethrin 5% (EC), Cypermethrin 3% + Quinalphos 20% (EC), Chlorpyriphos 16% + Alphacypermethrin 1% (EC), Acephate 25% + Fenvalerate 3% (EC), Streptomycin sulphate + Tetracycline hydrochloride (9 : 1, SP), Metalaxyl 8% + Mancozeb 63% (WP), Anilofos 24% + 2, 4-D Ethyl ester 32% (EC), Carbendazim 12% + Mancozeb 63% (WP), Ethion 40% + Cypermethrin 5% (EC), Endosulfan 29.75% + Deltamethrin 7.5% (EC) etc. For household purposes, natural pyrethrins in combination with either of cypermethrin, lindane, diazinon or piperonyl butoxide, deltamethrin with allethrin or allethrin and propoxur with cyfluthrin are registered. Herbicidal mixture of metsulfuron methyl and chlorimuron ethyl is also registered. In most cases, the proportion of one component is relatively higher than the other. It is easier to prepare products like D, WP, GR, WG, EC etc. while SC, EW, micro emulsions etc. are somewhat difficult to obtain.

Preparation of multiphase systems that often combine different formulations representing diverse phases is the most difficult preposition. Mosinski (1998) has reported some of the possibilities for making these formulations as follows:

A solid pesticide insoluble in organic solvent/ water (SP_1) with solid pesticide soluble in organic solvent (SP_2)	SC of (SP_1) in EC of (SP_2)
'SP_2' above with a solid pesticide insoluble in water (SP_3)	SC of (SP_3) in EW of (SP_2)
'SP_3' above with a solid pesticide soluble in water (SP_4)	SC of (SP_3) in SL of SP_4 (SP_3) solubilized in SL of (SP_4)
Liquid pesticide insoluble in water (LP_1) with 'SP_4'	EW of (LP_1) in SL of (SP_4) SL of (SP_4) in EC of (LP_1)
'SP_3' with 'SP_4' and 'LP_1'	SC of (SP_3) and EW of LP_1 in SL of SP_4

In multiphase systems, the product preparation, stability, performance etc. assume a special importance, necessitating due attention to be given to the formulation ingredients and the process.

Limitations

Analysis of active ingredients in multipesticide formulations is a difficult task. It is particularly so with multiphase formulations such as suspension concentrates, emulsions, suspo-emulsions etc. Quite often, multiple methods of analysis may be applicable in a given product.

Product toxicology is another matter of concern. Toxicology of the individual components and/or their mixtures has to be determined with no undesirable additive effect in mixture. Similarly, spectrum of mammalian toxicity should not get altered by the use of a mixture, as compared to the use of individual components.

A major difficulty in recommending mixtures is the non-availability of antidotes for the combination products. This is a serious bottleneck in promoting mixtures, as accidental intakes may become unmanageable.

REFERENCES

Anonymous (1989). *European Directory of Agrochemical Products.* 3rd Edn. (Hamish Kidd Ed.). Royal Society of Chemistry, U.K.

Mosinski, S. (1998). Pesticide mixtures. In: *Pesticide Formulation—Recent Developments and their Applications in Developing Countries*, (eds.), W. V. Valkenburg, B. Sugavanam and S. K. Khaitan. United Nations Industrial Development Organization, New Age International (P) Ltd. Publishers, New Delhi. pp. 30-43.

ANNEXURE I

DATA REQUIREMENTS FOR REGISTRATION OF COMBINATION OF PESTICIDES UNDER SECTION 9(3) OF THE INSECTICIDES ACT, 1968.

A. General Requirement

1. Proposed pesticides should be registered individually under Section 9(3) of the Act.

B. Chemistry

2. Source of supply
3. Chemical composition
4. Chemical identity
5. Physico-chemical properties
6. Specification
7. Method of analysis
8. Analytical test report
9.   (*i*) Shelf life claim
     (*ii*) Shelf life data
10.   (*i*) Process of manufacture
    (*ii*) Information about raw materials used
   (*iii*) Source of supply of raw materials
   (*iv*) Stepwise manufacturing process
    (*v*) Flow sheet diagram of process of manufacture
   (*vi*) Effluent treatment method

C. Bioefficacy and Residues

11.   (*i*) Bio-effectiveness
    (*ii*) Phytotoxicity
   (*iii*) Effect on parasites and predators
12. Persistence in soil
13. Persistence in water
14. Persistence in plant
15. Compatibility with other chemicals
16. Residues in plant
17. Residues in soil
18. Cost benefit ratio

D. Toxicity

19. Acute oral in rat and mice
20. Acute dermal
21. Acute inhalation
22. Primary skin irritation
23. Irritation to mucous membrane
24. Toxicity to birds
25. Toxicity to fish
26. Toxicity to honey bees
27. Medical data
28. Observation in man (health records of spray operators)
29. Toxicity to livestock (field trial and observation)

E. Packaging and Labelling

30. Labels and leaflets as per IR-1971 existing norms
31. Labels to contain

    (*i*)   Detailed chemical composition
    (*ii*) Purpose for import / manufacture
    (*iii*) Antidote
    (*iv*) Toxicity triangle
    (*v*)   Cautionary statement
    (*vi*) Brief direction concerning usages
    (*vii*) Restriction, if any.

32. Packs to contain detailed chemical composition on leaflets accompanying small labels.

    (*i*)   Introductory para about the pesticide
    (*ii*)   Detailed directions concerning usage(s)
    (*iii*) Time of application
    (*iv*)   Application equipment
    (*v*)   Waiting period
    (*vi*)   Symptoms of poisoning
    (*vii*) First aid measures
    (*viii*) Antidote and treatment
    (*ix*)   Restriction, if any
    (*x*)   Instruction for storage
    (*xi*)   Information regarding disposal of used pack.

33. Type of packaging (packaging material and compatibility with content)
34. Manner of packaging
35. Specification for primary package
36. Specification for secondary packaging
37. Specification for transport packaging
38. Manner of labelling.

CHAPTER 5

Laboratory and Principles of Analysis

LABORATORY

Role of a formulation laboratory may begin either with newly developed molecules, which need to be roughly formulated to enable their activity appraisal, or with the molecules of established bioactivity for which suitable conventional and/or newer formulations have to be developed. Apart from this, the laboratory may attend to quality control and quality assurance aspect of the formulations.

For convenience, a formulation laboratory may be divided into two sub-sections, addressing to the liquid and the solid formulations, having within each specialized unit(s) attending to different types or formulations (*viz.* wet flowables, dry flowables, emulsifiable concentrates, water dispersible powders, granules, controlled release products etc.) with specialised precision or attention.

Location

Pesticide formulation laboratories are generally located close to the technical material plant or near the administrative blocks. Ideally, these should be away from both, in an independent neat and clean environment, where it may be readily accessible to the above units.

Infrastructure
Buildings

Building of the laboratory should be well planned giving a due regard to needs of research and development and quality control functions. It should make a proper provision of sitting space and secretarial support needed for the unit. Separate units and sub-units need to be provided for the liquid and solid formulations and their subgroups. Ideally, the total complex should be well ventillated and provided with a controlled temperature and humidity system, dry, cold, explosion proof, sample and solvent storage space along with waste solvent and toxicant disposal place, washing room, rest room, toilets and allied facilities. Since solid and liquid carriers and diluents and various formulation auxiliary chemicals will be the back-bone of this unit, due space needs to be provided to create a sample bank for these materials. The whole building must be provided with efficient fume hoods, safety systems, electrical power and light points, lifts and so on. Round the clock, uninterrupted power supply needs to be ensured in the marked instrumentation rooms. Even in other rooms, emergency lights must be provided to cater to any emergency due to a power failure. The whole unit should be inter-linked for communication.

Manpower

Skilled and specialized scientific, technical and supporting staff is essential to man such laboratories. The job requirement and the job description for each position must be well defined. These will be governed by the ultimate information expected to come out of such laboratories e.g. diverse requirements for registration (refer Annexures I and III for guidance). Manpower needs, besides the laboratory work, must take into account the needs for running and maintenance of the equipment, stores procurement as well as data base of stores, equipment and research information, cleanliness, safety, medical aid and similar others.

Glassware and Equipment

Some of the common glassware, equipment and instruments needed for such a laboratory are listed in Annexure II. Precise need will be governed by nature of the studies/work to be carried out, the required precision, the tests to be performed, and so on. The tests commonly undertaken in a pesticide formulation laboratory are listed in Annexure III.

PRINCIPLES OF ANALYSIS

Traditionally, pesticide formulation analysis has been dominated by volumetric and gravimetric methods. Subsequently, infrared spectrophotometry became an important additional player in this analysis. During 1950s and 1960s, most analyses were based on either of IR, UV or visible spectrophotometric methods. Late 1960s saw the introduction of gas liquid chromatographic (GLC) methods, which became acceptable after an initial hesitation, probably because the GLC methods were extremely sensitive and formulation analysis had been characterized as primarily macro analysis. In contrast, high performance or high pressure liquid chromatographic methods which were introduced in late 1970s found a ready acceptance in pesticide formulation analysis, particularly because of their applicability in the analysis of thermo labile compounds.

General Characters
Moisture

Two methods commonly used are Karl Fisher Method and the Dean and Stark Method.

The former is applied in situations of less than 1% moisture content and the latter, for 1% and more of moisture content.

Karl Fischer Method: It employs a reagent prepared by action of sulphur dioxide upon a solution of iodine in a mixture of anhydrous pyridine and anhydrous methane. Water reacts with this reagent in a two-stage process consuming for each molecule one molecule of iodine. The end point is determined electrometrically. The current will decrease to zero (Null point), when the last trace of iodine has reacted.

Dean and Stark Method: The sample of pesticide dissolved in a suitable solvent is distilled to collect water in a glass trap containing a copper wire with one end twisted into a spiral. The water forms a layer at the bottom of the graduated portion of the receiver from which its volume is noted.

Density of Liquid Pesticides

Density of a liquid is given by:

$$\text{Density} = \frac{\text{Specific gravity and weight of given volume of liquid}}{\text{Weight of same volume of water}}$$

A specific gravity bottle (25 ml capacity) or a Pyknometer tube (25 ml capacity) is used for the purpose.

Melting Point

It is the temperature at which the solid pesticide just starts melting into liquid. It is generally determined by capillary method using liquid paraffin or sulphuric acid bath as heat assembly.

Softening Point

Softening point of a solid pesticide is defined as the temperature at which the compound becomes fluid. It is determined by first coating a molten solid on a thermometer by dipping it in the molten solid mass and then the solid coated thermometer is dipped in a test tube containing liquid paraffin that can be heated. Fall of first drop of liquid in the tube gives the softening point of the solid.

Setting Point

It is the temperature at which a solid pesticide solidifies from its molten state. The pesticide is first molten in a paraffin bath and the molten liquid is then immersed in another bath maintained at about 15°C below the expected setting point. The setting point is then noted.

Material Insoluble in Acetone

A known weight of pesticide is heated with acetone to dissolve the material. The solution is filtered and insoluble material is weighed.

Chemical Quantification

Volumetric and Gravimetric Methods

During the earlier days, pesticide analysis was predominantly based on volumetric and gravimetric methods. Analyses of chlorinated hydrocarbons by total and hydrolysable chlorine methods and those of nicotine, pyrethrins etc. by gravimetric methods can be cited as classical examples in this context. Such analyses were fairly accurate, and reliable with a good degree of reproducibility. However, these methods were applicable only for macro-analysis.

Spectrophotometric Methods

Need for undertaking analysis of micro-quantities of pesticides made spectro-photometric methods highly acceptable in pesticide analysis. These methods are very rapid, less cumbersome and fairly accurate. The UV, visible and infrared spectroscopy have found application in pesticide analysis. The colorimetry, though simple and widely applicable, suffers from the drawback that the transformation products of active ingredients, which are sensitive to a given chromogenic reagent and absorb in the region of the main compound, also get estimated along with. The analysis becomes precise if coupled with thin layer chromatography, TLC, which enables separation and isolation of specific compounds. The above drawback will hold good for UV-spectroscopic analyses too. In infrared spectrophotometry, specific bonds absorb specific wavelengths of infrared energy. The absorbance is directly proportional to the concentration of the chemical containing the specific chemical bond. However, the impurities or the transformation products containing the same chemical bond, if present, may interfere in the analysis. Coupling this analysis with preparative TLC will make the analysis more precise and reliable.

The analysis of elements such as sodium, potassium, calcium and others, when applicable, can be accomplished by flame photometry.

Chromatographic Methods

Beginning with thin layer chromatography, which served as a very useful tool for qualitative and semi-quantitative analysis and in combination with other techniques such as densitometry, and various spectrophotometric methods for quantitative analysis, the chromatographic methods have undergone revolutionary advances, particularly for quantitative analysis. Examples may be cited of gas liquid chromatography (GLC) for the analysis of heat insensitive compounds, high performance liquid chromatography (HPLC) for the analysis of thermo labile compounds and so on. Availability of an array of detectors with the above instruments has immensely enhanced their capabilities of analysis e.g. electron capture detector for chlorinated pesticides, flame ionization detector, alkaline flame ionization detector, flame photometer detector for organophosphates and carbamates etc. These instruments have also been clubbed with mass spectrometer to enhance the scope of application to the determination of molecular masses of the compounds or their transformation products. High performance thin layer chromatography (HPTLC) is another recent innovation in instrumentation that improves the resolution of compounds in a speedier way.

In view of their importance in modern analysis, a brief discussion on chromatographic methods is annexed (Annexure IV).

Radio Labeled Analysis

For studies involving pesticide metabolism, environmental fate, mode of action etc. radio labeled materials are used. Suitable functional group(s) that is(are) critical for a study is (are) labeled and the test compound obtained in radio labeled form. This material is used for planned study and the results interpreted appropriately by measuring the radioactivity using a scintilation counter.

Physico-chemical Characteristics
Solids
Powders

Particle Size Distribution: Particle size is normally determined by wet sieving method. A known weight of the powder formulation is taken on a BSS 200 mesh (approx. 74 μm) sieve and subjected to flushing with tap water at a fixed rate. The residue remaining on the sieve after the prescribed time is collected, dried and weighed. From the weight of the sample retained on the sieve, the material passing through the 200 mesh is determined.

Dry sieving in a 200-mesh sieve fitted on to a sieve shaker is also practiced. A rubber ball (roller) is introduced on the sieve to break lumps and facilitate passing of the material through the sieve. The material passing through the test sieve after the prescribed time is noted.

Acidity/Alkalinity: Acidity or alkalinity is determined volumetrically by titrating an aqueous extract of the sample with a standard alkali or acid solution respectively. Acidity is expressed in equivalents of H_2SO_4 and the alkalinity in equivalents of Na_2CO_3.

Bulk Density after Compaction: It is determined by first finding out the bulk density and then the same cylinder carrying the material is tapped by felling for a fixed number of times from a fixed height (usually 20 times from a 15 cm height). The latter volume for the originally taken weight provides bulk density after compaction..

Flowability: Flowability can be worked out by actually dusting the material out of application equipment. Distance travelled by different powders when subjected to a similar (constant) force provides an idea about their relative flowability.

The above determinations are common for all powders. Water dispersible powders, however, require additional determination of suspensibility and wettability.

Suspensibility: A suspension containing a prescribed value of the active ingredient in it is prepared and allowed to stand for a given time (30 minutes). After that, 9/10th of the suspension is sucked out from a pre-marked point with a minimum possible disturbance of the remaining 1/10. The amount of active ingredient in this 1/10th of the remainding surpension is determined and by back-calculation, suspensibility of the suspension is worked out.

Wettability: A known weight of the material is added to water and time of complete wetting of the material is noted. A good WP is expected to wet in as short a time as possible.

Accelerated Storage Stability (Shelf Life): A known weight of the powder sample is subjected to a uniform pressure of 25 g cm^{-2} and incubated at an accelerated temperature of 54 ± 2°C for a prescribed period (24 h, preferably for 14 days which is considered equivalent to 2 years shelf life). After the prescribed period, the sample is cooled to room temperature and subjected to the various physico-chemical tests prescribed for different powders as stated above.

Granules

Acidity or alkalinity, and moisture content are determined based on the above described principles. Storage stability is studied at 54 ± 2°C for 14 days at normal pressure.

Particle Size: The granule size is expressed in a range of sieve size e.g. 10/30, 30/60 etc. Therefore, sieves of both the size extremes are taken. Not less than 97% by mass is prescribed to pass through a test sieve having mesh size of the upper declared limit and not more than 5% shall pass through the sieve of lower declared limit. A known weight of the granules is taken on the sieve of the upper mesh limit. The sieve of the lower limit is fitted below it. The sieves are shaken and the material retained on the sieves is determined.

Dust Content: Not more than 1% by mass of the product to pass through a 75 μm (~200 mesh) sieve. The portion passing through the sieve should not contain more than the prescribed fraction (usually 8%) by mass of the declared nominal active ingredient content.

Liquid Holding Capacity: It is the capacity of blank granules to hold a particular quantity of the liquid without affecting free flow of the granules. Liquid used for the purpose is a mixture of naptha plus cyclohexanone or monochlorobenzene and xylene, so mixed as to give a relative density of 1.0.

Encapsulation and Encapsulated Material

Attrition: A known weight of the pesticide granule (100 g) is taken on the prescribed sieve (200 mesh) fitted to a rotary machine and rotated for a given time (6 hours).

Afterwards, the granules remaining on the sieve are analysed for the active ingredient content. Results are calculated on the basis of granules retained on the sieve.

Water Run off: A measured volume of water (50 ml) is passed through a known weight of the granules (10 g, taken in a glass column) after allowing some contact time (15 minutes). The drained off liquid is analysed for the active ingredient.

Wet Test for Encapsulation: In this test, the amount of active ingredient absorbed by a wet cloth when granules are rotated with it in a closed container, is determined. The per cent active ingredient released is then calculated.

Liquids
Emulsifiable Concentrate/Emulsion Concentrate

The principle of determination of acidity/alkalinity is same as given earlier.

Cold Test: The emulsifiable concentrate is subjected to low temperature (zero °C or as prescribed) for a given period (24 hours or as prescribed). During the period, the oil and solid separation and turbidity, if any, should be within the prescribed limits.

Flash Point: It is determined by employing either the Abel or the Pensky-Martin Flash point apparatus. The former is used for flash points from 24.5–49.0°C and the latter above 49.0°C. The flash point should preferably be above 24.5°C.

Emulsion Characteristics: A measured quantity of the concentrate to yield emulsion of the prescribed strength is added to a measuring cylinder (100 ml) containing water/standard hard water, shaken and the resultant emulsion allowed to stand at 30 ± 1°C for 24 hours. The emulsion characteristics in the form of formation of creamed layer, oil separation, solid settling and emulsion thinning etc. are noted. Normally, a creamed layer of maximum 2 ml is allowed. After 24 hours, the contents are re-emulsified and noted for the above characters after 30 minutes.

Heat Stability (Shelf Life): The sample is subjected to heat stability test at 54 ± 2°C for a period of 14 days or as prescribed. At the end of the period, the sample is cooled and re-subjected to the above stated prescribed tests for compliance.

ANNEXURE I

DATA REQUIREMENT FOR REGISTRATION OF PESTICIDES UNDER SECTION 9 OF THE INSECTICIDES ACT, 1968 (INDIA)

Particular	Requirement
General pesticides	
Chemical composition and allied data	(*i*) Source of supply
	(*ii*) Chemical composition (purity of technical material, identification of impurities and physico-chemical properties of active ingredient, chemical composition of formulation etc.)
	(*iii*) Specifications—The relevant I.S. (Indian Standard) number to be cited. If BIS (Bureau of Indian Standards) specifications are not prescribed, the information has to be supplied in the BIS format.
	(*iv*) Analytical test report—as per BIS specifications.
	(*v*) Shelf life data (accelerated storage data or one locational actual shelf life data) at the first instance followed by the other prescribed data.
	(*vi*) Methods of analysis. Give BIS specification (another method, if BIS specification not available). Preferably two methods to be given and the referee method to be indicated.
Packaging and labelling requirements	Packaging—type (as per BIS specifications or detailed specifications about primary, secondary and transportation package), to be given. Labelling (as per Chapter V of the Insecticides Rules, 1971). Leaflets, instructions for storage and use, first aid, precautions, disposal of used packages, surplus material and washing of pesticides, to be provided.
Bio-efficacy and residues	Laboratory and field data under the local conditions where the material is to be used should substantiate the claims on the label. Bio-efficacy in different agro-climatic conditions; phyto-toxicity on some of the key crops; translocation within the plant or animal being treated; persistence in soil, water and plant along with the nature of metabolites and their toxicities; compatibility with other chemicals; direction concerning the dosage, time of application, application equipment and manner of use. Residue data to be generated in the country. To provide methods of sampling and residue analysis in food and feed stuff, water, soil and wild life. Expected residue level in edible crops and soil, worker hazard. Residue data on insecticide formulations to be used indoor for public health programmes may be generated in the plastered/ mud walls of the homes. Methods of analysis to be given in case of combination products.
Safety/toxicology	Separate requirements are prescribed for the imported and indigenously made technical materials with active ingredient more or less than 90 per cent.
	For the imported technical pesticides in use in India prior to 31.12.1972, the requirements are:
	Acute toxicity studies on mammals (data to be submitted within 1 year for both technical material and formulations)

Oral: Rat or mice or any two species of mammals.
Dermal: Rabbit / rat
Inhalation: One species
Primary skin irritation: Rabbit
Irritation of mucous membrane: Any one species (data to be submitted within 2 years)
Birds (maximum lethal dose, MLD): Pigeon and chick
Fish: Fresh water fish
Bees: Field trials and observation

Observations on man
Field observation on the human population if technical is used as such.
Health records on industrial workers

Medical data
Record, signs and diagnosis of poisoning
Treatment of poisoning
First-aid measures
Medical treatment

For the technical pesticides imported for use in India after 1972
Acute toxicity studies in mammals (to be submitted within 1 year)
Oral: Rat or mice or any two species of mammals
Dermal: Rabbit/rat
Inhalation: One species
Primary skin irritation: Rabbit
Irritation of mucous membrane: Any one species

Sub-acute toxicity studies in mammals (on two laboratory animals rat and dog, to be submitted within 2 years)
Oral: Rat and dog for 3 months each
Dermal: Rabbit / rat for 21 days
Inhalation: Rat for 14 days

Supplementary toxicological data (to be submitted within 3 years)
Neurotoxicity: Chicken
Teratogenicity: Rat
Effects on reproduction: Rat
Carcinogenicity: Rat/mice
Synergism and potentiation: Rat
Metabolism: Rat/rabbit
Mutagenicity

Other information (to be submitted within 2 years)
Acute toxicity: Pigeon and chicken
MLD: Fresh water fish
Toxicity to beneficial insects: Field trials and observations
Toxicity to livestock: Field trials and stock observations

Medical data
Signs, diagnosis, chemical tests and record on poisoning
Treatment of poisoning
First aid measures
Medical treatment

Imported technical materials with active ingredient less than 90 per cent

Complete profile of the active ingredient and associated chemical constituents/impurities to be given. Toxicological data required to be given by the first registrant. Subsequent registrants need not give this data for the same product with the same chemical composition profile.

Indigenous technical pesticides in use in India prior to 31.12.1972

Acute toxicity studies in mammals (to be submitted within 1 year)
Oral: Rat and mice or any two species
Dermal: Rabbit/rat for 21 days
Inhalation: One species
Primary skin irritation: One species
Irritation of mucous membrane: Any one species
(Information to be submitted within 2 years)
Acute toxicity: MLD in pigeon and chicken (two birds)
Acute toxicity to fish: Fresh water fish
Toxicity to bee

Observations on man
Direct field observation on the population in case the technical material is used as such in the field
Health records of industrial workers

Medical data
Poisoning cases, signs of poisoning, diagnosis of poisoning, chemical tests for poisoning
Treatment of poisoning
First-aid measures
Medical treatment

Indigenous technical pesticides manufactured in India for the first time, not in use prior to 31.12.1972.

Acute toxicity studies in mammals (to be submitted within 1 year)
Same as above
Sub-acute toxicity studies in mammals on two laboratory animals- rat and dog (to be submitted within 2 years)
Oral: 3 Months each, rat and dog
Dermal: Rabbit/rat 21 days
Inhalation: Rat 14 days
Supplementary toxicological studies (to be submitted within 3 years)
Neurotoxicity: Chicken
Teratogenicity: Rat
Effects on reproduction: Rat
Carcinogenicity: Rat/mice
Synergism and potentiation: Rat
Metabolism and mutagenicity: Rat/rabbit
(Following information to be submitted within 2 years)
Acute toxicity to birds (MLD): Pigeon and chicken
Acute toxicity: Fresh water fish
Toxicity to bees

Medical data
As above.

Liquid formulations-emulsifiable concentrates, aerosols, aqueous and other solutions introduced in India before 31.12.1972

Acute toxicity studies in mammals (to be submitted within 1 year)
Same as above.
(Information to be submitted within 2 years)
Same as above

Observations on man
Direct field observation on the population exposed
Health records on industrial workers and/or pest control operators

Medical data
Same as above
Note: The following information on liquid formulations to be submitted.
Percent (w/w) solvent
Percent (w/w) emulsifier (if any)
Degree of emulsification as approximate particle size
e.g. macro-globules
Larger than 1 micron
Between 0.1 to 1 micron
Between 0.005 to 0.1 micron
Lower than 0.005 micron
Criteria for finding degree of emulsification

Emulsion appearance	Approx. particle size
Two phase	Macroglobules
Milky white	Larger than 1 micron
Blue white	1 to approx 0.1 micron
Grey semi-transparent	0.1 to 0.005 micron (translucent)
Transparent	0.005 micron and smaller

Liquid formulations-emulsifiable concentrates, aerosols, aqueous and other solutions introduced in India after 31.12.1972.

The following information in addition to those prescribed for formulations introduced before 31.12.1972, to be submitted.
Sub-acute toxicity studies in mammals on two laboratory animals-rat and dog (to be submitted within 2 years)
Oral: 3 Months each in rat and dog
Dermal: 21 days Rabbit / rat
Inhalation: Rat 14 days

Supplementary toxicological studies (to be submitted within 3 years)
Neurotoxicity: Chicken
Synergism and potentiation: Rats

Dusting powder formulation

If toxicological data for the technical grade pesticide, in reference, exists and provided that particle size conforms to the BIS specifications, only the following data is required.

Following information (to be submitted within 2 years)
Toxicity to other beneficial insects (field trials and observations)
Toxicity to livestock (field trials and observations)

Observations on man
Direct field observation on the population and livestock exposed
Health records of industrial workers and pest control operators

Medical data
Signs, diagnosis, chemical tests and records of poisoning
Treatment of poisoning
First aid measures
Medical treatment

Water dispersible powder formulation
Same as above
Note: If toxicological data is generated for EC preparation, as per guidelines, there is no need to generate data on wettable powder containing the same active ingredient. If data on toxicity to livestock (field trials and observations) and observations on man and medical data as per the guideliens exist with respect to EC preparation and/or dust, there is no need to generate data for wettable powder.

Granular pesticides
Medical data
Same as above
Note: If the granular formulation contains the technical grade of EC material for which the applicable toxicological data exists, no toxicological data is required. Also, none of the information on acute toxicity to bees, beneficial insects and livestock and data on observation on man be generated, if such data, applicable as per guidelines is already available.

Provisional registration
Data requirements for consideration for grant of provisional registration are detailed below:

Chemistry
- (*i*) Source of the technical material
- (*ii*) Minimum purity of the technical material
- (*iii*) Identity and properties of the active ingredient
- (*iv*) Specifications for the product quality
- (*v*) Chemical composition of the formulation

Packaging and labelling
- (*i*) Copies of the proposed/existing label and leaflet as per the Insecticides Act
- (*ii*) Manner of packaging
  - (*a*) Packaging specifications as approved by the Registration Committee or as per BIS specifications thereon
  - (*b*) If the packaging specifications for pesticides do not cover ii(*a*) above, specification for primary, secondary and tansportation packaging is required to be furnished
- (*iii*) Instructions for storage and use including first-aid and precautions for proposed labeling
- (*iv*) Information regarding disposal of used packages, surplus material and washing of pesticides

Efficacy	(i)	Effectiveness
	(a)	International data on bio-effectiveness
	(b)	Results of some field trials under different agro-climatic conditions of India
	(ii)	International data on translocation within the plant or animal

| Residues | (i) | Metabolism of the pesticide in plants, animals, soils and water conducted under foreign conditions with the nature of metabolites and their toxicities |
| | (ii) | Residue data on crops under foreign conditions |

Toxicity

(i) *Acute toxicity in mammals* (both for technical and formulation)
Oral: Rat and mice or any 2 species
Dermal: Rabbit/rat
Inhalation: One species
Primary skin irritation: Rabbit
Irritation of mucous membrane: Any one species

(ii) *Sub-acute toxicity in mammals*
Oral: Rat and dog, 3 months each
Dermal: Rabbit/rat, 21 days
Inhalation: Rat, 14 days

(iii) *Other information*
Toxicity to birds: MLD in pigeon and chicken
Toxicity to fish: In fresh water fish
Toxicity to other beneficial insects: Observations after field trials
Toxicity to livestock
Human toxicity data from foreign countries

Neem based pesticides

Chemistry

(i) Name of the plant(s) part to be used for extraction of the active ingredients/components

(ii) Outline of process of manufacture clearly identifying the chemicals as indicated at Sr. No. 3 below

(iii) Neem extract contains 'azadirachtin' as one of the major active constituents. The concentration of azadirachtin in the formulated neem extract should contain not less than 1500 ppm of azadirachtin active ingredient in 'kernel' based formulations and 300 ppm in 'neem oil' based formulations. When the insecticidal active ingredient is other than azadirachtin, its (their) name(s), quality and quantity to be indicated

(iv) Chemical identity of the ingredient at serial No. 3 above

(v) Physico-chemical properties

(vi) Specifications of ingredients as indicated at serial No. 3 above

(vii) Method of analysis of azadirachtin or other insecticidal a.i. other than azadirachtin

(viii) Analytical test report

(ix) Shelf life claim data

Bioefficacy

(i) Bio-effectiveness
(ii) Phytotoxicity
(iii) Compatibility with other chemicals
(iv) Purpose of manufacture
(v) Direction concerning dosage

(*vi*) Time of application
(*vii*) Waiting period
(*viii*) Application equipment
(*ix*) Information regarding registration in other countries, if any

Toxicity

Data on parameters (*i*) to (*iii*) are required for 9(3b) registration
(*i*) Acute oral in rat and mice
(*ii*) Acute dermal
(*iii*) Primary skin irritation, irritation to mucous membrane
(*iv*) long terms toxicity

    (a)   Neurobehavioural toxicity
    (b)   Reproductive toxicity
    (c)   Carcinogenicity
    (d)   Mutagenicity
    (e)   Effect on spray operators (health records)

Packaging and labelling

(*i*) Labels and leaflet as per Insecticides Rules 1971 existing norms
(*ii*) Type of packing
(*iii*) Manner of packing
(*iv*) Container-content compatibility
(*v*) Specification for package

(*a*) Primary package
(*b*) Secondary package
(*c*) Transport package

(*vi*) Manner of labelling
(*vii*) Instructions for storage and use
(*viii*) Information regarding disposal of used containers
(*ix*) Process of manufacturing indicating material, balance, generation of wastes

Note: In case of neem based products for export, the requirements at Sr.No. (*viii*) and (*ix*) under chemistry and (*iv*) to (*viii*) under toxicity are not prescribed. Under bio-efficacy, only information on registration status of tender or trade enquiry from the importing country and under packaging and labelling, only the packaging specifications of the importing country are prescribed.

Household pesticides

Definition

The insecticides which are used for direct domestic application in the form of different formulations in and around areas of all structures, including vehicles, areas associated with the household of human life, patient care areas of health related institutions or areas where children spend their time as in schools, garden etc. in order to kill/repel the insects and the vectors of diseases.

Classification

The household insecticides have been grouped into four categories.

Category I: Solids e.g. baits, glue etc.

Category II: Liquids e.g. ready to use sprays for space and spot application, EC and aerosols

Category III: Vapour/smoke generators etc. Household pesticides in technical form which emit vapours/gas/smoke, either as such or as a result of heating, burning or coming in contact with air/moisture or with the help of any other appliance/technology e.g. mats, coils, cakes etc.

Category IV: Miscellaneous products not covered under any of the above categories but used for the purpose

General requirement

(*i*) Household pesticides must be manufactured only from the technical grade pesticides, which are registered under the Insecticides Act 1968

(*ii*) Data on their formulations shall be considered along with the data on technical grade pesticides and not in isolation

Chemistry

The common use of the household insecticides which is to be printed on the labels and leaflets should be written as household pesticides containing 'X' % w/w and 'Z ...' % w/w etc.

Chemical composition and allied data requirement

(*i*) Source of supply of the technical grade material with minimum purity to be used in the formulation

(*ii*) Chemical composition

    (*a*) Complete chemical composition of the formulation

    (*b*) Complete data on the identity and physico-chemical characteristics of various ingredients of the formulation

    (*c*) The chemical composition for the formulation including the kind, name and percentage of all the ingredients

    (*d*) The number of active ingredients in the household combination product should not be more than two (However, if any applicant submits satisfactory method of analysis for more than two active ingredients in the combination product, the request can be considered on its merit)

(*iii*) Specification: As per national or international standards. Wherever no specifications are available, the Registration Secretariat may harmonize a procedure for evolving it

(*iv*) Analytical test report: The report of a particular batch giving the analysis of all the parameters in the specification shall be submitted from an independent reputed national/international laboratory

(*v*) Shelf life data: The expected shelf life claim for the product along with the supporting data, conforming to the guidelines prescribed by the Central Insecticides Board, to be submitted

(*vi*) Method of analysis:

    (*a*) Wherever available, IS methods of analysis to be followed

    (*b*) The methods for the analysis of various active ingredients used in a formulation to be provided. As far as possible, single multiple pesticide analysis method be suggested

    (*c*) For identification of active ingredients in the formulation, suitable quantitative methods be developed and provided following the requirement at *vi* (b) above

Bio-efficacy and residue

Bio-efficacy claims to be given on the labels

(*a*) A brief direction concerning the major usage of the pesticides

(*b*) Whenever a product has been approved for restricted use by the Registration Committee, it should be written clearly, in capital letters, on the label

(*c*) Instructions regarding insecticide, not to be used on any food crop to be given

Bio-efficacy claims to be given on the leaflets

(*a*) Detailed information on the usages of insecticide indicating the names of insects, method of application, dosage, places of treatments, plant protection equipments to be used etc. should be given. Common names of insects to be given

(*b*) If the product is approved for restricted use, it should be very clearly stated on the leaflet in bold letters

(*c*) Instructions, such as 'not to be used on any food crop' to be given

Bio-efficacy claims to be given on the labels and leaflets in case of technical grade material
The purpose of imported, manufactured technical grade material is required to be given on the labels and leaflets

Data requirements on bio-efficacy and residues for formulations of pesticides for provisional registration under section 9(3b) of the Act

(*a*) Published/cited Indian data on bio-effectiveness in support of the claims indicated on labels/leaflets. The data shall be produced from two national laboratories, based on minimum two repeated trials

(*b*) This should be further supported with any published information available from elsewhere (overseas data)

(*c*) Information on secondary pest outbreaks (particularly ticks and mites) should be given where synthetic pyrethroids are being used

(*d*) Data on residues: Data on the persistence of pesticides on different types of surfaces to be submitted/generated/obtained from foreign/ Indian conditions from two laboratories. This may also be supported by data generated elsewhere

Data requirement for the registration of formulations of pesticides on bio-efficacy and residues for regular registration under section 9(3)

(*a*) Minimum two trials in each of the three national laboratories

(*b*) Persistence of pesticides on different types of surfaces. Data to be generated in three national laboratories whenever applicable

(*c*) Information on secondary pest outbreaks particularly of ticks and mites to be given where synthetic pyrethroids are being used

Data requirement for the registration of new formulations of the approved pesticides
Data on bio-effectiveness and persistence on different surfaces as required in case of regular registration of formulations under section 9(3) above

Data required for combination products

Bio-efficacy data on combination products as well as individual products is to be submitted. Data on persistence on different types of surfaces should also be submitted. All data to be generated as required under section 9(3) of Insecticide Act.

Methodology

Flying/crawling insects

Residual films of insecticides are prepared by spraying insecticides on different types of surfaces such as glass, wood, mud and cemented surfaces. Insects are to be exposed for 30 minutes and then shifted to recovery chambers. The mortality count should be made after 24 hours. Satisfactory mortality would be more than 90%. The residual toxicity of insecticides should also be studied at different intervals. Evaluation of space sprays against insects should be conducted in Peet Grady Chamber as per IS: 1824. Mats/oils could also be evaluated inside the Peet Grady Chamber against caged mosquitoes and the knockdown effect recorded at different intervals. Aerosols are to be evaluated inside a standard room. The tests are to be conducted as per WHO Technical Report Series No. 206.

Guidelines on the data requirement for the packaging and labeling for the grant of registration of pesticide formulations of household products.

Type of formulation: (Physical state of formulations)

Solid—coil, cake, mat;

Liquid—Paste, aerosols, others

Proposed labels and leaflets: As per Insecticides Rules, 1971/ existing norms of the registration committee

Manner of packaging

(i) In general, as per IS: 8190 (Part III), 1979 and amendment thereof, if any

(ii) Glass bottles as primary packaging material are not permitted for household pesticides

(iii) If the packaging material does not conform to IS specification, the specification for primary and secondary packages and for transportation packages along with the history

Compatibility

If the container-pesticide compatibility is already established, the relevant Indian Standard for said packaging material may be quoted. In case compatibility is not established, detailed specifications for primary package are required to be furnished indicating dimension, material, construction, design etc., along with test reports with reference to physico-chemical requirements for field trials and studies for transport worthiness. The analysis report should be as per the relevant Indian Standards specifications on packaging and quality control or as per specification/test methodology approved by the Registration Committee. Instructions for storage and use including first aid and precautionary measures proposed for labelling are to be submitted.

Information regarding disposal of used package, surplus material and washing of insecticides proposed for labelling

Toxicity
The following data are to be generated and submitted for registration.
Category I (Solid)
Acute oral: Rat and mice, or any other two species
Acute dermal: Rabbit
     Primary skin irritation
     Irritation to mucous membrane
Toxicity to birds
Toxicity to fish
Toxicity to beneficial insects
Category II (Liquid): Same as above plus acute inhalation toxicity
Category III (Vapours): Same as Category II. Health monitoring study of the user by actually using the household pesticides. The symptoms and signs as a result of sub-acute chronic or delayed effects on man from single or multiple exposures to the product shall be observed and submitted.

Bacillus thuringiensis and *B. sphaericus* for pest and vector control

Chemistry
(Technical/Formulation)
(*i*) Physico-chemical specifications and composition
(*ii*) Systematic name and strain
(*iii*) Common name
(*iv*) Natural occurrence of the organism, its relationships to other species and history
(*v*) Manufacturing process
(*vi*) An appropriate test procedure and criteria used for identification, such as morphology, biochemistry and/or serology / immunology

Note: (*a*) Morphology description: Particle size, heat resistant spore count / protoxins per mg of dry material (formulation)

(*b*) Protein per mg dry weight of formulation. Protocols are required to be generated/developed for the assay, preferably Lowry Method

(*c*) Immunology assays: ELISA test or any other sensitive standard immunology test and Rocket Immunoassays are analytical methods. The ELISA test would have to be submitted by the firm while Rocket Immunoassay will be treated as referral method for insecticidal crystal proteins for *Bacillus thuringiensis* only

(*d*) Routine test: Referee method could be the basic tool for surveillance to the occurrence of the mutation, if any. Level of beta exotoxins to be identified, if expressed. The methodology of determination of the toxin content is by Dot Blot ELISA assay method.

(*vii*) Method of analysis:
(*a*) The protein content per mg (Dot Blot assay of B.t. toxin protein as alternate of bioassay)
(*b*) Viable spore counts
(*c*) Determination of the toxin content by Dot Blot ELISA assay method

(*d*)   Plasmid pattern—if any

(*e*)   A technique for the separation and purification of the crystals for raising antiserum against crystals (purification of crystals by sodium bromide gradient)

(*viii*)   Storage stability

      (*a*)   Data to be provided by bioassay method. Experimental details, calculations etc. are to be provided

      (*b*)   The importer/manufacturer has to provide the sample of technical material, which could be sent to Indian Institute of Microbial Technology, Chandigarh (Department of Biotechnology) for storage and future reference

Bio-efficacy

Technical

Laboratory test: For registration of microbial pesticides, LC_{95} or LD_{95} values for each testing insect species under laboratory conditions should be generated. The data are required from two independent laboratories recognised by DST/ICAR/CSIR/ICMR and State Universities etc. for the purpose.

Formulation

(*i*)   Laboratory test—as under technical

(*ii*)   Field test: The data on bio-efficacy based on two seasons/two trials under two different agro-climatic conditions in the form of published/authentic data is required to be provided. Information on phytotoxicity with the formulated material is also to be submitted along with bioefficacy report

(*iii*)   Data on persistence: Data on persistence in soil, water and plant should be submitted. Spore count method should be used for estimation

(*iv*)   Data on residues: Data on residues by the method used for persistence up to a period of 3 months is to be submitted

(*v*)   Compatibility: Data on compatibility will be required in case it is recommended to be used in mixture with some other pesticides. The data on interaction is desirable against most commonly used pesticides and fertilizers, under field conditions.

(*vi*)   Time of application: Information on time of application, equipment and the manner in which insecticides are used, is required to be submitted

(*vii*)   Data on non-target organisms: Information on toxicity to two species of parasites and predators is required to be submitted

Toxicity

(*a*)   *Single exposure studies*

	Technical	Formulation
(*i*) Oral toxicity/pathogenicity	Required	Required
(*ii*) Dermal toxicity/pathogenicity	- do -	- do -
(*iii*) Mucous membrane irritation	- do -	- do -
(*iv*) Inhalation toxicity/pathogenicity	- do -	- do -
(*v*) Allergy/sensitization/ immuno-suppression	- do -	- do -

Note: Allergy has to be conducted as per standard protocol prescribed by BIS-IS: 11601 (Part 2), 1990. Method of test for skin sensitization potential of synthetic detergents (Guinea pig maximization test).

(*b*)   *Repeated exposure studies*
(*i*)   Oral toxicity/pathogencity 90 days   Required   Required
(*ii*) Dermal toxicity/ pathogencity 20 days   - do -   - do -
(*iii*) Inhalation toxicity/ pathogencity 14 days   - do -   - do -
(c)   Supplementary toxicity/pathogencity
(*i*)   Mutagenicity   Required   Required
(*ii*) Teratogenicity   - do -   - do -
(*iii*) Carcinogenicity   - do -   - do -
(*d*)   Eco-toxicity
(*i*)   Toxicity to birds   Not required   Required
(*ii*) Toxicity to fish   - do -   - do -
(*iii*) Toxicity to honey bees   - do -   - do -
(*iv*) Toxicity to silk worm   - do-   - do -

Processing, Packaging,   Technical / Formulation

Labelling

(*a*)   *Manufacturing process / process of formulation*
(*i*)   Raw material / material balance
(*ii*) Plant and machinery
(*iii*) Process-unit operation / unit process
(*iv*) Output (finished product and generation of waste)

(*b*)   *Packaging*
(*i*)   Classification - solid, liquid or other types
(*ii*) Size - as per Indian Standards Specification
(*iii*) Compatibility - glass bottles are not recommended
(*iv*) Packaging specifications / systems of packaging

(c)   *Labels and leaflets*
       As per Insecticides Rules 1971 / as per existing norms indicating
       the common name, composition, antidote/storage statements etc.
       (the packaging material should also ensure freedom from
       contamination from handling, storage and transportation)

ANNEXURE II

LIST OF COMMON GLASSWARE, EQUIPMENT AND INSTRUMENTS FOR A FORMULATION LABORATORY

Glassware	Beakers, Buchner funnels, Buchner flasks, burettes, conical (Erlenmeyer) flasks, dessicators, distillation sets, round and flat bottomed flasks, funnels, separatory funnels, funnel stands, graduated cylinders with/without stoppers, volumetric flasks, graduated pipettes, single mark pipettes, reagent bottles, Soxhlet apparatus, test tubes with/without joints, standard male/female joints, chromatographic columns with stop-cocks, TLC-jars, single and double walled condensers, glass chromatographic sprayer, glass manifold evaporator, Kuderna-Danish evaporative assembly, pestle and mortar, reducing and enhancing adapters distillation heads, reciever adapters and thermometer pockets etc.
Equipment	Water baths, vacuum gauge (manometer), centrifuge, single pan analytical balance, physical/ordinary balance, pressure and vacuum pumps, heating mantles, energy regulators, laboratory jacks, Waring blender with glass/stainless steel cups, ovens, rotary evaporators, pressure assemblies for storage tests, stirrers, magnetic stirrers, sonicator/sonicator bath, refrigerator, hot plates, oil bath, sand bath, stop watch, granulator, ball mill, Wiley mill, emulsion homogenizer, constant temperature bath, distilled water preparing assembly, controlled temperature cum humidity chambers, shakers, sieves and sieve shaker, melting point block, grinding mills, wet grinding facility, thin layer chromatography apparatus, Karl Fischer apparatus for determination of water content, closed tester for flash point determination etc.
Analytical facilities	Gas liquid chromatograph (GLC), liquid chromatograph (HPLC), gas chromatograph-mass spectrometer (GCMS), nuclear magnetic resonance spectrometer, densitometer, flame photometer, atomic absorption spectrophotometer, UV-visible spectrophotometer, nephelometer, Geiger Muller Counter/ scintillation counter, particle size analyzer, image analyser, capillary electrophoresis, viscometer, rheometer, contact angle meter, tensiometer, flowmeter, pH meter, electrical conductivity meter etc.

KEY TESTS PRESCRIBED FOR THE TYPICAL PESTICIDE FORMULATIONS

Formulation	Test(s)
Solids	
Water dispersible powder	Active ingredient
	Particle size
	Acidity/alkalinity
	Suspensibility
	Wettability
	Repeat the above tests after accelerated storage
Dust	Active ingredient
	Particle size
	Acidity/alkalinity
	Bulk density (before and after compaction)
	Flowability
	Repeat the above tests after the storage stability test.
Granule	Active ingredient
	Size
	Dust content
	Moisture content
	Acidity / alkalinity
	Attrition
	Encapsulation
	Storage stability
Blank granules	Moisture content
	Bulk density
	Size
	Acidity
	Liquid holding capacity
Liquids	
Emulsifiable concentrate	Active ingredient
	Acidity/alkalinity
	Cold test
	Flash point
	Emulsion stability
	Persistent foam
	Repeat the above after the storage stability test
Solution concentrate	Active ingredient
	Acidity/alkalinity
	Cold test
	Flash point (for non-aqueous concentrates)
	Repeat the above after the storage stability test
Suspension concentrates	Active ingredient
	Suspension stability
	Dispersibility on dilution
	Viscosity
	Density
	Size
	Redispersion on settling

Note: Readers may also refer to Chapter 2, FAO (1999), ISI (1982) and CIPAC (1970).

Chromatographic Methods

Chromatography refers generally to the process of separation of components in a sample by distribution/partition between two phases, one that is stationary and the other that moves, usually but not necessarily in a column. Probably no other technique has been more valuable for a precise separation and analysis of highly complex mixtures (Table 5.1).

TABLE 5.1. COMPARATIVE ANALYSIS RANGE OF CHROMATOGRAPHY *VIS A VIS* OTHER METHODS

Method	Approx. range mol l^{-1}	Approx. precision(%)	Selectivity	Speed	Principal uses
Gravimetry	10^{-1}-10^{-2}	0.1	Poor	Slow	Inorganic
Titrimetry	10^{-1}-10^{-4}	0.1-1	Poor	Moderate	Inorganic, organic
Potentiometry	10^{-1}-10^{-4}	2	Good	Fast	Inorganic
Electro-gravimetry, Coulometry	10^{-1}-10^{-4}	0.01-2	Moderate	Slow	Inorganic
Voltametry	10^{-3}-10^{-10}	2-5	Good	Moderate	Inorganic, organic
Spectrophotometry	10^{-3}-10^{-6}	2	Good	Fast	Inorganic, organic
Fluorometry	10^{-6}-10^{-9}	2-5	Moderate	Moderate	Organic
Atomic absorption spectroscopy	10^{-3}-10^{-9}	2-10	Good	Fast	inorganic, multi-element
Chromatography	10^{-3}-10^{-9}	2-5	Good	Fast	Organic, multi-component
Kinetic methods	10^{-2}-10^{-10}	2-10	Good	Fast	Inorganic, organic

Principles of Chromatography

A solute equilibrates between a mobile and a stationary phase. The more it interacts with the stationary phase, the slower it moves along a column. The individual components interact to different degrees with the stationary/mobile phases.

The distribution equilibrium is described by the distribution coefficient.

$$K = [X_s]/[X_m]$$

where, $[X_s]$ is the concentration of component X in the stationary phase at equilibrium and $[X_m]$, its concentration in the mobile phase.

Classification of Chromatographic Techniques

Chromatographic processes can be classified according to the type of equilibration process involved, which is governed by the type of stationary phase. Various bases of equilibration are: (*i*) adsorption, (*ii*) partition, (*iii*) ion exchange and (*iv*) pore penetration.

Adsorption Chromatography: The stationary phase is a solid on which the sample components are adsorbed. The mobile phase may be liquid (liquid-solid chromatography); the components distribute between the two phases through a combination of sorption and desorption processes. Thin-layer chromatography (TLC) is a special example of sorption chromatography in which the stationary phase is a solid (cetite, silica gel, alumina etc.) supported on an inert plate (Stahl, 1969).

Partition Chromatography: The stationary phase of partition chromatography is a liquid supported on an inert solid. Again, the mobile phase may be a liquid (liquid-liquid partition chromatography) or a gas (gas-liquid chromatography, GLC). Paper chromatography is a type of partition chromatography in which the stationary phase is water adsorbed on cellulose in the form of a sheet of paper.

In the normal mode of operations of liquid-liquid partition, a polar stationary phase (e.g. alumina or silica) is used with a non-polar mobile phase (e.g. hexane). This favours retention of polar compounds and elution of non-polar compounds. If non-polar stationary phase is used with a polar mobile phase, then non-polar solutes are retained more and polar solutes are more readily eluted. This is called reversed-phase chromatography.

Ion Exchange and Size Exclusion Chromatography: Ion exchange chromatography uses an ion exchange resin as the stationary phase. The mechanism of separation is based on ion exchange equilibrium. In size exclusion chromatography, molecules in solution are separated according to their size, by their ability to penetrate a sieve like structure in the stationary phase.

Paper Chromatography

This is the simplest technique of partition chromatography that makes use of filter paper as the stationary support. The mobile phase is a liquid, which percolates within the porous structure of the paper, while water molecules present in the paper act as stationary phase for partitioning. A small amount, i.e. 1–10 μl of sample is placed and irrigated by solvent system. It is useful for qualitative and quantitative analysis of sample though the recovery is not necessarily in a purified form. Sometimes it takes long time for development and many a time, lack of the well-defined zones are limitations of this method (Hais and Macek, 1963).

The packing of fiber paper constitutes a porous medium for the retention of the stationary phase while space between fiber provide passage for mobile phase. The various varieties of paper, commercially available for this purpose, are Whatman No. 1, 2, 3, 31. Acetyl acid papers, Kieselguhr papers, silicon treated papers and ion exchange papers are also used. The papers made of pure cellulose or modified cellulose and glass fiber are also available.

Thin Layer Chromatography (TLC)

TLC is a method of chromatographic separation in which the adsorbent is coated on glass plate, which serves as stationary support. The mobile phase percolates and develops the chromatogram as in column chromatography. This method is simple, rapid in separation and sensitive. Speed of separation is fast and it is easy to recover the separated compounds.

Usually silica gel is used as coating material (stationary phase) but cellulose powder, Kieselguhr and diatomaceous earth can also be used. For hydrophilic stationary phase, plaster of paris is used as the binder. For spreading the slurry of the coating material with binder, an applicator is used in order to get a uniform layer. Ready to use thin layer plates are also available commercially, in the form of pre-coated glass/ aluminium plates, chromatotubes etc. Amount of moisture in plates should be controlled in order to get reproducible results.

TLC is useful for a large number of sample separation applications. It can analyze compounds, which are not volatile or too labile for gas liquid chromatography. It can also be used to check impurities in the solvent. The useful separations of plasticisers, antioxidants, ink and dye formulations have been accomplished by TLC. It has extensive applications in inorganic separations.

High Performance Thin Layer Chromatography (HPTLC)

It is comparatively newer version of TLC, which makes use of ultra violet scanner (detector) coupled with computer for uniform visualization of the developed plates. The technique is quicker because smaller, ready-made plates (5-10 cm long, glass or aluminium coated with finer particles of silica gel G or other suitable absorbent material are used). The technique is less sensitive than gas liquid chromatography (GLC) and even high performance liquid chromatography (HPLC).

Gas Chromatography (GC)

There are two types of gas chromatography: gas-solid (adsorption) chromatography and gas-liquid (partition) chromatography. The most important of the two is gas-liquid chromatography (GLC). Gas chromatography is widely used, particularly by organic chemists, and it undoubtedly ranks as one of the most important analytical techniques since its development in 1952. The separation of benzene and cyclohexane (b.p. 80.1 and 80.8°C) is extremely simple by gas chromatography, but is virtually impossible by conventional distillation. Very complex mixtures can be separated by this technique.

Principle: In gas chromatography the sample is converted to the vapour state (if it is not already a gas) and the eluant is a gas (the carrier gas). The stationary phase is generally a non-volatile liquid supported on an inert solid such a firebrick (Chromosorb-P or W etc.) or diatomaceous earth. There are a large number of liquid phases available, and it is by changing this phase rather than the mobile gas phase that different separations are accomplished.

Instrumentation: A good chromatograph consists of (*i*) pressure regulator (*ii*) sample injection port; (*iii*) stationary support column (*iv*) stationary phase (*v*) detector(s) (*vi*) signal recorder.

(*i*) *Pressure Regulater:* Gas pressure is adjusted in the range of 1-4 atmospheres while flow meter can measure 1-1000 litres of gas per minute. Flow valves are adjustable by means of a needle valve mounted on base, in flow meter. A trap containing molecular sieve is interposed to filter out the contaminating impurities. Carrier gas may be either of He, N_2, H_2, Ar. For thermal conductivity detector, He is preferred to others due to its high conductance.

(*ii*) *Sample Injection Port:* Samples are injected by a micro syringe through self-sealing silicon rubber septum into heated metal tube through a sample injector port. Electrical heaters heat the metal block. Sample size varies from 0.4-10 μl.

(*iii*) *Support Column:* The chromatographic column is made of glass tubing coiled into an open spiral. Copper is used to make columns for high temperature operation. Stainless steel is also used. Column diameter varies from 1/16 to 3/16 inch. Short columns (3-4 cm length) are also used but separation in such cases is not efficient. Usual size is around 2 meters. Longer capillary columns are used for better separation.

(*iv*) *Stationary Phase:* The structure and surface characteristics of the stationary phase are very important. The former contributes towards efficiency while the latter governs the degree to which it enters into separation. The surface holds immobile liquid phase as thin film. The commonly used supports are diatomaceous earth and kieselguhr.

The versatility of gas liquid chromatography lies in the availability of an infinite variety of liquid partition materials (stationary phases). Limitations are their volatility, thermal stability and ability to wet the support. Liquid phases can be divided into non-polar, intermediate polarity, polar carbowaxes and hydrogen bonding compounds like glycols. The maximum temperature of liquid phase is determined by its volatility. Excess volatility shortens column life. The loading of column is usually expressed as percent by weight (5% means 100 g column material has 5 grams stationary phase). Glass bead polymers have low specific surface area. The open tubular columns and support coated open tubular columns are generally used.

(*v*) *Detectors:* Ever since the introduction of gas chromatography, over 40 detectors have been developed. Some are designed to respond to most compounds in general, while others are designed to be selective for particular types of substances. Brief description of some of the more widely used detectors is given in Table. 5.2.

(vi) Signal Recorder: Receives signals from detector and records them as a chromatogram/read out (indicating experimental conditions, retention time and peak area % of various components).

Applications: GLC is used for identification of polymers and their structural analysis. In elemental analysis and class reactions, the pyrolysis is carried to extreme step when organic products break to CO_2 and H_2O. When a class of compounds is not readily chromatographed because of its thermal instability or lack of volatility, its conversion into derivatives that are more readily chromatographed is often possible, e.g. fatty acids converted to methyl esters by BF_3-methanol reaction. Similarly, the acetylation of fatty alcohol sterols and hydroxy compounds can also be accomplished with acetic anhydride and pyridine.

TABLE 5.2. APPLICATION AND SENSITIVITY OF VARIOUS DETECTOR SYSTEMS

Detector	Application	Sensitivity	Linearity	Remarks
Thermal conductivity	General response to all substances	Fair	Good	Sensitive to temperature, flow changes and concentration
Flame ionization	All organic substances; some oxygenated products	Very good 10^{-11} g ml^{-1} in carrier gas	Excellent	Requires very stable gas flow; response for water is 10^4-10^6 times weaker than for hydrocarbons; concentration sensitive (organics)
Flame thermionic	All nitrogen and phosphorus containing substances	Excellent	—	Needs recoating of sodium salts on screen; mass sensitive
Rubidium silicate bead	Specific for nitrogen, phosphorus containing substances	Excellent	—	Mass sensitive
Argon ionization (β-ray)	All organic substances; with ultra pure He carrier gas, also for inorganics and gases	Very good; 10^{-13} g ml^{-1} in carrier gas	Good	Very sensitive to impurities and water; needs very pure carrier gas; concentration sensitive
Electron capture	All substances that have affinity to capure electrons; no response for aliphatic and naphthenic hydrocarbons	Excellent for halogens	Poor	Very sensitive to impurities and temperature changes; quantitative analysis complicated; concentration sensitive

High Performance Liquid Chromatography (HPLC)

HPLC is the liquid carrier analogue of GC. The secret of its success is small uniform particles to give small Eddy diffusion and rapid mass transfer. HPLC allows separation and measurements to be made in a matter of minutes. Porous packing materials with particle size of 3-10 µm are usually used in modern instruments, with plate counts of 60,000-90,000 per meter.

Principle: The rate of distribution of solutes between the stationary and the mobile phases in traditional liquid chromatography is largely diffusion-controlled. Diffusion in liquids is extremely slow compared to that in gases. In order to minimize diffusion, the time required for the movement of sample components to and from the interaction sites in the column, two criteria should be met. First, the packing should be finely divided and have high spherical regularity to allow for optimum homogeneity and packing density; and second, the stationary liquid phase should be in the form of thin uniform film with no stagnant pools (Snyder *et. al.* 1997).

Instrumentation: HPLC consists of four principal parts.

Mobile Phase Supply System: The mobile phase supply contains a pump to provide the high pressure required and usually contains some means of providing gradient elution (i.e., changing concentrations of the eluant ions such as H^+ etc.). The solvent reservoirs can be filled with a range of solvents of different polarities, provided they are miscible, or they can be filled with solutions of different pH and mixed in the buffer volume. The solvents must be pure and be degassed.

Sample Injection System: This consists of a stainless steel ring with six different ports. A movable Teflon cone within the ring has three open segments, each of which connects a pair of external ports. Two of the ports are connected by an external sample loop of known or fixed volume. In one configuration, the cone permits direct flow of effluent into the column. Temperature control of column is usually not necessary for liquid-solid chromatography, unless it has to be operated at elevated temperatures, but is generally required for other forms of liquid chromatography (liquid-liquid, size exclusion or ion exchange).

Pump: The pump provides uniform pressure for flow of the solvent. It forms an integral part of the instrument for proper development of the chromatogram.

Detector: HPLC requires detectors with high sensitivity, usually with sensitivities in the microgram to nanogram range. Widely used detectors are refractometer and ultraviolet (UV) detectors. Some detectors, especially refractometer detectors, are very sensitive to temperature changes; and if the column is operated at higher than ambient temperature, a cooling jacket should be placed between the end of the column and the detector to bring the mobile phase back to ambient temperature.

Application: As far as the applications of HPLC with respect of adsorption chromatography are concerned, these are the same as for columnar methods but main use is in pharmaceutical and pesticide industries. In liquid-liquid partition chromatography using HPLC technique, separation of moderately polar to polar substances is possible.

Comparison between GLC and HPLC

It would be interesting at this stage to compare gas liquid chromatography (GLC) with high performance liquid chromatography (HPLC). From the viewpoint of speed and simplicity of equipment, GLC is better. For isolation of nonvolatile substances including inorganic ions and thermally unstable materials, HPLC is preferable. Both techniques, as a matter of fact, are complimentary, efficient, highly selective, need small sample, which can be handled by non-destructive testing and are applicable for quantitative analysis. The special feature of HPLC is that it can accommodate thermally unstable and non-volatile compounds and inorganic ions.

Gas Chromatography—Mass Spectrometry (GC—MS)

GC-MS is very powerful for positive identification. The appearance of a chromatographic peak at a particular retention time suggests (but not guarantees) the presence of a particular compound. The probability of positive identification will depend on factors like the type and complexity of the sample and sample preparation procedures employed. A gas chromatogram of an injected blood sample diluted with a solution of an internal standard (to verify retention time and relative peak area) that gives a large peak expected for alcohol, strongly suggests the presence of blood alcohol since there are few nontoxic compounds that are likely to interfere. Usually, there is indication of alcohol ingestion and the key legal question is what is the concentration? However, the appearance of a GC peak for cocaine may not be so straightforward in confirming the presence of this drug. Hence, confirmatory evidence such as infrared or ultraviolet spectrometry may be sought.

The integration of a capillary gas chromatograph with mass spectrometer provides an extremely powerful analytical tool. Capillary GC, with thousands of theoretical plates, can resolve hundreds of molecules into separate peaks and mass spectrometry can provide identification. Even if a peak contains two or more compounds, identifying peaks based on molecular ions and their fragmentation pattern, can provide positive identification, especially when combined with retention data. Mass spectrometry is a sophisticated instrumental technique that produces, separates and detects ions in the gas phase (McMaster and McMaster, 1998).

The basic components of a mass spectrometer are shown below:

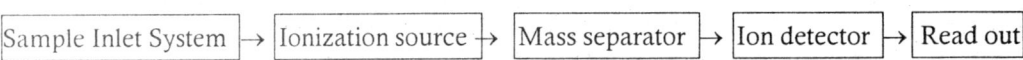

Sample Inlet System → Ionization source → Mass separator → Ion detector → Read out

A sample with a moderately high vapour pressure is introduced in an inlet system, operated under vacuum (10^{-4} to 10^{-7} torr) and at high temperature (up to 300°C). It vaporizes and is carried to the ionization source. Molecules to be analyzed are generally neutral and must be ionized. This is accomplished by various means but typically by bombarding the sample with high-energy electrons in an electron-impact source.

REFERENCES

CIPAC (1970). *CIPAC Handbook*, Vol. 1. *Analysis of Technical and Formulated Pesticide*, (ed. G.. R. Raw), Collaborative International Pesticides Analytical Council (CIPAC). Heffer and Sons Ltd., London.

FAO (1999) pesticide specifications. Manual on tea development and use of FAO specifications for plant protection products. 5th edition, FAO plant production and protection paper 149.

ISI (1982) Method of test for pesticides and their formulations. IS: 6940. Indian Standards Institute, New Delhi.

Hais, I. and Macek, K. (1963). *Paper Chromatography*, Academic Press, New York.

McMaster, M. C. and McMaster, C. (1998). *GC-MS: A Practical Users Guide*, John Wiley and Sons, New York.

Stahl, E. (1969). *TLC: A Laboratory Handbook*, Springer-Verlag, New York.

Snyder, L. R., Kirkland, J. J. and Glajch, L. G. (1997) *Practical HPLC Method Development*. Wiley Inter Science, New York.

CHAPTER 6

Machinery and Equipment

Each operation requires rather simple but precise equipment for uniformity of the end product. Machinery and equipment play a very important role in formulation.

Before grinding or milling, various ingredients of a formulation (e.g. carrier, surfactant, etc.) are mixed in suitable proportions. The equipment used for the purpose is a mixer or a blender. Proper kneading of the mixture of inerts and technical toxicant(s), which may be wax like in nature, can be thus accomplished. Impregnation of inert with liquid toxicant in pure form or in the form of solution can also be done. Solutions may be impregnated directly or sprayed under pressure on the solid carrier/diluent during mixing.

MIXERS/BLENDERS

Solid-mixing machines are generally of three types:

(*i*) Machines in which the container movement leads to mixing of ingredients.
(*ii*) Machines in which the container is stationary but a rotating device within the container causes mixing.
(*iii*) Machines in which a combination of rotating container and rotating internal baffles or blades is provided.

The major frameworks of these machines are outlined below.

Container Movement System

Tumbler Type Blender

Such blenders used for gentle blending, are suitable for dense powders and abrasive materials. The advantages include handling a large volume and easy cleaning of the blender.

Double Cone Type Blender

It is equipped with an agglomerate breaking device. Tumbler may be twin shell or Vee type in which the shell is in the form of a letter 'V' and is fitted with agglomerate breaking and liquid feeding devices. In this case, liquid feeding is not necessary as pin type agglomerate breaking device is used. Tumblers of this type are available in plain form or with other features. The mixers provided with baffles operate in such a way that the baffles lift the powder up and drop it during rotation of the drum in course of the mixing operation. Tip speeds of twin shell tumbler with pin type and with liquid feed pan are 500 and 1000 motions per minute respectively.

Horizontal Drum Mixer

It is a continuous barrel/drum like moving system with manual or automatic devices for charging and discharging. It is a dust proof machine in which mixing is done by use of specially shaped internal lifters, lifting the material and dispersing it amongst the main body of the charger in a series of cascades, falling in different directions. This machine can also be fitted with sprayers, thus permitting the addition of liquid additives during the process of mixing. Fig. 6.1 shows the horizontal cross section of a typical drum mixer.

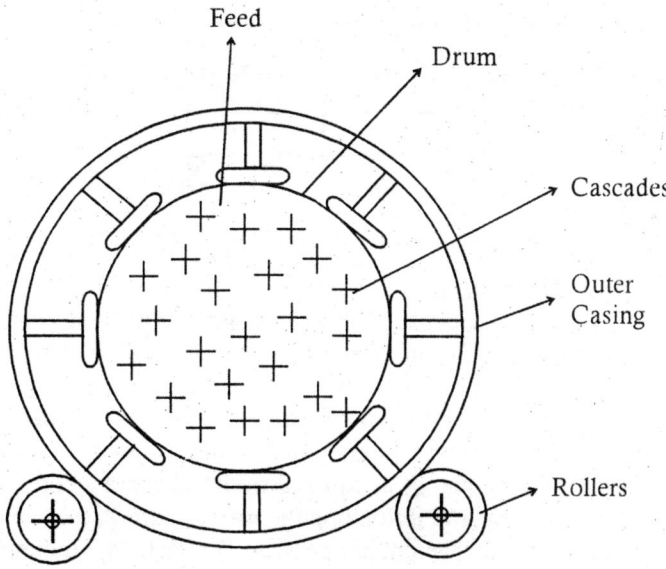

Fig. 6.1: Horizontal Drum Mixer

Stationary Container Mixer
Ribbon Blender

It falls under second category in which case the container is stationery and mixing or blending is accomplished by single or multiple rotating devices inside the container. The mixing devices used here are helical ribbons which help in thorough mixing of the material. The solids are fed from one end of the mixer and removed from the other end.

(a) ***Trough Type Ribbon Mixer:*** It is probably the most simple device. It utilises semi-continuous ribbon elements, having a central horizontal shaft supported at each end with the flights fixed to the shaft by a series of arms.

(b) ***Ribbon Mixer with Double Continuous Ribbons:*** The blender contains left and right hand spirals, which cause a controlled and uni-directional flow of the material. It can also be fitted with suitable sprayers to add liquid additives during the process of mixing (Fig. 6.2). Ribbon cross section and pitch clearances between outer ribbon and shell and number of spirals on the ribbon are some features which can be varied to accommodate materials ranging from low density, finely divided, that move rapidly, to fibrous or sticky products which require positive discharge aid. Tip speed of ribbon blender is approximately 90 meter per minute. A broad ribbon can be used for lifting and conveying while a narrow one will cut through the material during conveying.

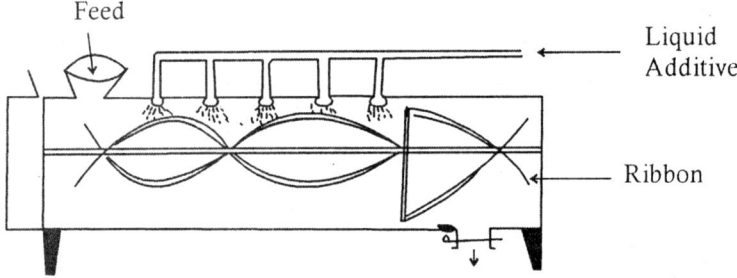

Fig. 6.2: Outline of a Ribbon Blender

The selection of mixing or blending machine is important. Ribbon blender can be used both as primary and final mixer. This is capable of primary mixing of most of the formulations in less than 15 minutes. The mixing of finished fine products can be carried out in a comparably lesser time.

During mixing, care should be taken to minimise dust formation by proper venting, so that the displaced air can be filtered and unregulated loss of dust laden air is avoided. Another consideration in selection is the frequency at which formulations are going to be charged by duct like arrangements for loading and unloading the mixer. Care should be taken to avoid powerful suction on the mixer or the weight hopper from which the ingredients feed into the mixer. If suction of the dust collection on the mixer is too strong, then vital ingredients may be sucked out.

The third category of machines *viz.* those containing a combination of rotating container and rotating internal baffles or blades is seldom utilized in pesticide formulation industry owing to faster wear and tear of gaskets and hazards associated with the leakage of the toxicants.

CRUSHERS

Crushing is another important operation in the formulation of pesticides. The feed materials in the form of lumps, varying in size from a quarter of inch to sixty inches are crushed. The materials having smaller size (1/4") require pulverising and disintegration.

Three types of crushers generally used are (*i*) Jaw crusher, (*ii*) Gyratory crusher and (*iii*) Hammer crusher. Their principle and working are described below in brief.

Jaw Crusher

Jaw crushers are mainly used for primary crushing and are usually followed by other types of crushers. There are two main types: (*a*) The Blake, (*b*) The Dodge. In smaller sizes, these are used as single-stage machines but the Blake design is more popular. These are available in cast iron and fabricated steel body. The crushing is done by to and fro motion of the swing jaw in the jaw stock.

The Blake

It is provided with movable jaw at the top giving greatest movement to the smaller lumps. It has a replaceable crushing plate, usually corrugated and fixed in a vertical position at the front end of a hollow rectangular frame. A similar plate at a suitable angle, is attached to a swinging lever (movable jaw) suspended from a shaft resting in the sides of the frame. Movement is accomplished through a knuckle action by the rising and falling of a second lever (Pitman) carried by an eccentric shaft. The vertical movement is communicated horizontally to the jaw by two plates called toggles.

The Dodge

It is provided with the moveable jaw pivoted at the bottom, giving greatest movement to the bigger lumps. It has the advantage of larger feed opening for the same cost as the blake and is useful for low production intermittant service to produce a uniform product in sizes, smaller than 11 by 15 inch. The setting of a jaw crusher is important. The close or wide opening is set between the moving jaws at the outlet end. The reciprocating motion of the jaws controls the opening to vary between close and wide. Specifications are usually based on close adjustable settings.

For steady running, flywheels made of cast iron are provided. Jaw stock is made of heavy duty cast iron. Other jaws are made of manganese alloy, iron or steel and are replaceable. A typical jaw crusher is shown in Fig. 6.3.

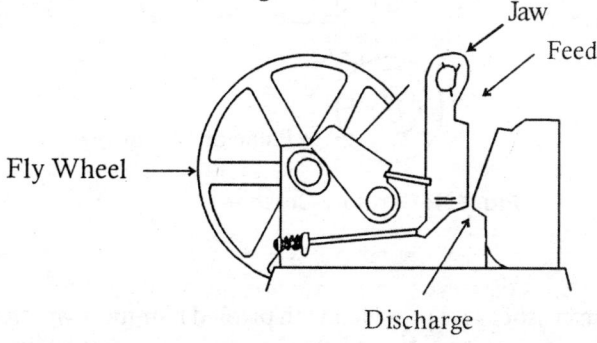

Fig. 6.3: A Typical Jaw Cursher

Gyratory Crushers

A typical gyratory crusher is shown in Fig. 6.4. Two types of gyratory crushers are available.

Primary Gyratory Crusher

It has the shape of the frustums of two cones placed together with the narrow sections in the centre. It has two halves: the upper half crushes and the lower half houses the draining mechanisms, eccentric etc.

Secondary Gyratory Crusher

In this case both the concavering and the mantle are generally curved to minimise wear and tear and to eliminate packing between them. The eccentricity at the lower end gives to the shaft a gyratory motion. This causes the head and its mantle to approach each other from the concave surfaces of the common gyratory, breaking the feed in the downward path. It is provided with a close or a wide opening between the mantle and the concave ring at the outlet end. The close opening is known as the close setting or the close side setting while the wide opening is known as the wide side or open-side setting. The close setting is adjustable.

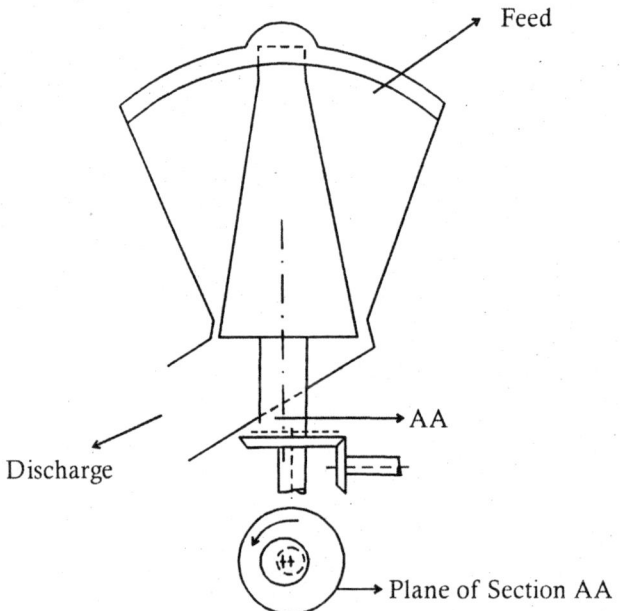

Fig. 6.4: Gyratory Crusher

Hammer Crusher

As the name indicates, such crushers are provided with pivoted hammers which are mounted on horizontal shafts, and crushing takes place by the impacts between the hammers and the plates. A cylindrical grating may be positioned beneath the rotor. It retains material until reduced to a size small enough to pass between the bars of the grating. The direction of rotation of hammer crushers can also be reversed to distribute wears evenly on the hammers and breaker plates. The size of the product can be regulated by changing the spacing of the grate bars and also by manipulation of the hammers to make it longer or shorter. Speed varies from 500 to 1800 rpm depending on the size of the machine. Fig. 6.5 depicts an outline of a hammer crusher.

The crushers are used for high capacity primary crushing or as follow up to crushers of jaw type. These are generally not used in pesticide formulation industry.

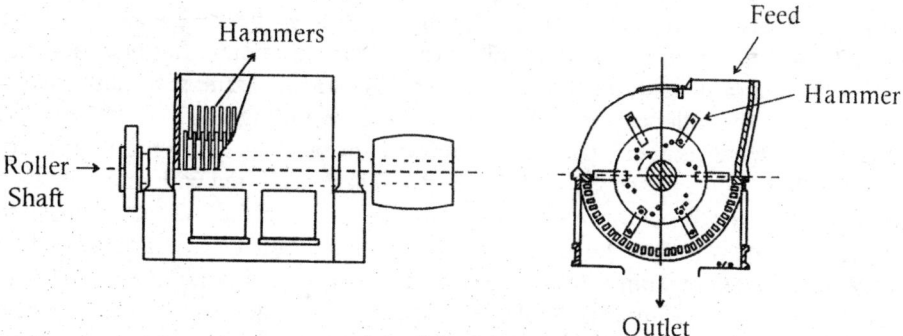

Fig. 6.5: Outline of Hammer Crushers

GRINDERS/PULVERIZERS

Grinding can be divided into two classes:

(*i*) Milling of mixed ingredients for the production of dusts or dust concentrates for use on the crop.

(*ii*) Pre-milling of ingredients for the production of WP and seed dressings before they are ground to the required particles size. The inerts should be selected carefully to avoid choking of the mill.

In the milling of mixtures of filler and technical material, difficulty is generally associated with the low softening point of the latter and their tendency to soften and clog the equipment during milling. This problem leads to an increase in the frictional heat and power consumption. It is also a factor in judging the suitability of the fillers used in the formulation of the mixtures. To minimise this problem, dry ice or liquid nitrogen is generally added. In the case of fluid energy micronizer type mills and special hammer mills, providing suitable rubber lining in the mills can reduce this problem. This problem is acute in hot, humid climates. In case, this problem takes place, cleaning of the equipment becomes necessary, resulting in stopping the production temporarily. For milling to fineness of the order of 200 mesh, conventional mills like hammer mill, disk mill with or without classifier can be used. Some of these utilise screens or grates while others are provided with screenless type classifiers. In general, they are invariably the hammer or pulverising type machines.

Ring/Roller Mills

These are equipped with rollers that operate in conjunction with the grinding rings. Grinding takes place between the surface of the grinding elements i.e. rings and rollers. Pressure may be applied with heavy springs or by centrifugal force of the rollers against the ring. The grinding rings may be in a vertical or horizontal position. These are also called roll ringing mills or roller mills. The ball and ring and bowl mills belong to the types of ring roller mills.

The base of an internal air classification type mill carries the grinding ring, rigidly fixed to the base and lying in the horizontal plane. Underneath the grinding rings are tangential air blowing parts through which the air enters the grinding chamber. A vertical shaft with a bevel gear near the bottom and resting on a thrust bearing is driven by a horizontal shaft through a pinion. Keyed rigidly to the shaft near the top is a spider which carries the mollen journals. These journals have rollers on the bottom rotating on their own bearings while travelling around the ring. Two or more journals are pivotally suspended by triennions (fastened at the top of the journal) and supported in the arms of the spider. They are held almost vertically so that when the mill is at rest the rollers exert a low pressure on the grinding ring.

Internal whizzer classifier is used to get uniform fine product. The whizzer disks rotate in a horizontal plane. They may consist of one or two disks each fitted with multiwhizzer blades. The whizzer is driven through a variable drive and the fineness of the finished product is regulated by the speed of the whizzer. The fineness of the finished product varies directly with the speed (rpm) of the whizzer machine. Before passing out of the machine at the top, the final product passes between the rotating whizzer blades. The coarse particles are returned to the periphery and dropped back for further grinding while the particles of required size pass through the whizzer blades into the duct leading to the cyclone collector.

Ball Mill

It consists of a horizontally rotating bowl or drum which contains, a number of grinding balls. The materials to be mixed or ground, along with the balls are taken in the bowl and rotated. As the bowl rotates, the balls fall to wear the material, leading to crushing and grinding effect. The fineness of the material depends on the hardness and weight of the balls and the grinding period (Fig. 6.6).

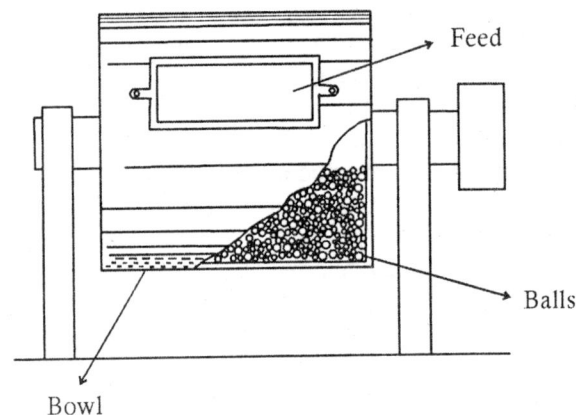

Fig. 6.6: Ball Mill

Planetary Ball Mill

This ball mill consists of two rotating bowls mounted on a moving platform, both of which are rotating on vertical axis but not co-axially. The bowls are loaded with the material and the required number of grinding balls added. The machine creates a planetary motion on the balls. The material is ground faster here than in the conventional ball mills. This machine can be used both for dry and wet grinding. The size of the feed should be 5 mm or less in diameter. A fineness of < 1 mm can be achieved. Friction forces and impact play a very important role in the grinding of the material in the mill (Fig. 6.7).

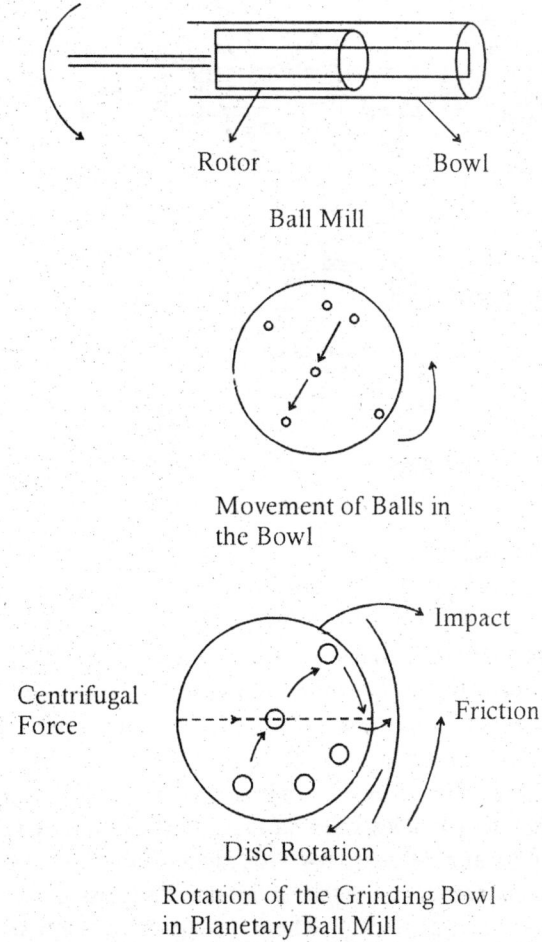

Ball Mill

Movement of Balls in
the Bowl

Rotation of the Grinding Bowl
in Planetary Ball Mill

Fig. 6.7: Principle of a Planetary Ball Mill

Hammer Mills

These mills work on the principle of rotor and stator. It can grind particles of 5-0.5 mm size.

Hammer Mill with Internal Air Classifiers

In the high-speed hammer pulveriser (Fig. 6.8), the materials are fed by means of screws into the mill through the path of the rotating hammers. On entering the mill, the materials are reduced in size by their impact against the hammers and are ultimately discharged through the grates in the bottom section of the grinding chamber, after attaning the required size. The machines are available in various sizes. These are equipped with a hopper below, which serves as the star feeder, activated by a plow and sachel mechanism. A vent from the top of the return air pipe passes through a tubular dust collector, so that the system does not become contaminated and remains dust free.

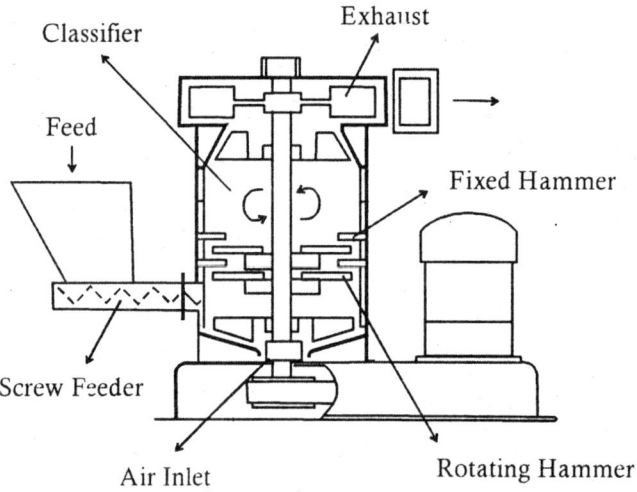

Fig. 6.8: Hammer Mill

A fan is placed on one end of the hammer shaft. Between the fan and the hammers in the whizzer, are fitted two or more thin blades with tips tapered to conform to the housing distance between blades. The motion of the whizzer along the shaft, regulates the housing. The whizzer moves towards the hammers and a coarse product results. Action of the whizzer is similar to that of a fan wheel opposing the motion of the main fan. The space between the blades and housing is cleaved and a maximum counter current is set up at the periphery in the direction of the arrows. An air current accompanies the feed when it enters the pulverising chamber along with carrier and the pulverised material then passes through to the collector. As the centrifugal force is greater on the coarser particles, their radial velocity will exceed the lateral; hence they are thrown to the periphery and deflected towards the hammers by the counter current, while the finer particles are discharged through the fan intake. The classified product passes through the fan and is blown to a cyclone collector where it is discharged into bins or containers. The air current goes back to the pulverizer completing the cycle. The mill is provided with spot which acts as an air vent. It is connected to a suitable filtering medium, which is necessary for the disposal of the air displaced by the rotor.

Automatic Pulverizer

The automatic pulverizer is a hammer type machine equipped with air classifier of the double whizzer multivane type. It has a horizontal shaft on which one or more disks fitted with hammer can be mounted. On the door of the pulverizing chamber there is a mechanical arrangement, which functions to remove coarse material contained in the feed (such as sand and gravel from clay). A span-feeder with plow and sachel mechanism receives the raw material from a stock bin and drops it into the pulverizing chamber on the top of which is mounted the air classifier. The air enters the pulverizing chamber at the rear end, removes the pulverized material. Particles of proper fineness are discharged in the bins or containers, while oversized ones are returned to the pulveriser through the bottom valve of the inner cone.

The oversize impurities accumulate in the grinding chamber until the rapidly revolving hammers pick them up. These are then pushed through the slot on the door into the throw out chamber where they are finally rejected by the flap valve. The slide damper on the top of the trough may be adjusted to admit air from the atmosphere, which circulates into the pulveriser through the door slot. In its movement in the throw out, the air leaves the rejected material and blows back fine particles into the pulverising chamber.

Victoria Hammer Mill

It is a screenless machine in which charge is fed by a screw type feeder. There is a variable speed control but the machine does not utilise any grates or screws. The material passing into the mill is broken by the impact of the hammers. The outgoing air stream directs it to a classifier where the oversized particles are separated from the product and thrown back to the mill for crushing again. The outgoing air stream passes from the classifier to a filter collector where it is discharged. From the filter, through bottom outlet is conveyed the product of required particle size. The control of particle size is achieved by movement of a selector valve that can be varied during the process of milling. Fig. 6.9 shows an outline of a Victoria Hammer Mill.

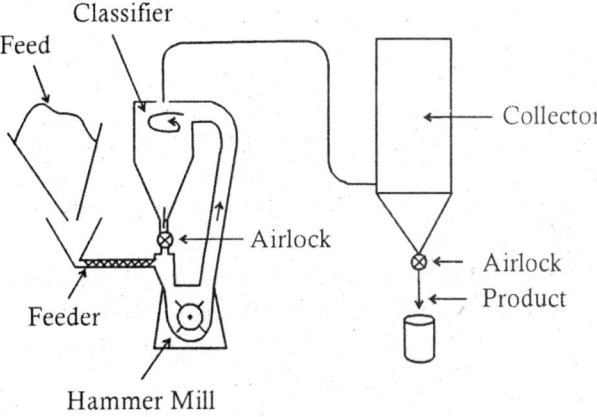

Fig. 6.9: Victoria Hammer Mill

Disintegrator

It is essentially a type of hammer mill suited for coarse grinding of soft and medium hard minerals. Shaft is only revolving part and it runs on balls/roller bearings. It is fitted with four or more hammers installed on a hub. The screens control the fineness of the desired product and can be replaced as and when needed. Portions of the body of the disintegrator, which are expected to resist wear and abrasion, are lined with replaceable hard alloy liners.

Pin Disc Mill

It is a high speed mill having pin breakers in the grinding circuits. The material to be ground is fed into the centre of the machine and directed on to a rotating disc fitted with a series of pins projecting upwards and passing through a similar arrangement of pins fixed to a static cover pointing downwards. The material moving on the rapidly revolving bottom disc is flung outwards against the pins by centrifugal force. It then passes over the outer edge of the rotating disc through the narrow pin clearing until it is discharged into the collecting hopper. The fineness of product is achieved by variation of the feed rate and speed of the machine.

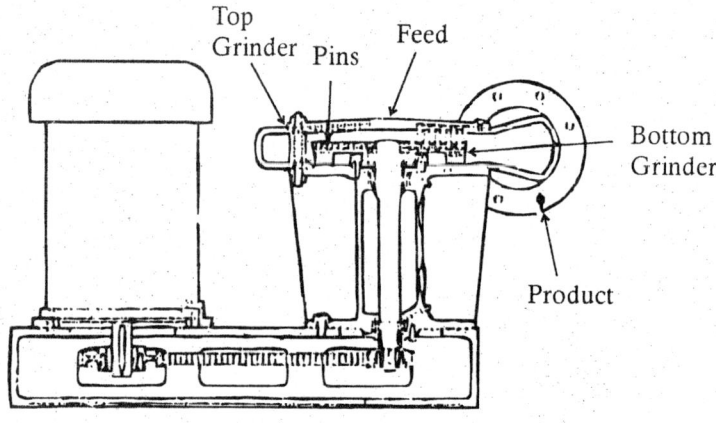

Fig. 6.10: Pin Disk Mill

Disc Attrition Mill

It is a modified version of early Buhr-stone mill used for milling grain to obtain flour (**Fig.** 6.11). Stones are replaced by steel discs, inter changeable metal or abrasive grinding plates rotating at higher speeds, permitting a much broader range of application. Grinding takes place between the plates, which may rotate in a vertical or horizontal plane. Generally, the vertical plane is more popular. The two discs may rotate in same or opposite directions. The assembly, consisting of a shaft, disc and grinding plate is known as runner. It contains a rotor fitted with a horizontal disc. In the centre of the disc is fitted a 2" long metal piece with sharp edges. The material falls on the moving discs and gets ground.

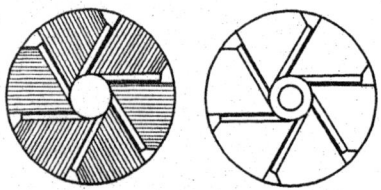

Fig. 6.11: Grinding Plates of a Buhr-Stone Mill

Salient features of different mills are given in Table 6.1.

TABLE 6.1. SALIENT FEATURES OF DIFFERENT MILLS

Type of mill (Principle)	Fineness achieved (microns)	Built-in advantage/disadvantage
Fluid energy mill (air attrition)	below 10	Suitable for heat sensitive and low melting point materials No moving part
Pin mill (disk grinder)	below 10	Heat generated during grinding. Not suitable for highly abrasive materials
Raymond mill (Impact/extrusion)	30	Heat generated during grinding
Hammer mill (Impact/extrusion)	30	Heat generated during grinding
Micropulveriser (impact/extrusion)	20	Heat generated during grinding

Milling of Wettable Powders

The type of mills described above yield a product of about 200 mesh size but wettable powders and seed dressings require a product of particle size around 10 μm. For this purpose, special machines are required. The well known mills generally used for this purpose are called fluid energy mills.

Fluid Energy Mills

These machines utilise the energy of compressed air or a steam of air for milling operation. In the grinding chamber, the materials are reduced in size by the action of supersonic air jets causing violent turbulence and consequent particle attrition. In the case of insecticide units, only compressed air can be used. In such type of mills, the temperature of the outgoing air is almost the same as that of the incoming air. It restricts the appreciable rise in temperature of the material preventing its softening. For milling of heat sensitive insecticides, the inlet air can be cooled before entering the unit. Nitrogen or carbon dioxide gas can replace air during processing of compounds that are easily oxidisable or perhaps form explosive mixtures with air. Filter collector is necessary for each unit. Before the final filter collector, high efficiency cyclone separators in series are incorporated to reduce the heavy load of dust. In fluid energy mills, the air stream conveys the materials around inside the mill body until they are of the desired size such that they can be drawn by the viscous drag of the outgoing air to a product collector. The cross section of a fluid mill is depicted in Fig. 6.12.

The fluid energy mills can be classified into two types in terms of the nature of operation of the mill.

(*a*) In first category of such mills the fluid energy is admitted in fine high velocity streams at an angle around a portion or all of the peripherry of a grinding and classifying chamber. The micronizer, jet pulveriser, reductionizer, Jet-O-Mizer and similar other machines are examples of this class.

(*b*) In the other class the different fluid streams convey the particles at high velocity into a chamber where the two streaks of gas/air strike upon each other. The Majac and other mills fall under this category.

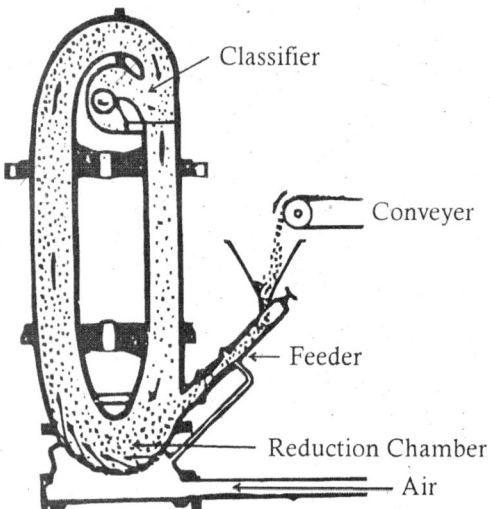

Fig. 6.12: Cross Section of a Fluid Energy Mill

The particles are conveyed with the jet or these may be intercepted by angular jets as they travel around the periphery of the grinding-classifying chamber. There is higher energy release and a high order of turbulence which causes the particles to rupture and grind amongst themselves. As all the particles are not ground, it is necessary to classify them and return the coarser particles for further grinding. Most of these mills utilise the centrifugal energy of the flowing fluid. The Majac mill differs in construction and uses a mechanical arrangement.

Micronizer

It consists of a shallow circular grinding chamber in which the material is acted upon by a number of gaseous fluid jets, which are forced through orifices spaced around the periphery of the chamber (Fig. 6.13). The diameters of orifices may vary up to one fourth of an inch. Jet orifices are drilled through the peripheral wall of the grinding chamber and vary in number from 3 to 16, equally spaced around the chamber. They are generally placed in a tangential position (the streams coming in a tangential fashion on an imaginary circle inside the chamber), so that entering stream or air will promote the rotation of the material to be pulverised in one direction. The fluid pressure is converted to approximately atmospheric pressure in the velocity head due to expansion. This causes a high-speed rotation of the contents of the grinding chamber. The centrifugal force of rotation causes the material to concentrate at the periphery where the jets are introduced. Since the energy of the fluid jets gets dissipated near the point of entry, intense local velocity gradient and interactions are set up within the circulating material. A great reduction in size of the material takes place by impinging of the particles upon one another. This also reduces the wear of the pulveriser housing.

The gaseous fluid supplying the grinding energy is withdrawn at an inward point, tending to cause the dust laden gas to travel spirally. The coarse particles, which are thrown to the periphery, are subjected to further reduction. The grinding chamber serves as an internal classifier also.

The air pressure at the outlet from the mill being usually slightly above one atmosphere, the velocity of air is sufficient to pneumatically convey the finely ground product to locally placed cyclone/filter collectors.

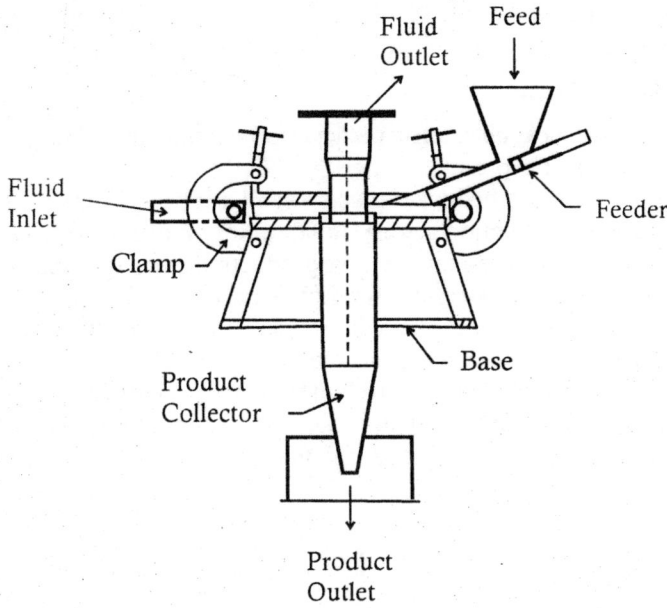

Fig. 6.13: Micronizer Mill

The outlet from the grinding chamber leads directly into a concentric centrifugal collector. This collector receives the material as it is rotating in a high velocity rotary motion, which is conducive to the separation of the material from the fluid, so that about 85-95% of the product is collected in the concentric collector.

Jet-O-Mizer

The equipment consists of a hollow elongated tower, which is placed vertically. The feed is injected to the outer periphery of the base. Feeding is done through a vent, which also has a navigating conditioner at the entry point. The removable jets of required size are placed tangentially to the curvature of the grinding chamber, so as to permit entry of high velocity air on to and into the incoming feed material. The fluid stream creates a rapidly circulating flow in the hollow doughnut shaped mill casing. Grinding is achieved by the pressurized air entering the grinding chamber and the entrainment of the feed material with air, causing the particles to break by impact and attrition against another. The upper turn of the casing is the classifying zone where larger particles concentrate at the periphery while the fluid which is being displaced by fluid from the jets passes to the outlet by a reversal of flow from the inner part of the ring. It carries with it the particles of acceptable fineness. Some fluid and coarser particles pass down the stack for further grinding.

GRANULATORS

Granulator

It consists of a rotor comprising of four rectangular prismatic bars fixed between two circular plates. A wire mesh is fitted below the prismatic bars. The rotor while rocking on the wire mesh cuts the matter to be granulated. By fixing the size of the wire mesh, suitable mesh size of the granules is obtained.

Fluidised Bed Granulator

The equipment is versatile for application to pesticide formulations including granulation, drying, coating and mixing of the final products.

     The formation of a fluidised bed takes place when air is passed upward through a bed of solid particles resting on a retention screen. In laboratory models, air flow is necessary for fluidization which is generated by a fan. In large production units, fluidisation is generated by suction fans that create negative pressure. A useful feature of the process is the ability to heat the air or pass warm air to dry the product while the bed is being made (Woodford 1998). The material to be processed can be placed into the product container or it can be part of the granulating liquid. The perforated plate at the bottom of the container has a fine mesh screen to prevent the solid particles from falling into the machine. On top, it has exhaust filters that allow air to escape while retaining small particles and dust (Fig. 6.14).

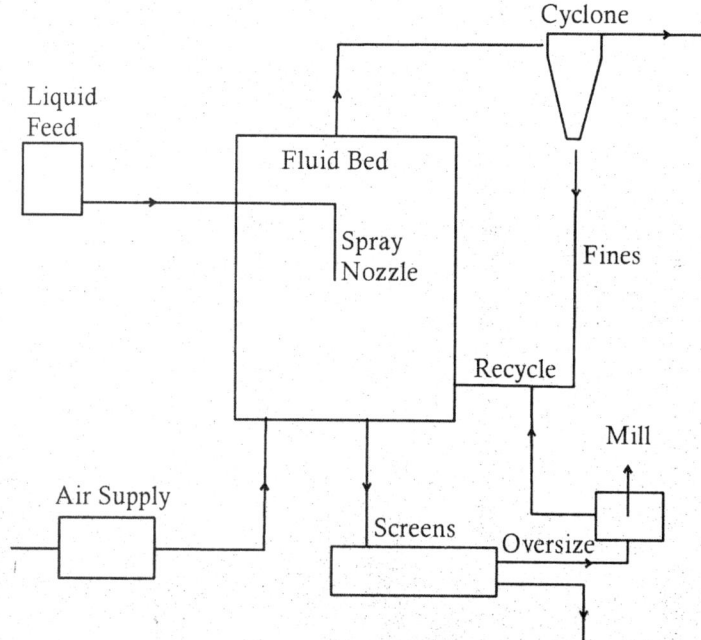

Fig. 6.14: Outline of a Fluid Bed Spray Granulation Plant

EQUIPMENT FOR EMULSIFICATION

The purpose of the emulsification equipment, whether simple or complex, is to break up or disperse the internal phase in the external phase such that the particle size of the resulting emulsion is sufficiently small to withstand coalescence and the consequent breakdown of the emulsion during the required time of stability. The choice of emulsifying equipment is governed by (*i*) the apparent viscosity in all stages of manufacture, (*ii*) the amount of mechanical energy input required and (*iii*) heat exchange demand. Formation and stability of the emulsion is greatly affected by time and type of agitation.

Propeller Agitator

In this equipment one or more propellers are mounted on a common shaft in a mixing tank. Modifications include variation in the location of propellers in the tank and the use of two or more propeller shafts and complex propellers. It is most suited for generation of emulsions of low and medium viscosity.

Turbine Agitator

The equipment contains fixed baffles either on the tank wall or adjacent to the propellers. A turbine rotor and stator increase the efficiency of agitation considerably. Turbine agitators are preferred since baffle plates in a tank frequently result in areas of little or no agitation. These are available in various sizes, speeds and rotor-stator clearances and in many modified designs. Turbine type systems may be designed to give a very high degree of shearing action and may be used with fluids of higher viscosities. It is not very efficient for emulsions of higher viscosities.

Colloid Mill

It can be considered as a modification of a turbine agitator, although in this case the clearance between the rotor and stator is of the order of a few micrometers. With such small clearance, an extremely high shearing action occurs. The product from a colloid mill usually has a uniform particle size because of the fixed clearance between the rotor and stator. Owing to the tremendous shearing force applied to the emulsion, a significant rise in the temperature occurs during emulsification and in most cases external cooling has to be employed.

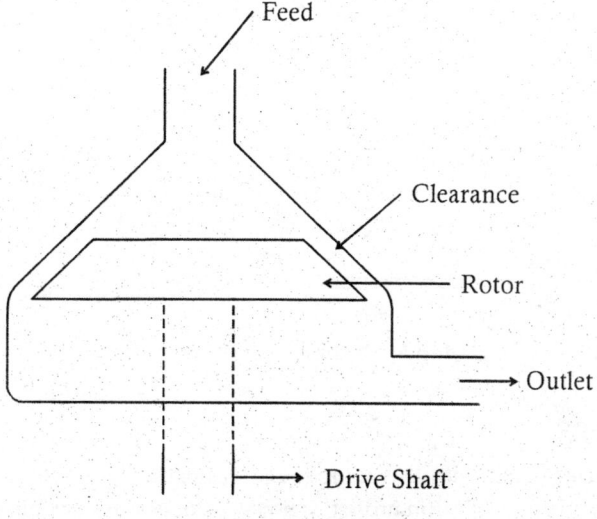

Fig. 6.15: Colloid Mill

Homogenizer

In a homogenizer, emulsification is affected by forcing the two phases past one or two spring seated valves at pressure of 3.45-34.5 MPa (500-5000 psi). Emulsification occurs not only while the components pass under the valve seat but also when the emulsion impinges against the wall that surrounds the valve. These are also built with more than one stages of emulsification i.e. with successive relief valves. This is of value when high pressure homogenisation promotes clumping of the fine particles of emulsion. At the second stage of homogenisation, at a lower pressure, clumps are broken up and produce product of lower viscosity using comparable ingredients. The homogeniser usually gives an emulsion of finer average particle size generally finer than colloid mills, though the particle size is not as uniform. The rise in temperature during homogenisation is only between 5-10 °C. However, the actual temperature rise throughout the homogenizer and the pump may be 5-20°C or as much as 25-50°C depending upon the type of pump employed. A piston pump generates lower heat than a gear pump. Owing to clearances in the gear pump a certain quantity of liquid continuously bypasses the pump and is partially homoginised before reaching the homogenizer head. Homogenizers can handle liquids or pastes and the rate of homogenization is not much affected by viscosity.

Ultrasonic Homogenizers

The high frequency ultrasonic emulsification is better suited for low viscosity liquids, although it has also been successful with systems having high viscosities. Ultrasonic energy is developed either mechanically or electrically. In the former case a pump forces the combined phases to pass through a tuned vane which vibrates and produces energy via cavitation. Pressure ranges from 1 to 3.5 MPa (150-500 psi) in the chamber surrounding the tuned vane.

Electrically, an oscillator causes ultrasonic vibration in either a magnetostriction device or a piezoelectric crystal. Major difficulties in the transfer of this energy to the liquid phase have restricted extensive commercialization of these oscillators in emulsification of pesticides.

Pohlman whistle (Fig. 6.16) is an example of ultrasonic homogenizer. The nodal supports of the blade are separated by a distance equal to half the wavelength of the characteristic vibration of the system. By this, ultrasonic energy is developed mechanically.

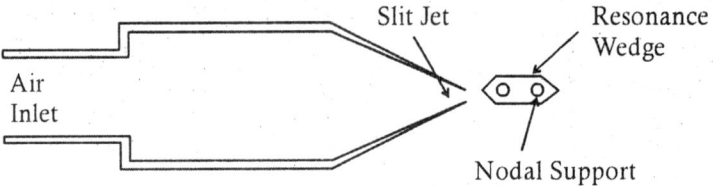

Fig. 6.16: Pohlman Whistle

DUST COLLECTORS

Dust collection is concerned with the removal or collection of solids suspended in gases. Particularly, in pesticide formulation industry, it is very necessary to collect and measure the dust from surrounding environment and to take precaution so that the generation of dust in the process is kept within the prescribed limit. Dust collection/elimination increases the running as well as the fixed cost of the plant. A proper balance between the cost and the dust collection efficiency has to be arrived at.

The selection of system of dust collection depends mainly on plant layout and capacity. Small plants have dust collection equipment with each dust producing unit and no more care is necessary regarding dust in the atmosphere. For example, the pulverisers and blenders can be provided with separate dust collection equipment at a relatively low cost. In bigger plants, each equipment is provided with dust collection system and in addition all points, where dust nuisance can occur (changing, discharging and packing places etc.) are also provided with hoods and canopies to exhaust the dust laden air which is again subjected to dust separation process before discharging. In special cases, plants possess central unit in the form of central vacuum clearing plant to separate dust from the atmosphere.

The performance of a dust collector is commonly termed as collection efficiency. It is the weight ratio of percentage of dust collected to that entering the equipment. The different types of equipments used for dust collection are given below.

Gravity Settling Chamber

It is probably the simplest and earliest type of dust collection equipment, consisting of only a suitable chamber in which gas velocity is reduced to enable the dust to settle out by action of gravity. Its simplicity leads it to almost any type of construction, but the industrial utility is limited. It is capable of removing the particles having size larger than 325 mesh (43 μm diameter) only. For removing smaller particles, the required chamber size is generaly bigger. Gravity collectors are usually built in the form of long, empty horizontal/rectangular chambers, with an inlet at one end and an outlet at the top of the other end. Gas velocity in the chamber should not exceed about one meter per second.

Horizontal plates arranged as shelves within the chamber give a marked improvement in collection. The disadvantage of this arrangement is the difficulty of cleaning due to the close shelf spacing. The pressure drop through a settling chamber is small. These are now being replaced by more compact dust collectors like cyclone collector etc.

Cyclone Collector

The most widely used type of dust collection equipment is the cyclone collector in which dust laden gas enters a cylindrical or conical chamber tangentially at one or more points and leaves through a central opening. The dust particles by virtue of their inertia tend to move towards the outside separator wall from which they are led into a receiver. A cyclone is essentially a settling chamber in which gravitational acceleration is replaced by centrifugal acceleration. Under the operating conditions commonly employed, the centrifugal separating force or acceleration may range from 5 times of the gravity in case of very large diameter, low resistance units to 2500 times gravity in very small, high resistance cyclones. The immediate entrance to a cyclone is usually rectangular. Cyclone separators offer one of the least expensive means of dust collection from both an operating and investment point of view. The cyclones for removing solids from gases are generally applicable when particles over 5 μm are involved, although some of the multiple tube parallel units attain 80-85% efficiencies on the particles of 3 μm diameter. To collect the particles of over 200 µm, cyclones with gravity settlers are usually satisfactory. In special cases where there is high dust concentrations (73 kg meter^{-3}) are involved, cyclones will remove dusts having a much smaller particle size. In certain cases, efficiencies as high as 98% have been realised on dusts having an ultimate particle size of 0.1 to 2.0 µm because of the predominant effect of agglomeration.

In a cyclone collector the gas path involves a double vortex with the gas spiralling downward at the outer side and upward at the inner part (Fig. 6.17). For the collection of heavy dusts and where the efficiency and the degree of dust exhaust is not a prime factor, the cyclone units are convenient equipment. In pesticide formulation industry these are usually employed as precollectors, prior to filters for reduction of the final dust loading on the filtering media. In case of pulverisers and blenders, they are added to reduce dust nuisance. One example of the operation of such a unit follows:

The blower delivers the pulverised powder in the cyclone air separators, the powder is discharged from the cyclone spout, and dust free air then returns to the grinding chamber to complete the air cycle. The excess air is blown off through the dust collector filter bags to collect the extra fine particles.

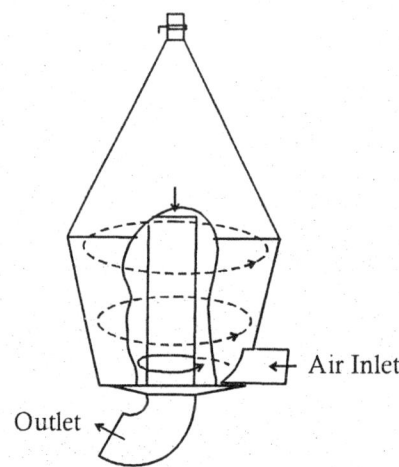

Fig. 6.17: Cyclone Collector

Filter Bag Collector

In pesticide formulation industry handling of dusts of low particle size is a common feature. After collection of coarse dust with the help of cyclone separators, the finer ones are collected within the filter bag type collectors which are generally of two types: the first and the older type employs a relatively thin woven fabric as the filter medium and the second uses felts. Both have collection efficiencies upward of 99% and are suitable for fully automatic operation. In woven fabric filters, the dust-laden gases are passed through a woven fabric which filters out the dust allowing the gases to pass on. Actually the separation is not a simple filtration, since the pores in the cloth are usually many times the size of the particles separated. When the dust laden gases first pass through the cloth, the efficiency of separation will be low until enough particles have been removed to build up that corresponds to a precoat in the fabric pores. This initial deposition of dust takes place because of interception and impingement on the cloth fibres and by gravity settling and Brownian movement in the pores. With the type and amount of dust normally encountered in industrial processes, this precoat layer will form in a few minutes. Once this layer has formed, the efficiency of separation becomes usually well over 99 percent.

Two types of mechanical bag dust filters are available as standard commercial units. One of them utilises cloth envelopes supported by screens and the other uses either oval or round vertically mounted bags, usually 5 to 8 inches in diameter and 8 to 17 feet in length (Fig. 6.18). Excess platforms are provided on the clear-air side. These units may be shaken manually or with motor. Mechanical filters generally have a pressure of 2 to 6 inches water and are rated at 0.3-2.5 M^3 min^{-1} M^{-2} cloth area. For very fine dusts or high dust loading, the rating should not exceed 1M^3 min^{-1} M^{-2} cloth and it may be desirable to reduce the rating to one-half. For very fine talc dust, bag type filters are better than screen or envelope type because of more effective sharing provisions. When dust loading is high, as usually is the case in commercial units, cyclone pre-collectors are employed to reduce the load on the filter. When enough dust is collected, the filters are freed of the dust by shaking and removing the dust from blast gate (Fig. 6.18). Filter fabrics are usually made of cotton satin because of low cost.

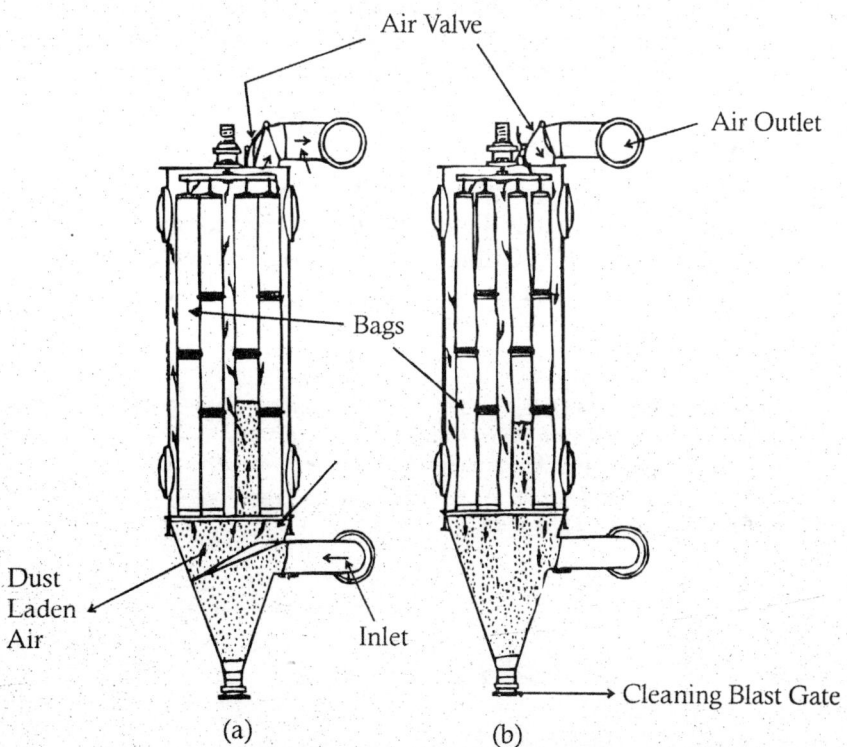

Fig. 6.18: Bag Dust Filters (*a*) In operation (*b*) Cleaning

SCREENS

Screening is the process by which a mixture of various grains is separated into two or more portions by means of a screening surface. The screening surface may consist of woven wire mesh, plastic cloth or silk for fine screening.

Vibrating screens can be used for screening. Such screens may be (*i*) Mechanically vibrated screens or (*ii*) Electrically vibrated screens.

Mechanically Vibrated Screens

These are used for particles of medium to coarse sizes. An eccentric or unbalanced shaft produces a vertical circle. One such type of screen is a balanced circle throw machine mounted on a base frame, having a full floating mount on shear rubber mounting units which absorb the shock of heavy material and allow the shaft to revolve around its own natural centre of rotation. Mechanically vibrated screens may be of two bearing and four bearing types.

In the two bearing type, the two outer bearings and the base frame are absent. The gyrating motion is caused by eccentric weights on the shaft and the screen itself is supported by overhead cables or springs on the floor. Screening machines actuated by rotating unbalanced weights have a symmetrical shaft through the screen body with an unbalanced flywheel on each end. Counter weights on each flywheel, which may be moved in relation to the shaft, permit adjustment of the amplitude of vibration. The complete shaft assembly is contained in a unit bottled to the top of the screen body. The horizontal-type screen is actuated by an enclosed mechanism consisting of off-centre weights geared together on a pair of short horizontal shaft. The mechanism is usually mounted between the side plates and above the screen body.

Electrically Vibrated Screens

These are used mainly in pesticide formulation industry. They handle successfully many light, fine dry materials from approximately 4 mesh to as fine as 325 mesh. Most of these screens have an intense, high speed (1500-7200 vibrations min^{-1}) with low-spaced amplitude vibration supplied by means of an electro-magnet.

REFERENCES

Faulkner, B. P. and Rimmer, W. H. (1983). Size reduction. *In: Kirk-Othmer Encyclopaedia of Chemical Technology* Vol. 21,. (ex. ed. M. Grayson), John Wiley and Sons, New York pp. 132-162.

Woodford, A. R. (1998). Water-dispersible granules. *In: Pesticide Formulation, Recent Developments and their Applications in Developing Countries.* (eds. W. V. Valkenburg, B. Sugavanam, and S. K. Khaitan). United Nations Industrial Development Organisation, Vienna pp. 203-231.

CHAPTER 7

Quality Control and Assurance

FOR all quality control/quality assurance needs good laboratory practice (GLP) has to be followed as per the guidelines of Organization of Economic Co-operation and Development's 'Principles of Good Laboratory Practices' (OECD, 1999). GLP is a quality system concerned with the organizational process and conditions under which non-clinical health and environmental studies are planned, performed, monitored, recorded, achieved and reported.

QUALITY

The quality of a product is its degree of possession of those characteristics, designed and manufactured into it, which contribute to the performance of an intended function when the product is used, as directed.

Quality Control

An effective system for coordinating the quality maintenance and quality improvement efforts of various groups in an organization so as to enable production at the most economical levels with full customer satisfaction (For a general reading, refer Feigenbaum, 1950).

Quality must be built into the product during research and development and production. Awareness of product quality should be a guiding principle from product conception through the various stages of development to final delivery of the product to the customer.

Key Features

Development of quality products and a system to maintain quality must bear in mind that:

(*i*) Additional costs are carefully considered. The quality of the products has to be maintained in an economic and efficient manner.

(*ii*) The quality control incharge has a proper status in the organizational hierarchy. In no case should he be a subordinate to sales or manufacturing activities.

(*iii*) The quality control cannot be an end in itself. The freedom of quality control incharge must not demand a degree of perfection that cannot be achieved.

(*iv*) Proper and regular controls are a key to quality maintenance.

Responsibilities of the Section

The following responsibilities are usually entrusted to the quality control section.

(*i*) Develop methods of analysis and specifications for raw materials and finished products.

(*ii*) Develop specifications for packaging materials.

(*iii*) Check equipment design and insist upon changes when equipment has been found to cause a poor quality production.

(*iv*) Run a continual check on production practices and report any condition found contributing to poor product quality.

(*v*) Check incoming raw materials for conformity to specifications.

(*vi*) Check all finished products for conformity to specifications, after which lots meeting all specifications are released for sale and those found defective are rejected with instructions to rework or destroy.

(*vii*) Handle customer complaints.

(*viii*) Inspect finished materials and check the contents of containers.

(*ix*) Collect representative samples for analysis.

The key parameters concerning the area of quality control section may thus be identified as: inspection, sampling, customer complaints, specifications, analysis or test methods, shelf-life and cross contamination, if any.

Both the industry and the government play a role in maintaining/enforcing quality control.

Role of Industry

Quality Cost: Internal failure costs like scrap, rework, retest, down time, yield losses and disposition would disappear if no defects existed in the final product. In the same manner, external failure costs arising out of detection of defects by the end user such as complaint, adjustment, returned material, warrant charges, allowances etc. would also be eliminated. Therefore, the production of quality product could lead to an overall reduction of production cost and industry has a major role to play in reducing it by achieving quality standards.

New Product Quality: Newly developed/introduced products are required to meet tough quality demands. In India, if development cycle curtails environmental and field testing, the product may not pass the tough tropical conditions, leading to serious avoidable consequences. Accelerated testing is an inexpensive and reliable test.

Mathematical models enable production of actual shelf life from these short time tests conducted under drastic stress conditions.

Quality Control During Manufacture: 'Process capability' is measured by control chart, frequency distribution and lot plot methods. To achieve required quality standards, set up for approval, running approval and quality approval is necessary.

Procurement of Materials and Components: Quality of raw materials and packaging materials received from vendors affects quality. Here the relationships work more than the other factors. Financial soundness, prices, ability to stick to delivery schedules etc. affect the overall supply position and indirectly, the quality.

Quality Control Inspection: 'Incoming inspection' for control of vendor quality, 'process inspection' for process variable effect on manufactured product, 'finished goods inspection' are the key inspections. The data are submitted to management for information and necessary action.

Quality Control Measurement: Proper record of analysis, calibration checks on measuring instruments, preparation and use of primary and secondary and reference standards are maintained for proper quality control through accuracy of the measuring instruments and analysis.

Marketing of Quality: In the product label providing product information to users, instructions for safe use of pesticides etc. are clearly given. The label conforms to government regulation in all respects. It also gives information about the inherent nature of the product such as net weights of package, ingredients used, instructions for use and warnings of danger. The information is provided either in the form of a small tag, or label or printed on the container, placed as accompanying leaflet etc. The campliance of the products to the national/international specifications is indicated. All there paramateres enable marketing of quality.

Field Performance Quality Complaints: To minimize deterioration of the product including its degradation during storage, sale or use, the following actions will help in reducing complaints.

(*a*) Establishing shelf life of the product.
(*b*) Establishing standards to assign limits on time in storage.
(*c*) Marking of the manufacturing and expiry dates conspicuously in order to make easy identification of the age of the product in stock.
(*d*) Designing the package and control the environment to minimize degradation.
(*e*) The feed back on storage and transport is also obtained by conducting 'depot audits'.
(*f*) The manufacturer must take up quality complaints very seriously. It would be better to set up a 'registration desk' to which copies of complaints are sent. This would assist in:

(*a*) registering all complaints

(*b*) to assign serial numbers to carry out tracing progress

(*c*) route complaints to different departments

(*d*) assist in identifying complaint cases requiring fundamental studies

(*e*) follow up and guaranttee that adequate analyses and disposition are made

(*f*) help in summarized reporting.

Thus the pesticide industry will be in a position to offer quality goods to consumer by adoption of total quality control employing, wherever required, statistical methods of evaluation.

Role of Government

Quality Controls: With industrialization, quality controls are widening fast. They could be broadly classified as under:

(**a**) **Extension of Standardization:** Official bodies like Bureau of Indian Standards (BIS) in India establish standards.

(**b**) **Certification of Products:** Independent laboratories are established to certify conformation to standards. For example, certification by BIS is mandatory in India for certain key products involving human health and safety. In case of pesticides, certification in majority of cases is voluntary but some of the state governments insist on buying ISI marked formulations.

(**c**) **Import Quality Control:** Brought about by Governments for both quality reasons and to protect use of foreign exchange.

(**d**) **Currency Controls:** Fiscal in nature, affect quality. Also used to restrict import of competing goods in the country.

(**e**) **Service to Manufacturing Units:** Various institutions provide consultancy assistance in quality control. In India, institutions such as BIS, Indian Statistical Institute and other similar bodies do so.

(**f**) **Quality Evaluation and Marks:** The state provides quality evaluation and certification through quality marks. In India, BIS do it.

(**g**) **Specification AVP 92:** In U.K., the Ministry of Technology issued in 1968 the above specification to which all industrial undertakings approved for government contracts, should comply. Its key stipulations are:

Quality Management: A qualified top man should cover all aspects of quality.

The Quality Manual: Should be prepared and updated periodically. Should show how quality is managed and those responsible for it.

Design and Development: These personnel must be completely aware of all quality and reliability requirements.

Bought out Items: Their compliance to quality must be ensured.

Pre-production and Production Quality Control: Involves satisfactory material, plant, equipment, methods, shop floor quality control working conditions, etc.

Statistical Methods: Should be employed wherever appropriate.

Inspection, Measuring and Test Equipment: Adequate provision should exist to obtain, periodically inspect and maintain the above equipment.

Audit of Quality Control: Quality audits should be conducted frequently by impartial persons.

Requirements and Assistance from other Bodies: The provisions of British Standards, are not always mandatory. However, the users of thses provisions realize that their products are easy to market with less likelihood of confusion between the seller and the consumer, about the quality demanded.

The specification AVP 92, though applicable to British trade, has similar implication in other countries.

The Insecticides Act, 1968 and Insecticides Rules 1971: Enacted in India to enforce quality standards of all pesticides, imported, manufactured and sold in India. Similar acts exist in other countries of the world.

The Standards of Weights and Measures Act, 1976 and the Standards of Weights and Measured (Packaged Commodities) Rules, 1977: In India, the customer interest is safeguarded by proper definition of packages, manner of labelling and variations in qualities of the contents packed. Under 2(5), 6(1), 7(1), 9(1), 24(1-6), 27(1-4), 27(1-4, 33) various provisions applicable to pesticides are detailed.

Under second Schedule, Rule (i), (ii), maximum permissible errors in relation to packaged commodities are specified. Liquid packs fall in class A and solid packs under class B, Section 35 (Chapter IV) mentions about registration of manufacturers and packers with the Directorate of Weights and Measures.

Most countries have enacted similar laws to ensure supply of proper quality and quantity of the goods to their consumers.

Indian Standards Institution (Bureau of Indian Standards): The Indian Standards Organization was set up in 1947 under the Ministry of Industry and Civil Supplies, with the following aims:

(*a*) Formulation of Standards relating to products, commodities, materials and processes.

(*b*) Promotion and general adoption of standards at national and international level.

(*c*) Operation of Indian Standard Institution's Certification mark scheme.

Indian Standards set recognized quality levels. These also provide the framework for mass production processes and enhance labour efficiency. The standards incorporate latest results of research and development.

Certification marks scheme gives ISI license and mark for the goods produced as per IS standards. The scheme is governed by ISI (Certification Marks) Act 1952 and the Rules and Regulations framed thereunder. ISI has its network of laboratories at several places.

ISI standards on technicals, EC, WSC, D, WP, GR, mosquito larvicidal oil, lime sulphur, solution, smoke generators, tablets, etc., packaging materials, methods of tests for pesticides, guides for handling pesticide poisoning, pest control and pesticide application equipment exist. Standards are added with time.

Quality Assurance (QA)

Quality and quality assurance seem simple and self-explanatory concepts, but in fact the terms are frequently misunderstood. Quality assurance is often confused with "quality control". It involves all those systematic actions, which provide adequate confidence that a product will satisfy given requirements for quality, e.g. an adequate standard, causes for failures to meet the standard, correction of the causes of the failure. Quality control, on the other hand, addresses to the operational techniques and activities that are used to fulfill the requirements for quality.

A quality assurance programme as per revised OECD principles of good laboratory practice is "a defined system, including personnel, which is independent of study or conduct and is designed to assure test facility management of compliance with these principles of good laboratory practice" [Section 1.2.2(8)]. The responsibilities of the management of a test facility include ensuring "that there is a quality assurance programme with designated personnel and assure that the quality assurance responsibility is being performed in compliance with the principles of good laboratory practice" [Section II.1.1(2f)]. In addition to the test facility, the management should ensure "that the Study Director has made the approved study plan available to the quality assurance personnel" [Section II.1.1(2j))] and the responsibility of the Study Director should include ensuring "that the quality assurance personnel have a copy of the study plan and any amendments in a timely manner and communicate effectively with the quality assurance personnel as required during the conduct of the study" [Section I.1.2(2b)]. The test facility management should also ensure that for a multi-site study, clear lines of communication exist between the Study Director, Principal Investigator(s), the quality assurance programme(s) and study personnel [Section II.1.(2o)].

General

(*i*) The test facility should have a documented quality assurance programme to assure that studies performed are in compliance with the principles of good laboratory practice.

(*ii*) The quality assurance programme should be carried out only by an individual or by individuals designated by and directly responsible to management and who are familiar with the test procedures.

(*iii*) The individual(s) should not be involved in the conduct of the study being assured.

Responsibilities of the Personnel

The responsibilities of the quality assurance personnel include, but are not limited to, the following functions. They should:

(*a*) maintain copies of all approved study plans and standard operating procedures in use in the test facility and have access to an up to date copy of the master schedule;

(*b*) verify that the study plan contains the information required for compliance with these principles of good laboratory practice. The verification should be documented;

(*c*) conduct inspections to determine if all studies are conducted in compliance with these principles of good laboratory practice. Inspections should also determine that the study plans and standard operating procedures have been made available to study personnel and are being followed.

Inspections can be of three types as specified by quality assurance programme standardizing operation procedures as mentioned below:

– Study-based inspections
– Facility-based inspections
– Process-based inspections

(*d*) inspect the final reports to confirm that the methods, procedures and observations are accurately and completely described, and that the reported results accurately and completely reflect the raw data of the studies;

(*e*) promptly report any inspection results in writing to management and to the Study Director, and to the Principal Investigator(s) and the respective management, when applicable;

(*f*) prepare and sign statement, to be included with the final report, which specifies types of inspections and their dates, including the phase(s) of the study inspected, and the dates inspection results were reported to management and the Study Director and Principal Investigator(s), if applicable. This statement would also serve to confirm that the final report reflects the raw data.

The QA-management Link

Management of a test facility has the ultimate responsibility for ensuring that the facility as a whole operates in compliance with GLP principles. Management may delegate designated control activities through the line management organization, but always retains overall responsibility. An essential management responsibility is the appointment and effective organisation of an adequate number of appropriately qualified and experienced staff throughout the facility, including those specifically required to perform quality assurance functions.

The manager ultimately responsible for GLP should be clearly identified. Thus person's responsibilities include the appointment of appropriately qualified personnel for both the experimental programme and conduct of an independent quality assurance function. Delegation to assurance group of tasks which are attributed to management in the GLP principles must not compromise the independence of the quality assurance operation, and must not entail any involvement of these personnel in the conduct of the study other than in a monitoring role. The person appointed to be responsible for quality assurance must have direct access to the different levels of management, particularly to top level management of the test facility.

Qualifications of QA Personnel

The personnel should have the training, expertise and experience necessary to fulfill their responsibilities. They must be familiar with the test procedures, standards and systems operated at or on behalf of the test facility.

Individual(s) appointed should have the ability to understand the basic concepts underlying the activities being monitored. They should also have a thorough understanding of the principles of GLP.

In case of lack of specialized knowledge, or the need for a second opinion, it is recommended that the quality assurance operations ask for specialist support. Management should also ensure that there is a documented training programme encompassing all aspects of quality assurance. The training programme should, where possible, include on the job experience under the supervision of competent and trained staff. Attendance at in-house and external seminars and courses may also be relevant. For example, training in communication techniques and conflict handling is advisable. Training should be continuous and subject to periodic review.

The training of personnel must be documented and their competence evaluated. These records should be kept up to date and be retained.

The QA Statement

The principles of GLP require that a signed quality assurance statement be included in the final report, which specifies types of inspections and their dates, including the phase(s) of study inspected, and the dates inspection results were reported to management, the Study Director and the Principal Investigator(s), if applicable [Sections II.2.2(1f) and II.9.2(4)]. Procedures to ensure that this statement reflects quality assurance's acceptance of the Study Director's GLP compliance statement and is relevant to the final study report as issued for the responsibility of management.

The format of the quality assurance statement will be specific to the nature of the report. It is required that the statement includes full study identification and the dates and phases of relevant quality assurance monitoring activities. Where individual study-based inspections have not been part of the scheduled assurance programme, a statement detailing the monitoring inspections that did take place must be included, for example, in the case of short-term studies where repeated inspections for each study are inefficient or impractical.

It is recommended that the quality assurance statement only be completed if the Study Director's claim to GLP compliance can be supported. This statement would also serve to confirm that the final report reflects raw data. It remains the Study Director's responsibility to ensure that any areas of non-compliance with the GLP principles are identified in the final report.

QA and Non-regulatory Studies

Compliance with GLP is a regulatory requirement for the acceptance of certain studies. However, some test facilities conduct in the same area studies, which are and which may or may not be intended for submission to regulatory authorities. If the non-regulatory studies are not conducted in accordance with standards comparable to GLP, this will usually have a negative impact on GLP compliance of regulatory studies.

Lists of studies kept by quality assurance should identify both regulatory and non-regulatory studies to allow a proper assessment of workload, availability of facilities and possible interferences. Quality assurance should have access to an up to date copy of the master schedule to assist them in this task. It is not acceptable to claim GLP compliance for a non-GLP study after it has started. If a GLP-designated study is continued as a non-GLP study, this must be clearly documented.

QA at Small Test Facilities

At small test facilities it may not be practicable for management to maintain personnel dedicated solely to quality assurance. However, management must give at least one individual permanent, even if part-time, responsibility for co-ordination of the quality assurance function. Some continuity in the assurance staff is desirable to allow the accumulation of expertise and ensure consistent interpretation. It is acceptable for individuals involved in studies that comply with GLP to perform the quality assurance function for GLP studies conducted in other departments within the test facility. It is also acceptable for personnel from outside the test facility to undertake quality assurance functions if the necessary effectiveness required to comply with the GLP principles can be ensured.

The concept may be additionally applied to multi-site studies, for example field studies, on the condition that overall responsibility for co-ordination is clearly established. The major features of quality assurance that have to be practiced even by a small set up are being described below.

Major Features

The quality assurance programme covers all aspects of the laboratory work, which can affect the quality of its output. There is no universal quality assurance programme, suitable for all control laboratories. The emphasis given to each aspect of quality assurance will reflect the laboratory's work.

Sample Identity and Integrity: It is recommended that each sample should have an individual identity and contain all the information required for reporting it. If a choice of analytical methods is available to the analyst, the identity card must clearly and unequivocally mention the analytical method used.

The analyst of a pesticide sample almost always has to remove a portion of the entire sample for analysis and keep the remnant of the sample as close to its original condition as possible, for future use in case further analysis is necessary. The analysis can sometimes lead to testimony in court. Sample integrity may be lost by contamination. Improper sub-sampling due to lack of homogeneity can also lose it. In many cases sample procedures are laid down and these are given in the methods. Many samples are difficult to homogenize, or contain constituents of a radically different particle size that tend to separate from the rest.

Analysis Required: The analysis to be carried out on a sample may be predetermined as a part of planned sampling programme or by legal requirements. The purpose of the analysis must be borne firmly in mind. When samples are examined under the general and specific provisions of laws pertaining to consumer protection and food quality, the analysis must be designed to answer certain questions such as: Does the formulation meet established minimum quality and safety standards? Is it correctly labelled? Are the claims on the label justifiable and legal? Does it contain non-permitted additives or permitted ones in excess? Are any contaminants present at unacceptable levels?

Choice of Method: The analyst must establish that reliable results can be obtained with the method used. Generally, it is preferable to use a method that has been subjected to collaborative study. It is important to follow the procedures for validating a method. If a method has been in use and found to give reliable results, it should not be changed for another until the new method has been shown to be of equal or better reliability in that particular laboratory irrespective of its collaborative status. The ability of the analyst to use the method is the essential requirement and the person taking the responsibility for the result, whether the analyst himself or his superior, must be confident of this.

Qualitative and Quantitative Analysis of the Active Ingredient: The qualitative analysis deals with identification of the elements, ions, or compounds present in a sample, while the quantitative analysis deals with determination of how much of one or more constituents is present. The sample may be solid, liquid or gas. A quantitative analysis involves several steps and procedures. The analytical process may be defined as a sequence of events such as: defining the problem, obtaining a representative sample, preparing the sample for analysis, performing necessary chemical separations, performing the measurement and calculating the results and presenting the data.

Quality Control of Laboratory Results: The analyst has to be as certain as possible that a result does not vary by more than an acceptable amount from the right answer.

It is usual to expect small variations in the results on carrying out the same test on the same sample, either by the same analyst, or a selection of analysts. The important criterion is that the variations in results are not so large as to represent significant errors, utilizing statistical concepts to test the variation between results. It is, therefore, necessary to monitor the quality of results by the introduction of various check procedures.

Such monitoring takes a considerable amount of time and resource. On the other hand, a wrong result casts doubts on the credibility of other results and may have very serious economic or social repercussions. In practice, the laboratory head must exercise careful judgement in the amount of resources put into monitoring the quality of the work done under his or her charge. Occasional checks may be adequate. Every possible check, including blanks, recovery experiments, instruments, re-calibration and check analysis by other laboratories may be justified for a sample which is to be reported adversely. The following checks can be used:

- One test in every ten should be on a standard or blank.
- Ten to fifteen per cent of laboratory time should be devoted to quality assurance of results.
- Periodic analysis of sample for which the analytical level is known.

Practices such as analysis of spiked sample, inter-laboratory proficiency testing, re-validation of the method of analysis of sample of known composition, etc. may help.

Certification/accredition of the laboratory/test facilities by various national/international agencies would go a long way in achieving the goals.

Accuracy and Precision in Quality Control/Assurance

Accuracy is the degree of agreement between the measured value and the true value. An absolute true value is seldom known.

Precision is defined as the degree of agreement between replicate measurements of the same quantity. That is, it is the repeatability of a result.

Good precision does not assure good accuracy. Accuracy is how close you get to the bull's eye. Precision is how close the shots are to one another. It is nearly impossible to have accuracy without good precision.

Accuracy/precision of any analysis must be expressed in statistical terms. Some basic parameters of help are stated below:

Significant Figures: The number of significant figures can be defined as the number of digits necessary to express the results of a measurement consistent with the measured values. The last digit of a measurement has some uncertainty. Therefore, it is not desirable to include more digits than necessary.

Errors: Two main classes of errors can affect the accuracy or precision of a measured quantity. Determinate or systematic errors are non-random and occur when something is wrong with the measurement. Indeterminate errors are random and cannot be avoided. Absolute error is the difference between the true and the measured value, with regard to the sign, and it is reported in the same units as the measurement. Relative error is the absolute or mean error expressed as a percentage of the true value.

Standard Deviation: Each set of analytical results should be accompanied by an indication of the precision of the analysis. Various ways of indicating precision are acceptable. Standard deviation is one such indicator. It is defined as the square root of the squared deviations of individual values from the mean (Rangaswamy, 1995).

Control Charts: A quality control chart is a time plot of a measured quantity that is assumed to be constant for the purpose of ascertaining that the measurement remains within a statistically acceptable range. A control chart is constructed by periodically running a known control sample.

Tests of Significance: In developing a new analytical method, it is often desirable to compare the results of that method with those of an accepted (perhaps standard) method. This way one can tell if there is a significant difference between the new method and the accepted one?

The F Test: The F test is used to determine if two variances are statistically different.

The Student t Test: The t test is used to determine if two sets of measurements are statistically different.

Rejection of a Result—The Q Test: The Q test is used to determine if an outlier is due to a determinate error. If it is not, then whether it falls within the expected error and should be retained.

Detection Limit: The concentration that gives a signal equal to three times the standard deviation of the background is generally taken as the detection limit.

Analysis by an Alternative Method: In practice, due to pressure of work, determinations are often done in duplicate and a third analysis carried out only if the first two disagree. Although this may be an acceptable practice, it is better that the first two determinations are done at different times or by different analyst or by different methods.

Unsatisfactory Samples: If repeat analyses by the original analyst show that the sub-sample does not comply with the legal requirements or other acceptable standards, then whole sample should be considered unsatisfactory. Check analyses must be carried out and attention paid to the integrity of the sample with utmost rigor, and all results must be meticulously noted on the sample card, bearing in mind that these records could be required in court if legal proceedings took place.

Writing Reports: This can only be learnt by experience. Reports for the use of magistrates and others must be written in non-technical language, or if this is not possible (which is unusual), technical terms must be carefully explained. Results must always be reported only to a number of significant figures justified by the accuracy and precision of the method. The report must be clear and confined to factual information, and statements, which cannot be substantiated, must not be made.

Report of Analysis: The wording of results of analysis should be chosen with great care, particularly information about the sample, analyst, analytical details etc.

The information must be enough to allow reviewers to review the analysis to draw their own conclusions.

House Keeping

Maintenance of sample integrity requires that any contamination and decomposition of the sample is prevented. Storage in freezers, refrigerators or metal cupboards is satisfactory, depending on the nature of the sample. Sometimes it is necessary to add a preservative such as formaldehyde or mercuric chloride. A preservative must be chosen which does not interfere in the analysis, as even samples already reported may have to be analyzed again in case of dispute. The cleanliness of the laboratory is very important in preventing loss of sample by infestation and in minimizing contamination. Laboratory must be duly certified.

Training

Staff training is an important activity and must be pursued. Periodic human resource development programmes need to be formulated keeping in view the need of the unit or the personnel involved. Training in test methods, instrumentation, accredition procedures, data requirements by different agencies for the diverse purposes, data storage/retrieval, software use and development, laboratory upkeep and upgradation, international and national developments in standards/specifications, specific on the job trainings in identified fields etc. are some of the areas suggested for this purpose.

REFERENCES

Feigenbaum, A.W. (1950). *Quality Control: Principles, Practice and Administration*. McGraw Hill, New York, USA.

OECD (1999). Quality assurance and GLP. OECD Series on Good Laboratory Practice and Compliance Monitoring No. 4 (Revised) ENV/JM/MONO (99)20. Organization for Economic Cooperation and Development 83306, Department of Science and Technology, New Delhi.

Rangaswamy, R. (1995). *A Text Book of Agricultural Statistics*. New Age International (P) Limited Publishers, New Delhi.

CHAPTER 8

Packaging and Labelling

Pesticides, being poisonous, ought to be handled with extreme care and precaution in order to avoid hazards associated with them. The guidelines for safe handling of a pesticide are printed on the label pasted on the container, and the detailed information is provided in the leaflets enclosed with the package, which have to satisfy certain requirements depending upon the type of the pesticide. Hence, effective labelling and suitable design of pesticide containers are the two major factors in ensuring their safe handling to minimize health hazards and environmental pollution. Manner of packaging, labelling and content of the label and leaflets are statutory requirements for registration of pesticides under the Central Insecticides Act of India and various similar acts in other countries.

LABELLING

The label pasted on the package containers includes, besides name of the active ingredient, its weight and price, any written, printed or graphic matter accompanying the pesticide, generally essential for its safe handling. The label and leaflets found inside the container contain very important information as required under the Indian Insecticides Act, 1968, section 18 and 19 and/or similar acts in other countries of the world. They are not only very important but also have been termed as the most expensive literature in the world. Before registering a pesticide, the label and leaflet are to be approved by the Central Insecticides Board in India or the other similar regulatory body of a country where the material is to be used.

Every pesticide label contains the pertinent safety information. Hence, before using the pesticide, a warning must be given on the label as "stop" sign (Fig. 8.1) in an appropriate position. The use of this system by every manufacturer or formulator will help the user to know the precautions before opening the container.

Besides the information regarding amount, price, date of manufacture and expiry date for the active ingredient, the label has two distinct functions associated with the pesticide:

(*i*) It describes where, when and how to use the product

(*ii*) It conveys warnings and instructions for its safe handling.

Fig. 8.1: Warning Sign for Pesticide Containers (Reynolds, 1983)

Therefore, the label should be so designed as to convey information, which forms an integral part of the container. The format of a typical pesticide label is given in Fig. 8.2.

Information/Content of the Label

The following information should be contained on a typical label.

(*i*) Name of the product, including brand names, trade names and common name

(*ii*) Name and address of the manufacturer

(*iii*) Registration number of the pesticide

(*iv*) Description of the formulation/ingredients

(*v*) Purpose of the product

(*vi*) Directions for use

(*vii*) Warnings, warning symbols and signal words (Fig. 8.4) or pictograms (GIFAP, 1988)

(*viii*) Toxicity: Grading of the product or the class or category to which it belongs in the classification of pesticides based on toxicity (Tables. 8.1 and 8.2).

(*ix*) Precautionary statements and antidotes

(*x*) Type of formulation

(*xi*) Content of the package

(*xii*) Batch number

(*xiii*) Date of manufactur

(*xiv*) Expiry date

(*xv*) Storage and disposal

(*xvi*) Re-entry statement-period after which it is safe for people or for animals to re-enter an area which has been treated with pesticide for normal farm use.

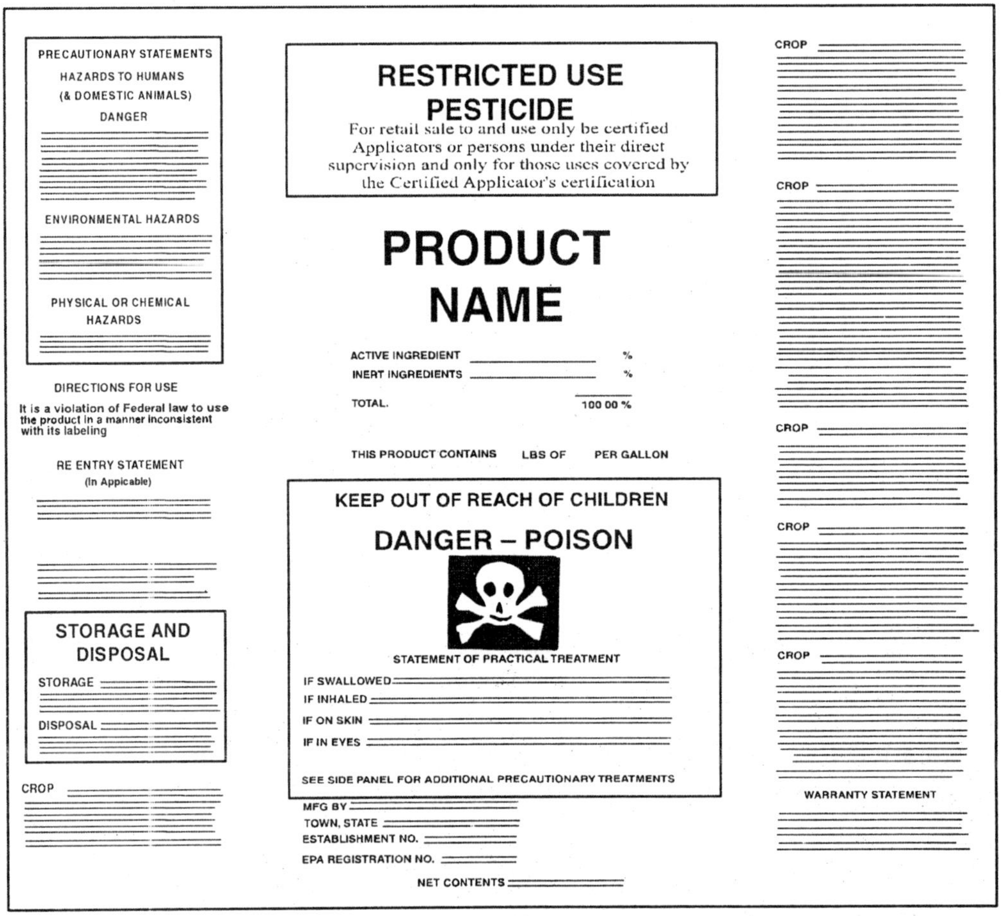

Fig. 8.2: Format of a Typical Pesticide Label (Reynolds, 1983)

In India, the labels contain all the information in three languages namely Hindi, English and a regional language.

The label has to be so affixed to the container that it cannot be removed ordinarily. It shall be located on a prominent place and occupy not less than 1/16 of the total area of the face of the container. Within the label, there is a square highlighting toxicity hazard. The dimension of the said square shall depend on the size of the package. It can be divided into two equal triangles. The upper portion shall contain the symbol indicating the hazard category and a signal word or phrase e.g. 'Poison' or 'Toxic' etc. (Fig. 8.3).

"KEEP OUT OF THE REACH OF CHILDREN IF SWALLOWED OR IF SYMPTOMS OF POISONING OCCUR, CALL PHYSICIAN IMMEDIATELY"

CATEGORY 1	LD_{50} (Oral) (mg/kg)	LD_{50} (Dermal) (mg/kg)
	1 to 50	1 to 200

(Extremely Toxic)

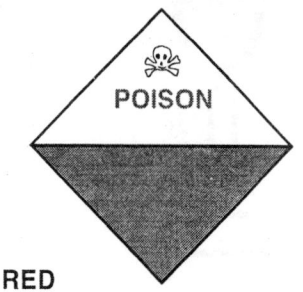

RED

"KEEP OUT OF THE REACH OF CHILDREN"

CATEGORY 2	LD_{50} (Oral) (mg/kg)	LD_{50} (Dermal) (mg/kg)
	51 to 500	201 to 2,000

(HIGHLY Toxic)

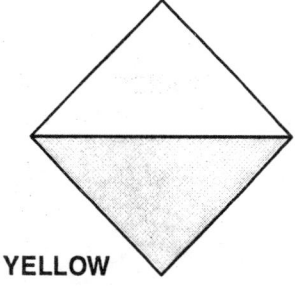

YELLOW

"KEEP OUT OF THE REACH OF CHILDREN"

CATEGORY 3	LD_{50} (Oral) (mg/kg)	LD_{50} (Dermal) (mg/kg)
	501 to 5,000	2001 to 20,000

(Moderately Toxic)

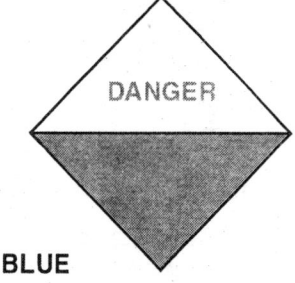

BLUE

CATEGORY 4	LD_{50} (Oral) (mg/kg)	LD_{50} (Dermal) (mg/kg)
	More than 5,000	More than 20,000

(Slightly Toxic)

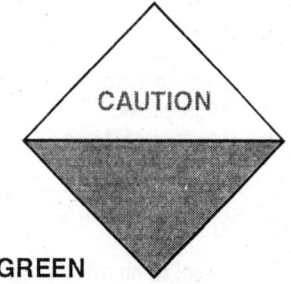

GREEN

Fig. 8.3: Categories of Warning Symbols (PAI, 1989) (see colour view of this Fig. at the end of this book)

Fig. 8.4: Signal Words

TABLE 8.1. BIS CLASSIFICATION OF PESTICIDES BASED ON MEDIAN LETHAL DOSE ALONG WITH THE COLOUR OF IDENTIFICATION

	Category/ Class	Median lethal dose by the oral route (Acute toxicity) LD_{50} mg kg^{-1} body wt. of the test animal	Median lethal dose by the dermal route (Dermal toxicity) LD_{50} mg kg^{-1} body wt. of the test animal	Colour for the identification based on the class*
1.	Extremely toxic	1-50	1-200	Bright red
2.	Highly toxic	51-500	201-2000	Bright yellow
3.	Moderately toxic	501-5000	2001-20000	Bright Blue
4.	Slightly toxic	> 5000	> 20000	Bright green

BIS = Bureau of Indian Standards, *Fig. 8.3.

The label to be affixed on the package containing highly flammable insecticides shall indicate that it is inflammable or that the insecticide should be kept away from heat and open flame. In addition, in the United States of America, pesticides with low flash points are required to carry the following warning statements.

(*i*) *Flash point is 20°F (–6.7°C) or less:* Danger-extremely flammable. Keep away from fire, sparks and heated surface.

(*ii*) *Flash point between 20°F and 80°F (–6.7-26.7°C):* Warning—flammable, keep away from heat and open flame.

(*iii*) *Flash point is between 80°F and 150°F (26.7-96.7°C):* Do not use or store near heat or open flame.

TABLE 8.2. WHO CLASSIFICATION OF PESTICIDES BASED ON LETHAL DOSE

Hazard Category/ Class	LD_{50} rat (mg kg^{-1})			
	Oral		Dermal	
	Solids	Liquids	Solids	Liquids
la Extremely hazardous	5 or less	20 or less	10 or less	40 or less
lb- Highly hazardous	5-50	20-200	10-100	40-400
II - Moderately hazardous	50-500	200-2000	100-1000	400-4000
III - Slightly hazardous	> 500	> 2000	> 1000	> 4000

Note: As per the above WHO (World Health Organization) classification, the liquid formulations of the same pesticide are more hazardous than the solid ones. It has been worked out that oral dose of less than one teaspoon (0.5 ml) of category la and lb would be enough to kill a 70 kg man whereas, for the same result, 0.5-30 ml and 30-500 ml respectively of categories II and III will be required.

Label must not bear any unwarranted claims for the safety of the products or its ingredients i.e. statement like 'safe', 'non-poisonous', 'non-injurious' or 'harmless', etc. should be avoided.

Several countries of the world have modified label details to suit local requirements. For example, in Malaysia, coloured strip covering 20 per cent of the area of the label are to be printed on all the packages along with the cautionary note and symbol. In Coasta Rica, many posters illustrating the safe handling of pesticides showing symbols for skin, stomach, respiratory poisoning, etc. are displayed.

In order to avoid overcrowding of the label, it may be convenient to provide directions for use including crop information, etc. on a separate leaflet that accompanies each package, reserving the label for those items only, which are essential for immediate safety.

Label Specifications

Labels may fade or get detached. It necessitates specifications as to the quality of adhesives and also of the colour fastness and weather resistance of inks. Internationally accepted test methods such as ISO 1050 and BS 5609 may be used. Such tests involving special equipment should be carried out for 65 hours. Samples of ink painted on the substrate should be observed under worst conditions for at least 21 days, which is sufficient to indicate stability.

Labels for Regions of Low Literacy

In regions of low literacy, use of graphics, cartoons, posters etc. on the label has been reported to be effective. The message should be conveyed in a simple and recognizable form that explains the level of hazard to non-target pests and crops and method of application, etc.

For illiterate labour, pictorial markings are used to avoid use of language (Fig. 8.5). This ensures protection to handler as well as packaged goods.

Storage Pictogram

Activity pictograms

Advice Pictograms

Warning Pictograms

Fig. 8.5: Pictorial Markings

PACKAGING

Package means a box, bottle, casket, tin, barrel, case, receptacle, sack, bag, wrapper or other material in which an insecticide or its formulation is placed or packed. Packages of pesticides should be leak proof and exhibit good design.

Successful packaging needs good quality packing material because of the very toxic nature of the pesticide. Therefore, the pesticide formulator may guide the packaging manufacturer about the requisite packaging specifications. Defective packaging results in negative effects on sales due to additional labour cost, disposal problems and hazards from pesticide exposure.

Package Standards

International committees of experts have made following useful suggestions on the standards and specifications for pesticide package/container:

Package

A structure that contains or limits its contents. Packaging extends the life span of products by its protective action. It reduces the losses of the product at different levels of distribution. The faulty packaging would invalidate even a most stable formulation. In designing a package, one will have to consider the needs and requirements of the society, and ensure that these are met.

Functions of a Package

Major functions of a package are to contain, carry and dispense material. It should contain the elements of aesthetics, structure, technology, etc. Testing the package under vibrations for three-quarters of an hour in a shaker is considered equivalent to one thousand jerks in a truck ride.

Laws and Regulations

There are three broad areas of regulation

(*i*) *Weights and Measures*: Ensure the buyer the full value for his money.
(*ii*) *Adulteration*: Material and construction of the package should not affect the product in an adverse manner.
(*iii*) *Public Safety*: Ensure safety while transporting.

Existing rules cover six categories *viz.* product quality, public health and hygiene, safety, export promotion, transportation and consumer protection.

General Package Requirements

(*a*) Package should comply with the national standards and regulations in force and where required, with the international transport regulations such as those recommended by United Nations Committee of Experts on the Transport of Dangerous Goods (Reynolds, 1983).
(*b*) The shelf life of two years for a container and the product should be established. In the case of shorter life than this, it should be clearly shown on the package.

(c) Pesticide should be packed only in clean, dry containers designed to provide protection against deterioration, compaction, weather change or often spillage. Containers must withstand all anticipated levels of handling, storage, stacking, loading and unloading conditions and should not be adversely affected by change of atmospheric pressure, temperature and humidity.

(d) The inner surface of container or closure may be coated or lined with a corrosion resistant substance.

(e) Outer surface of the pesticide container must be constructed or coated with such materials, which resist corrosion or deterioration, and which will accept printed labels.

(f) Labelling should be positioned so as to be readily identified and should remain legible and fixed throughout the shelf life.

(g) Containers of specific design, which have met the required standards through tests performed for one specific product, must be retested if they are to be used with another product, or with a new formulation of the existing product. Similarly, used or reconditioned packaging should be used only when it can pass the test of being equal in performance to the original packaging.

(h) Inspection procedures should be strictly followed at the filling site.

(i) All liquid containers should have an ullage of at least 5% i.e. they should not be filled above 95% of their capacity.

(j) Closures and seems must be thoroughly tested for leakage and corrosion resistance. Containers with their closure in place must undergo standard drop tests such as ASTM D-775 of American Society for Testing Materials.

(k) Design of closure and containers must be tamper proof (by children).

Specifications of Packages for Different Groups of Products

Solid Products

Powder, dust or granule package up to 3 kg can be selected from ready made packages such as bags, pouches, canisters, cans, glass/plastic jars, etc. Bags and pouches should be leak proof through the bottom and sides. Generally, they are made of more than one ply of material. Inner ply will be a polyethylene film of 0.02 - 0.05 mm thickness that offers excellent moisture barrier and is resistant to attack of most chemicals.

Cans and canisters should be manufactured with leak proof bottoms and tops. Canisters are made with layers of paper to form body, which can be embedded with polyethylene or any other material like aluminium foil to develop necessary barrier qualities. Cans are usually manufactured from tin plates, which provide good environmental and chemical protection.

Glass or plastic jars are one piece containers comprising bottom and body. Although glass is not corroded by pesticide formulations, yet it is not commonly used because it requires special care during handling and transport. The plastic jars made of polyethylene are the most commonly used because they act as moisture barrier and possess shatter proof characteristics.

Large packages with 10-30 kg capacity may be selected from sacs, fibreboard or plastic steel drums or corrugated boxes. Sacs must be manufactured entirely from polyethylene film or from layers of papers and film to make them leak proof.

In case of fibreboard, plastic or steel drums of standard size, linings are used in the form of polyethylene bags either for protection from moisture or to reduce the contamination of drums.

Fibreboard drums are manufactured from layers of paper and have polyethylene or other barriers such as embedded aluminium foils. However, plastic or steel drums are more effective.

Packing requirements for specific solid pesticides are given in Table 8.3 (IS 8190, 1980 a).

Indian Standards Institution (ISI) Requirement of Packaging of Solid Pesticides (IS 8190, 1980a)

General Requirements

1. Solid pesticides, technical materials and their formulations shall be packed in a container in the quantities of 100, 250, 500 g, 1, 5, 20, 25 and 50 kg each.
2. Retail packs up to 500 g shall be further packed in corrugated fibreboard boxes or wooden cases when they require transportation.
3. High Density Polyethylene (HDPE) used in fabrication of HDPE containers shall be of designation 43 BA or 52 BA or 52 BB having melted flow index of not more than 1.8 g per 10 minutes.
4. Containers made of mild steel and tin plate shall be painted or layered externally to prevent rusting, if not lithographed. In case these are lithographed, the top and bottom components shall be lacquered externally.
5. Heat sealing of containers, machine stitching of bags shall be such that no powder comes out from the seams during normal handling.
6. Closure, whenever provided, shall be leak and tamper proof. The stitching threads on bags shall be sealed with metallic seals at both ends.
7. Double Hessian Jute Bags: Double walled tarpaulin laminated jute bags and HDPE woven sacs are used for packing of pesticides. Wherever allowed, these shall be provided with an internal loose polythene liner of not less than 0.062 mm in a tubular sheath, · bottom sealed. The length and breadth of loose liner shall be 10 and 5% more respectively than the outer bag.
8. The hand gloves should also be provided with the packs of 5 kg and above in case of encapsulated granules.

Safe Container for Overseas Shipment of WP Formulations

Water dispersible powder formulations require special packaging depending on bulk and nature of ingredients. Several containers ranging from the paper bags to fibre cases were tried for shipment of DDT, an extensively used insecticide globally, but its transportation to various countries was a problem. On the recommendation by World Health Organization (WHO) and the Agency for International Development (AID) shipment in bags was discontinued and only fibre drums were used. The difficulty in handling and high shipment cost led to the development of rectangular boxes which were strong enough to stack at least up to 8 boxes high. The boxes could fit together without any space in between them and withstand storage without degradation in humid climate. WHO and AID recommend similar boxes for the shipment of all WP formulations including those of malathion (Miles and Churchill, II, 1983).

TABLE 8.3. PACKING REQUIREMENTS OF SOME SOLID PESTICIDES/FORMULATIONS

S.No.	Pesticide	Requirement
Technical grade materials		
1.	BHC (Technical and refined), binapacryl, carbendazim, $CuSO_4$, mancozeb, DDT, endosulfan, phosalone, quintozene, thiram	Mild steel, tin plate or fibreboard containers of double hessian jute bag or high density polyethylene woven bags
2.	γ-BHC, lindane, dalapon sodium	Mild steel, tin plate or fibreboard containers with polyethylene liner of thickness not less than 0.062 mm
3.	Aldrin, endrin, diquat, propoxur	Mild steel, tin plate or fibreboard containers with polyethylene liner of thickness of not less than 0.062 mm
4.	Toxaphene	Galvanized iron container
5.	Heptachlor	Fibreboard drums
6.	2, 4-D sodium, 2, 4-D, MCPA	Mild steel drums, double hessian jute bag or double walled tarpaulin jute bags, high density polyethylene woven bags
7.	Ziram	Mild steel/fibreboard drums, lined with polythene of not less than 0.062 mm thickness
8.	Copper oxychloride, CuO (cupric oxide), Zineb	Double hessian jute bag, double walled tarpaulin, laminated jute bags or high density polyethylene bags
9.	Zinc phosphide	Galvanized steel sheet, tin plate or high density polyethylene and mild steel air tight container, mild steel drum lacquered with polyethylene lining of not less than 0.125 mm thickness
10.	Chlorpyrifos	Mild steel drums lacquered and with a polyethylene liner of not less than 0.125 mm thickness
11.	Coumafural, warfarin, warfarin-sodium	Galvanized steel sheet/tin plate/mild steel air tight container
12.	Carbaryl	Mild steel container, multi walled polythene coated paper bag or double walled jute bag or double walled tarpaulin, laminated jute bag or high-density polyethylene woven bags
13.	Dimethoate	Mild steel container lacquered with a polyethylene liner not less than 0.062 mm thickness
14.	Alachlor and chlormequat chloride	Air tight mild steel/jute bag containers lined with polyethylene of thickness less than 0.062 mm
15.	Sodium cyanide	Mild steel air tight receptacle drums with an opening, big enough to allow the contents to be easily removed, and having high density polyethylene lining of thickness not less than 0.062 mm
16.	Nitrofen	Mild steel container
Dusting powders		
1.	Aldrin, BHC, carbaryl, chlordane, DDT, endosulfan, fenitrothion,	Mild steel/tin plate/fibreboard containers, double walled jute bags or double walled tarpaulin, laminated jute bags or high

heptachlor, malathion, methyl parathion, phosalone, quinalphos, quintozene, toxaphenes — density polyethylene woven sacks

2. Copper oxychloride, Cu_2O (cuprous oxide) — Double hessian jute bag, double walled tarpaulin, laminated jute bag, high density polyethylene woven sacks

3. Pyrethrum, sulphur — Mild steel/tin plate/fibreboard container/double hessian jute bag or double walled tarpaulin, laminated jute bags or high density polyethylene woven bags

Water dispersible powder

1. BHC, DDT, diuron, mancozeb, quintozene, zineb, ziram — Mild steel/tin plate/fibreboard lined with polyethylene of not less than 0.062 mm thickness

2. Endosulfan, malathion — Mild steel/tin plate container suitably lacquered with polyethylene liner of not less than 0.062 mm thickness

3. Diazinon — Mild steel/tin plate containers

4. Thiram — Galvanized iron/mild steel/tin plated/fibreboard containers with polyethylene liner of thickness not less than 0.062 mm

5. Carbaryl — Polyethylene bag of thickness 0.062 mm inserted in suitable cardboard containers which are further packed in corrugated boxes

6. Copper oxychloride, Cu_2O (cuprous oxide) — Mild steel/tin plate/fibreboard container with polyethylene liner of not less than 0.062 mm thickness

7. Sulphur — Galvanized iron/mild steel/tin plated container with polyethylene liner of not less than 0.062 mm thickness

Granules

1. Blank — Single hessian jute bag/high density polyethylene bag

2. Alachlor, aldicarb, encapsulated butachlor, carbaryl, carbofuran, fenitrothion, fenthion, lindane, phorate, encapsulated quinalphos, trichlorfon — Jute bag, mild steel or composite container suitably and properly lacquered or lined with high density polyethylene of not less than 0.062 mm thickness

Other solid pesticides

1. Esters of 2, 4-D — Mild steel drum, fibreglass container may be used for packing upto 10 kg material

2. Thiram seed dressing formulation — Mild steel, tin plated or fibreglass board containers, double hessian bags or high density polyethylene box

3. Warfarin bait concentrate — Galvanized steel sheet, tin plated, mild steel, plastic or high density polyethylene air tight container

4. Warfarin Na salt WSP — Galvanized sheet, tin plated or high density polyethylene containers

5. Formulation based on stabilized ethoxyethyl mercury chloride concentrate — Mild steel or tin plated container having polyethylene liner of thickness more than 0.062 mm

6. Mercurous chloride, concentrate dalapon sodium salt — Metal steel or fibreglass board containers

Liquid Products

Indian Standards Specification for Packaging of Liquid Pesticides (IS 8190, 1980 b)

General Requirements

1. Technical liquid pesticides, and their liquid formulations, shall be packed in containers of capacity 100, 250, 500 ml, 1, 5, 10, 20, 25 or 50 litre. Technical grades and emulsifiable concentrates meant for bulk consumption may be packed in quantities of 200 litre also.
2. Retail packs can be further packed in corrugated fibreboard or wooden case if they require transportation (ISI, 1979c).
3. High density polyethylene used in fabrication of containers or its components shall be of designation HDPE 43 BA or 52 BB having melt flow index not more than 1.8 g per 10 minutes.
4. Aluminium used in fabrication of container should conform to grade 19,500 of 1.9 of Indian Standards.
5. Mild steel or tin plated containers shall also be painted internally to prevent rusting.
6. Closure provided shall be such as not to allow any material to pass through it.
7. Container holding five litres or more of liquid pesticide shall be provided with a pouring device.
8. The container shall be of sturdy design and conform to Indian Standards/BIS specifications.

Some requirements for packing of specific liquid pesticides are listed in Table 8.4 (IS 8190, 1980 b).

TABLE 8.4. PACKING REQUIREMENT OF SOME LIQUID PESTICIDES/FORMULATIONS

S.No.	Pesticide	Requirement
Technical grade pesticides		
1.	Diazinon, disulfan, nitrothion, fenthion formothion, oxydimeton, phenthoate, phorate, quinalphos, temephos, triometon	Mild steel drums, suitably and preferably lacquered from outside
2.	Parathion methyl	Mild steel drums
3.	Chlordane	Al/Fe container or phenolic enamel lined metal container
4.	Dichlorvos	High density polyethylene drums of 1.75-2.25 mm thickness
5.	Phosphamidon	High density polyethylene drums of 1.75-2.25 mm thickness
6.	Dicofol	Mild steel containers suitably lacquered, or anodized aluminium drums
7.	Monocrotophos	Mild steel containers suitably lacquered or high density polyethylene containers
8.	Butachlor, edifenphos, fluchloralin concentrate, tridemorph	Mild steel containers properly lacquered or aluminium containers

Emulsifiable concentrates

1. Alachlor, aldrin, BHC, butachlor, chlordane, DDT, diazinon, endosulfan, fluchoralin, heptachlor, malathion, toxaphene, triallate

    Tin plated mild steel containers properly lacquered from inside

2. Parathion methyl

    Leak proof aluminium containers with an ullage of 10% for 100-500 ml. One or 5 litre containers with an ullage of 5 per cent

3. Chlorpyrifos, dimethoate, fenthion, oxydemeton, phosalone, quinalphos, thiometon, tridemorph

    Mild steel, aluminum or tin-plated container, suitably and properly lacquered

4. Binapacryl

    Aluminium or tin plated lacquered containers

5. Dichlorvos, phosphamidon

    Mild steel or tin-plated or high density polyethylene containers which can be sealed with high density polyethylene plugs

6. Dicofol, propanil

    Mild steel or tin plated, properly lacquered, or anodised aluminium drum

7. Pyrethrum

    Aluminium, mild steel or tin plated containers

8. Phenthoate

    Mild steel or lacquered tin plated containers or aluminium containers

9. Paraquat dichloride

    Mild steel suitably and properly lacquered or high density polyethylene containers

10. Nitrofen

    Mild steel, aluminium or high density polyethylene containers

11. Ediphenphos

    Mild steel or properly lacquered from inside or aluminium containers

Other liquid formulations

1. Lime sulphur

    Mild steel container

2. Larvicidal oil EC, pyrethrum extract

    Mild steel or tin plated. Galvanized steel not used for packing pyrethrum extracts

3. Nicotine sulphate solution

    Properly lacquered mild steel or tin plated containers

4. Liquid amine salts of mono chlorophenoxy acids

    Properly lacquered mild steel, tin plated

5. Liquid amine salts of 2, 4-D, chlormequat chloride aqueous solution, monocrotophos SL

    Mild steel or tin plated or high density polyethylene containers

Note: If aromatic hydrocarbon solvents are used in making the formulation, high density polyethylene containers shall not be used for packing the formulation. Some liquid formulations such as acids will corrode the base of tin plate in storage. To prevent it, its inorganic nitrite salts are added as corrosion inhibitors. However, dimethylamine formulation of 2, 4-D has been shown to produce the potential human carcinogen, dimethyl nitrosamine, which can be prevented by the use of plastic lined metal containers.

The packing requirements for household pesticides as per Bureau of Indian Standards are given in Table 8.5 (IS 8190, 1979a).

TABLE 8.5. PACKING REQUIREMENTS FOR HOUSEHOLD PESTICIDES

S. No.	Pesticide	Requirement
1.	Zinc phosphide (Tech. 10 g)	Hermetically sealed trilaminated pouch, further packed in paper envelope
	(50 g)	Leak proof seamless, extruded rigid aluminium tubes packed in tin containers
2.	Ethylene dibromide (EDB) (0.5 ml)	Plain neck clear glass ampules either in absorbent media like cotton/blotting paper weighing 0.5 g, and then further packed in cloth and paper, fob weighing 1.0 g plugged with cotton at both ends, weighing 0.1 g and stapled at both ends. For 1 ml, the above weights are respectively 1.0 g, 1.5 g and 0.2 g. For 3 and 6 ml, 2 g, 2.5 g and 0.2 g. For 10 ml, 3 g, 4 g and 0.3 g
3.	Propoxur (2% bait)	Laminated pouch of 40 g glossy paper/0.02 mm or 0.0375 mm high density polyethylene and mouth sealed. Suitable number of smaller packs up to 100 g packed in fluted cartons

General requirements

For solids (up to 1 kg)	Polyethylene bags of thickness not less than 0.06 mm, and the mouth heat sealed and packed in cardboard cartons or high density polyethylene
For solids (more than 1 kg)	Mild steel, tin plated, fibreboard container, double hessian jute bag, or double walled tarpaulin lined jute bag or high density polyethylene woven sacs
Pastes (15, 25 g)	Collapsible aluminium tubes, internally plain or suitably lacquered and externally plain or lithographed
Liquids (25, 50, 100 ml)	Mild steel tin plated/high density polyethylene or glass containers
(Above 100 ml)	Mild steel tin plated, high density polyethylene, aluminium containers
Aerosols (100-500 ml)	Mild steel tin plated, high density polyethylene, aluminium containers. Hermetically sealed, monoblock, seamless aluminium rigid containers
Fumigants (IS: 8190, 1979 b)	Packs up to 1 kg/1 litre shall be further packed in corrugated board boxes or wooden cans. Only high density polyethylene of grade HDPE 43 BA, melt flow index 1.8 g per 10 minute shall be used. Mild steel, tin plated and externally lacquered or aluminium (grade 19500 of Indian Standards) containers and pilfer proof and leak proof closures to be used. Containers containing 5 kg or more of liquid fumigants shall be provided with a pouring device

Some specific requirements for fumigants are given below:

S.No.	Pesticide	Requirements
1.	Ethylene dibromide-Carbon tetrachloride (EDCT) (3 : 1 v/v) - (100, 250, 500 ml, 1, 5, 20, 25, 50 litre)	Galvanized iron or mild steel containers
2.	Ethylene dibromide (EDB) (100, 250, 500 ml, 1, 5, 20, 50 litre)	Mild steel containers, suitably lacquered from inside
3.	Methyl bromide	Welded low carbon steel, gas cylinder conforming to IS: 7682-1975, fitted with appropriate valve conforming to IS: 3224-2002.
4.	Aluminium phosphide	Leak proof, air tight, seamless, rigid extruded aluminium bottles containing either 10 or 20 tablets (3 g each) or 50 tablets (0.6 g each)

Packing Materials

Aluminium bottles for packing of liquid pesticides (IS 9503, 1980 C)

Aluminium bottles of the size 100, 250, 500 and 1000 ml come under this specification.

Requirements

(*i*) Ullage of 10 per cent for 100 and 250 ml and 5 per cent for all other capacities shall be provided over the normal capacity.

(*ii*) Neck finish: Bottle neck finish should be such as to provide a leak and pilfer proof closure system with flexible sealing plug and a sealing ring.

(*iii*) Sealing ring of polyvinyl chloride (PVC), rubber or any other material compatible with the contents of the bottle. The plug shall be made of polyethylene, PVC or other flexible material.

(*iv*) Wad material: Wads can be made of corkboard, pulpboard, rubber or any other suitable material. The wad thickness shall be such as to provide effective sealing. Wad facing shall be compatible with the product packed.

(*v*) Dimensions: Dimensions of different aluminium bottles are given in Table 8.6.

TABLE 8.6. KEY DIMENSIONS FOR ALUMINIUM BOTTLES OF DIFFERENT CAPACITIES

Capacity (ml)	Brimful capacity (ml)		Diameter	Height	Thickness
	Max.	Min.	(mm ± 2)	(mm)	(mm)
100	130	120	45.0	108	0.35
250	290	275	54.0	158	0.35
			58.0	135	0.35
500	600	550	63.5	225	0.45
			74.5	168	0.45
1000	1220	1080	80.0	250	0.45
			88.5	230	0.45

(*vi*) Tests

Leakage tests: Fill with coloured solution, close and keep it upside down for 24 hours on a white paper and examine for any leakage.

Stress, cracking, resistance test: Fill to 2/3 of its capacity with surface active agents like lissapol-DX, Teepol-TS, etc. Then fit with polyethylene plug and screw it tightly with the cap and keep it in the inverted position at 50°C for 24 hours. The plug shall not show any stress cracking or any other permanent deformation after the test is over.

Bottles are to be marked with manufacturers' name, initials or trade name.

Polyethylene Containers for Transport of Materials (IS 6312, 1980 d)

Polyethylene containers are used for transport of solid, semi solid or liquid materials and these are available in a wide range of capacities.

Note: Guidelines on the requirements for polythene containers may be obtained by consulting the following regulations:

(*a*) Ministry of Transport and Shipping regulations related to road transport and sea transport including the IMCO code for the transport of dangerous goods by road and sea.

(*b*) Restricted article regulations of the International Air Transport Association (IATA)

(*c*) Ministry of Railways Regulation

Classification (Type of Containers)

Type   I: Containers intended for transport of hazardous goods by air

Type   II: Multi-trip free standing containers. These are subdivided as:

      Type II A: Containers for surface transport of hazardous goods

      Type II B: Containers for transport of non-hazardous goods

Type III: Single trip free standing container

Type IV: Multi-trip container with protective outer layer. These are subdivided as:

      Type IV A: Containers for the transport of hazardous goods

      Type IV B: Containers for non-hazardous goods

Type V: Single trip container with protective outer surface. These containers are further sub-classified as:

      Class L: Container suitable only for liquids and semi solids

      Class S: Containers suitable for solids only

Requirements

(*i*) Closure: The closure shall be a screw type made from a material which is inert to the contents to prevent the leakage.

(*ii*) Neck size and thread form: The internal diameter of the neck of the principal closure shall not be less than 25 mm.

(*iii*) Handle: It should suitably be located in such a position that it may be used for both carrying and pouring.

(*iv*) Spout: It should permit the contents to be dispensed without leakage or undue loss due to spillage.

Units having capacity of more than 10 litre for transport of flammable liquids shall include a flame arrestor to be incorporated with the spout. It shall be a perforated metal screen of not coarser than one millimeter aperture.

(*v*) Tests: The various types of containers must be subjected to the tests as given hereunder:

Stacking test

Closure leakage test

Dry test, which includes:

(*a*) Drop test at ambient temperature for containers of Types 1, 2A, 3, 4B and 5

(*b*) Drop test at 0°C for containers of types 2A and 4A to be used at low temperatures

(*c*) Drop test at 18°C (for type I only).

Pressure-tightness test (class 'L' only)

Hydrostatic tests (class L only)

Handle strength test

Top load resistance test

Environmental stress cracks resistance.

(*vi*) Marking: Each container shall be legibly marked with the following information.

(*a*) Manufacturer's name or registered trade mark

(*b*) Nominal capacity of the container in litre(s)

(*c*) Classification i.e. type and class

Disposal of Used Pesticide Containers and Surplus Pesticides

In order to avoid various hazards, due care is essential in disposal of used pesticide containers and surplus pesticide. Decontamination of unutilized toxicant and proper disposal of the once used pesticide containers is important. Disposal of such materials without taking precautions may affect wild life, fish and other life forms. They also cause contamination of water, soil or air and damage crops, plants, etc. Therefore, safe disposal of surplus pesticides and empty pesticide containers is very important.

The pesticide containers on one hand should be very durable to offer adequate protection to the pesticide product during transportation over long distances and vigorous handling and storage under high moisture and humid conditions. On the other hand, these should be easily disposable after taking out the material.

Disposal of Toxicant

Deteriorated Chemical

The ways to dispose off a deteriorated chemical should be such as no danger to humans or contamination of the environment arises. The major means of disposing off chemicals include chemical destruction, incineration, soil disposal, etc.

Chemical Destruction: It utilizes specific chemical reactions to destroy the chemical. This may be oxidation by another chemical, like permanganate or chlorate or reaction with environmental forces of light and water, or alkaline hydrolysis, but disposal of the final waste product after the reaction is difficult. Use of safer, effective and economical methods involves photochemical and microbiological degradation in the environment. Use of costly xenobiotic chemicals to achieve the above effects may be avoided.

Incineration: Incineration or burning refers to use of specially constructed burners (furnace) that cause a flame temperature of over 900°C with adequate air intake. The chemical should be kept for a long time in the combustion chamber to ensure its destruction.

Soil Disposal: It is an effective method of disposing off limited amounts of certain organic pesticides. Here a combination of chemical and microbial degradations is involved. The site should be carefully selected so as to avoid problem of water contamination.

Surplus Pesticides

Disposal of large quantities of obsolete/surplus pesticide is problematic. The methods, which can be adopted, are as follows:

Incineration: It is burning and destruction of the chemical under controlled time and temperature conditions. It is to be carried out in suitably designed incinerators (fume hood or muffle furnace) as detailed above. It is, however, an energy intensive process.

Soil disposal: There are two different practices:

(a) Which utilize sanitary land fill where a pit is dug in soil and the chemical is disposed off in the pit. Some chemical additives such as lime and charcoal are used to accelerate the break down of the chemical in the pit. The practice is to add lime and the chemical in alternate layers and finally a layer of soil on the top. Hence the site for sanitary land pit must be so selected that the break down of the chemical would not be very rapid and that there will be no chance of leakage from the pit, which may contaminate the ground water. Also, the pits should be such that relatively small quantities of pesticide are disposed in each pit. They should be 10-30 cm apart, having depth of not more than two to two and a half meter and have a covering of half a millimeter of soil.

(b) Distributing the material over a large area incorporating into the soil to a depth of 5 or 6 inches to bring about microbial breakdown. Management of such an area by adding fertilizer and maintaining soil moisture accelerates the rate of break down.

Liming: Addition of lime provides an alkaline medium, which will hydrolyse the organophosphorus and carbamate pesticides to less toxic compounds. It also encourages microbial activity, which brings about further degradation of the metabolites. In case of chlorinated hydrocarbons, the alkali dehydrochlorinates the toxicants into less toxic products. If rapid break down of chlorinated hydrocarbons is desired, addition of animal manure or good organic compost and some fertilizers to create aerobic conditions is desirable.

Note: Other possible methods of disposal are carbon treatment and sand filtration systems, enzymatic degradation by the enzymes produced by microorganisms, disposal pit and evaporation beds, disposal in rivers sea and/or in deep wells, etc. However, these are not accepted due to the risk of environmental pollution.

Disposal of Pesticide Containers

In the case of bulk containers for transport of technical or formulated materials, it is not possible to clean the container efficiently, for reuse for other chemicals. Hence they should be dedicated to the transport of one or more closely related pesticides of the same group. If the drums are not to be used again, these should be returned to the manufacturer for reuse.

These are rinsed three to four times with appropriate solvent at least with one tenth to one fourth of the volume capacity. If it were emulsifiable concentrate (EC), then washing with water and pouring into the spray tank is possible. In case these processes do not work, then the container can be disposed off as per the guidelines given in the U.S. Federal Register Vol. 39 No. 85, Part IV, pp. 15234, 15241. It is suggested that the properly rinsed containers, if they cannot be reused or recycled as metal, should be disposed off in suitably designed sanitary landfills.

Pesticides and Container Disposal by the End User: Disposal of empty pesticide containers and prevention of their misuse, is one of the responsibilities of the person, enterprise or agency that receives and uses the pesticide. Also, the used pesticide containers are never completely empty and, therefore, a source of serious hazard to the people, animals and environment. Both the containers and the residues are dangerous to persons, especially to children and other livestock, pets and other animals. They may contaminate food, feed and the environment. Thus for the reasons of safety, pesticide containers should be cleaned carefully.

In case of containers of liquid pesticides, drain the container into spray tank in a vertical position for additional 30 seconds after the contents are poured out. Then rinse three times and drain into spray tank in a vertical position for 30 seconds. Use one fourth of the diluent for each rinse. This rinse and drain procedure is simple, quick and economic method which reduces the potential hazard of used pesticide cans and drums. This procedure will render the container ready for proper disposal, but not for reuse in any other process. No amount of washing can render the pesticide container sufficiently safe for storing edible commodities for any other safe use.

Lightweight single trip containers, which have been thoroughly drained and rinsed, should be crushed and buried at a safe location at the farm. Burial pits should be dug in clay or loam soil, in an isolated level spot at least 150-200 meter away from livestock feeding area, wells and streams so that water supply may not get contaminated. Pits should be deep enough to provide a soil cover not less than three fourth of a meter.

Combustible pesticide containers except those used for carrying hormone type herbicides can best be disposed off by burning, provided it is permissible under local laws and regulations. While burning pesticide bags, boxes, fibreboard drums, etc. care must be taken not to expose people and animals to the smoke or fumes from the fire. Combustible containers in which 2, 4-D, 2, 4, 5-T and other hormone type herbicides have been transported must not be burnt as the fumes may affect crops or plants grown nearby. Hence it is advisable to bury them in fallow pits.

Decontamination of Application Equipment

Sprayers may be washed immediately after use, otherwise the nozzle can get choked. The sprayer contaminated with pesticide should be decontaminated suitably depending on the pesticide in use.

(*i*) For organophosphorus compounds, a decontaminating mixture containing 100 g washing soda, 20 g detergent powder, and 50 g bleaching powder in 10 litres of water, should be taken in spray tank and shaken thoroughly. Allow mixture to stand for one hour and then wash with water.

(*ii*) For organochlorine compounds, it is slightly more difficult to clean them with water. The above mentioned cleaning mixture may also not be suitable. Hence substitute bleaching powder with ammonium hydroxide.

(*iii*) Decontamination of equipment after application of carbamates should be carried out in a manner similar to that described for organophosphate compounds.

(*iv*) Oil soluble formulations are best removed by treatment with half a litre of kerosene followed by washing with some detergent and treatment as above.

REFERENCES

BS (1986). Specification for printed pressure sensitive adhesive coated labels for marine use, including requirement for label base material. BS 5609, British Standards Organization, U.K.

Copplestone, J. F. (1982). *Education and Safe Handling in Envvironmental Science, 18.* Elsevier Scientific Publishing, Amesterdam.

Insecticide Act (No. 46 or 1968). Gazette of India, Part II Section 1, dated 2nd September 1978, p. 579.

PAI (1989). *Manual for Pesticide Users.* Pesticide Association of India, New Delhi, pp. 66-90.

ISI (1980 a). *Requirements for Packing of Solid Pesticides.* IS 8190, Part I, p. 12, Indian Standards Institution.

ISI (1980 b). *Requirements for Packing of Liquid Pesticides.* IS 8190, Part II, p. 10, Indian Standards Institution, New Delhi.

ISI (1979 a). *Requirements for Packing of Household Pesticides.* IS 8190, Part III, p. 8, Indian Standards Institution, New Delhi.

ISI (1979 b). *Requirements for Packing of Fumigants.* IS 8190, Part IV, p. 5, Indian Standards Institution, New Delhi.

ISI (1979 c). *Specification of Wooden Packaging Cases.* IS 1503, p. 14, Indian Standards Institution New Delhi.

ISI (1980 c). *Indian Standard Specification for Aluminium Bottles for Packing of Liquid Pesticides.* IS 9503, p. 15, Indian Standards Institution, New Delhi.

ISI (1980 d). *Indian Standard Specification for Polyethylene Containers for the Transport of Materials.* IS 6312, p. 18, Indian Standards Institution, New Delhi.

ISI (1985). *Specification for Welded Low Carbon Steel Gas Cylinders for Methyl Bromide Gas.* IS 7682, Indian Standards Institution, New Delhi

ISI (2002). *Valve Fittings for Compressed Gas Cylinders Excuding Liquid Petroleum Gas (LPG) Cylinders—Specifications.* IS 3224, Indian Standards Institution, New Delhi.

ISO (1975). *Continuous Mechanical Handling Equipment for Loose Bulk Materials—Screw Conveyors.* ISO 1050, International Organization for Standardization, Geneva, Switzerland.

Pesticide Association of India (1989). *Manual for Pesticide Users.* pp. 66-90.

GIFAP (1988). *Pictograms for Agrochemical Labels.* Groupment International des Association Nationales de Fabricants de Products Agrochimiques (GIFAP), Belgium.

Miles, J. W. and Churchill, II, F. C. (1983). Development of a safe container for overscas shipment of water-dispersible power formulations. In: *Psticide Chemistry, Human Welfare and Environment.* Vol. 4, *Pesticide Residues and Formulation Chemistry.* (eds. J. Miyamoto and P. C. Kearney). Pergamon Press, Oxford pp. 385-391.

Reynolds, Richard (1983). Labelling and packaging: The formulator's responsibility to the user. *In*: *Formulation of Pesticides in Developing Countries*. United Nations Industrial Development Organisation (UNIDO), New York, pp. 189-206.

World Health Organization, "Guidelines to the Use of WHO Recommended Pesticide by Hazard" (VBC/78.1 Rev. 2) WHO CH 1211, Geneva, Switzerland.

CHAPTER 9

Application

PRINCIPLES

Application implies distribution of the toxicant, as uniformly as possible, in a given environment, with a view to keep the pest population under check. The key factors governing the success of pesticide application on crops are (*i*) quality of the pesticide, (*ii*) timing of its application and (*iii*) quality of application and coverage.

A uniform pesticide application is vital for successful pest control. It depends generally on: (*i*) dosage, (*ii*) technique of dispersal, (*iii*) droplet size or spectrum, (*iv*) density of droplets, (*v*) crop density and allied aspects.

The pesticides are dispersed employing various types of nozzles such as hydraulic, pneumatic, centrifugal, thermal etc. Choice of droplet size depends on the type of pest, its location and weather conditions. Both wide and narrow spectrum of droplets can be used for pesticide application. When pesticides are applied on the plant surface in the form of droplets, they should be deposited in an effective form to tackle the problem at hand. Each of insecticides, herbicides, fungicides, nematicides etc. require a specific method and form of delivery for best results.

Choice of Pest-Control Chemical

For the control of insects having biting or chewing mouth parts, a stomach poison is usually recommended which may be distributed over the surface upon which the insect feeds. The materials need to be insoluble in water and capable of withstanding reasonably long periods of exposure to environmental factors without decomposition. In case of insects, which are not surface feeders, such as aphids, it is necessary to apply the insecticide on the surface on which they move. These materials are called contact insecticides, and may be water or oil soluble. Insect infestations in the stored products and in other similar situations in which direct application of toxicant is difficult, are usually treated by fumigation. Fungicides are recognized as eradicant or protective. The purpose of an eradicant fungicide is to destroy fungal organisms already established in a given location. Materials for this purpose are characterized by their ability to destroy fungi quickly and efficiently and are usually water-soluble. Protective fungicides, on the other hand, are applied in anticipation of fungus attack, and for this reason must be more resistant to natural forces. In case of rodent control with chemicals, the choice lies between fumigants and baits. Usually fumigants are preferred in areas that can be closed, while baits are useful in open areas.

Depending on mode of action, the toxicant has different requirements with respect to persistence. In all cases, however, an universal requirement is stability of the compound on or in the vicinity of the target. The persistence of the compound introduced into the soil should be higher but not exceed one season for herbicides or two seasons for agents to control soil-inhabiting pests. In such cases the toxicant should neither be absorbed by the plants nor accumulate in fruit or other parts of such plants which are edible by man or domestic animals. If treatment of the crop is carried out shortly before harvest, pesticide should break down completely at the time of harvest. The nature of breakdown products of the pesticide in soil, plant etc. is also very important to ensure the safety of men and animals.

Formulation *vis-a-vis* Application

The formulation usually consists of a solid or liquid carrier diluent combined with the active and other ingredients to provide requisite properties for application. The type and condition of application, target organism, the nature of pesticide itself etc. govern the requirements. Among the factors that must be considered in preparation of formulation are the method of application, factor of dilution of the active ingredient, safety in handling, reduction or loss through drift, vaporization or degradation and adherence to the crop or soil being treated.

Diversity in Formulations

Chapters one and two provide information on possibilities and types of different solid or liquid formulations. The vast diversity among the available products underlines a need to pick up the right formulation for a given situation.

Choice of Formulation

Formulation choice is mostly dictated by user convenience. Farmers with large tractor mounted sprayers fitted with hydraulic agitation prefer emulsifiable concentrates which can be poured into the tank straight from the can, particularly as the volume of such concentrates is easier to measure than to weigh out a powder. Nevertheless, in many parts of the world, the less expensive wettable powder is used, despite the need to form a thin paste before dilution. Pre-packing of required weights of wettable powders for knapsack or tractor equipment has helped to reduce the problem of weighing powders on the farm.

With powder formulations, particle size is an important factor. In general, micronisation of a formulation provides finer particles, which are more effective than coarse particles as contact poison. When stomach poisons are applied, surface deposits are effective against leaf chewing insects, but less so against borers which often do not ingest their first few bites of plant tissue. The effectiveness of stomach poison can be improved by addition of feeding stimulant, e.g. molasses, to the sprays.

Wettable powder formulations of systemic insecticides, fungicides and many herbicides are often used. Their uptake is sometimes increased by a suitable solvent or oil carrier, which can penetrate leaf and seed cuticles more rapidly. Control of scale insects on citrus and other crops is an example where addition of suitable oil improves the effectiveness. Addition of an emulsifier also enhances wetting of certain types of foliage and redistribution of pesticides.

Choice of formulation has often been dictated by availability of equipment in developing countries. Low percentage concentration dusts and granules can often be applied by hand or shaken from a tin with a few holes punched in it. On the other hand, the farmers may be reluctant to use granules where neither labour nor specialized equipment for spreading are available. Shortage of water in many areas has dictated the use of dust or granules, but higher transport costs have favoured highly concentrated formulations.

The losses due to drift, particularly with aerial application, are controlled by formulations containing thickening agents e.g. polysachharide gum, alginate derivatives, hydroxyethyl cellulose and various other polymers. Using larger droplets or granules can more effectively reduce drift.

Choice of formulation may also be determined by phytotoxicity consideration. Some plants or indeed individual varieties are susceptible to certain solvents and other ingredients or merely to the impurities present in cheap solvents. Persistence of a formulation can be improved by adding stickers but care must be taken to avoid protecting the deposit so much that its availability to pest is reduced. Persistence to rain washing can be improved in the formulations that are effective on foliage application. Rain fastness can also be achieved with fine particles, which are not readily washed off by rain. Advantages of small size and slow release are now being combined with the new micro-encapsulated pesticide formulations.

Timing of Application

Pesticides are frequently applied as a prophylactic treatment on a fixed calendar schedule irrespective of the occurrence or level of pest infestation. Such a policy is favoured, since forward planning is so much easier. The user knows exactly which chemical to purchase. The supplier knows when the delivery is required so that the stock is not carried over to the following year. Fewer applications are needed if they are timed more accurately and this will reduce selection pressure for resistance. A routine pest management is required, preferably aided by a pest forecast of the probable level of infestation to avoid fixed schedules.

Forecasting

Accurate forecasting of pest incidence will depend on collection of a vast amount of appropriate biological data and integrating meteorological data. At present, the amount of appropriate data is limited. The probability of infestation of *Aphis fabae* is predicted for various regions of the United Kingdom to assist the farmers to decide whether or not to spray field beans. In East Africa, data from light traps have been used to forecast outbreaks of the armyworm, *Spodoptera exempta* in relation to the movement of the inter tropical convergence zone. The success of such schemes depends on giving sufficient time margin to the farmers.

Economic Threshold

The population density at which control measures should be applied to prevent an increasing pest population from reaching the economic injury level *i.e.* the lowest population density that will cause economic damage. Such an approach requires considerable skill in interpreting data.

Sprays have been applied in relation to data obtained by monitoring a number of crops including cotton. Further use of the system has been limited due to shortage of data on damage by particular pest population at various stages of the crop development. Relatively simple techniques of monitoring pest populations and damage are needed if spraying according to an economic threshold is to be widely used. Various trapping techniques have been used for sampling populations. The use of pheromone traps as a means of timing sprays may be particularly important, owing to their selectivity and effectiveness when pest densities are low, but such traps may only indicate the likelihood of an infestation and scouting within the crop may still be necessary when most activity has been noted.

The pesticide should be applied early at the start of an infestation of first instar larvae to ensure effective control. A larger dose is required to kill later instars, even if the amount of active ingredient per unit weight of insect remains constant. Change in behaviour of morphology of the latter instars of an insect necessitates an increase in the dosage. Any natural resistance to the pest may enhance chemical control of the first instar larvae if applications are timed properly.

Application Sites and Placement

Scouting for a pest may show up particular foci of infestations in a crop and these can often be treated separately to reduce the spread of the pest and avoid the cost of applying pesticide to the whole area. In Egypt, pink bollworm (*Pactinophora gossypiella*) infestation is observed on the edges of fields near the villages where cotton stalks are stored for fuel. Red spider mite outbreaks are sometimes close to isolated trees within a field, partly because the treatment effects are often lower in such areas, especially after aerial application. These sections of a field marked out by scouts can be successfully treated on the spot with an acaricide. Sprays directed at the bottoms and apical foliage of chrysanthemums have been shown to give control of leaf miners without affecting the parasitoids and their hosts on the lower part of the plant and leaves, where they are protected by dense canopy of foliage.

Contact poisons can reduce natural enemies of the pest, immediately after application of these chemicals unless the dose applied is just sufficient or selective to kill the pest alone without affecting the predator. Systemic insecticides are less likely to affect natural enemies once they have penetrated plant tissues. When foliar sprays are needed, the effect of a non-persistent insecticide on predators and parasites can be reduced in plantations of perennial crops e.g. citrus and coffee by application to different sections on separate occasions. Natural enemies that survive in the untreated sections are given sufficient time to disperse into the area that is to be left unsprayed, when the remainder of the untreated trees are sprayed.

Dose Distribution and Coverage

The biological efficacy of a pesticide is influenced by the mean level of deposit (dose), distribution of the deposit and coverage of the target. Dose refers to total amount of pesticide used in treating a unit area of the target. There can be different targets for controlling different pests, which may be the pest itself (flying or settled, mobile or immobile, hidden or exposed), the plant or its specific part(s), the soil or another surface on which the pest rests or crawls. The nature of the pest to be controlled and that of the crop or animal to be protected determine the target to be treated. Coverage refers to the percentage of the target area receiving the pesticide, and distribution refers to the evenness or otherwise of the pesticide deposit on the target surface.

In most pest control situations, a discontinuous but an adequate coverage of plants is desirable. Adequate coverage implies a reasonable distribution of a certain minimum level of deposit. The density of droplets per unit area, the droplet size and amount of chemical are all important. Both the nature of the pest and the mode of action of the pesticide determine the requirements for the degree of coverage needed for an effective pest control.

Drop Diameter

Drops of various sizes (expressed as volume median diameter) are formed when spray liquid is dispersed through nozzles. None of the nozzles generates uniform sized droplets. A spectrum of varying sizes is created. Some of the droplets are large and others small.

TABLE 9.1. CLASSIFICATION OF SPRAY ON THE BASIS OF DROPLET SIZE

Droplet size (Microns)	Spray classification	Equipment required
400-1000	Coarse	Hydraulic sprayer
100-400	Fine	Mist blower
50-100	Mist	Micron sprayer
1-50	Fog	Fogging machine
0.001-1	Smoke	Smoke generator
Less than 0.001	Vapour	Vapour generator

Droplet Spectrum

The size range that defines formation of droplets of varying sizes is known as droplet spectrum. In order to ensure proper coverage for adequate pest control, a suitable droplet spectrum is a pre-requisite. Largest volume of spray is found in bigger droplets, which are few in number. Smaller droplets are proportionately more in number than larger droplets in any spectrum. Too large droplets roll down the sprayed leaf surface and drain as run off while very small droplets drift away with air current and cause pollution hazards.

Classification

The droplet spectra are classified as under (Table 9.2)

TABLE 9.2. KEY CLASSES OF DROPLET SPECTRA

Spectrum class	Droplet size (Microns)
Wide droplet range spectrum	500-1000
Medium droplet range spectrum	100-500
Narrow droplet range spectrum	50-100
Micro droplet range spectrum	1-50

Coverage

The droplet spectrum governs the coverage of various targets. The type of coverage in turn determines the quality of pest control.

(*a*) Entire coverage		Entire plant surface (ideal condition). Not possible in practice
(*b*) Specific coverage		
	(*i*) Contact insecticide	Full coverage, coarse droplets
	(*ii*) Fungicides	Full coverage, coarse droplets
	(*iii*) Herbicides	Coverage to the need, coarse droplets Preferably no drift. Dosage is exact
	(*iv*) Systemic pesticides	Fine droplets to be absorbed by the plants. Dosage is exact. Drift problem
	(*v*) Granular pesticides	Specific coverage, dosage is exact, placement at the target

Effect of Droplet Size on Coverage:

Large droplets—Placement on one side of the surface of leaves, run off losses
Fine droplets—Placement on both sides of leaves, drift losses and pollution hazard

Droplet Size and Number of Drops:

50-100 μm multra low volume pesticides	50-70 droplets cm^{-2}
100-300 μm low volume pesticides	30-50 droplets cm^{-2}
400-100 μm high volume pesticides	20-30 droplets cm^{-2}

Penetration/Deposition of Droplets in Air

The following relationship holds good to determine the distance(s) to which a droplet can penetrate through the air:

$$S = \frac{D^2 \rho V}{18N}$$

where, S is distance travelled in centimeters, D is diameter of droplet, ρ is density of droplet, V is the velocity of droplet and N the viscosity of air.

Suppose, $D = 50$ microns, $\rho = 1$ g ml^{-1}, $N = 1.8 \times 10^{-4}$ poises and $V = 5000$ cm sec^{-1}

$$S = \frac{50 \times 50 \times 5000 \times 1250}{18 \times 1.8 \times 10^{-4}} = 38.6 \text{ cm}$$

Let us take another example:

Suppose, D, the diameter of other spray droplets $= 200$ microns, $N = 1.8 \times 10^{-4}$ poises, $V = 2000$ cm sec^{-1}

$$S = \frac{200 \times 200 \times 2000}{18 \times 1.8 \times 10^{-4}} = 246.9 \text{ cm}$$

Hence the penetration distance $= 246.9$ cm

This has been concluded that the spraying distance is increased with a reduction in the spray angle and with an increase in the liquid output. The atomization of spray by compressed air is influenced by (*i*) low volume of air at its high velocity, (*ii*) low rate of output of the liquid, (*iii*) low pressure on the liquid. Higher deposits of insecticidal spray occur on a rough surface.

In order to increase the deposit on the target, a simultaneous combination of spray and dust is often blended. Driving a stream of dust at around 150-300 miles per hour carries out this operation. The dust and spray liquid are driven against each other with great force in a narrow outlet. It is advised to apply one gallon of liquid per 0.5 to 2 kg of dust depending on the efficiency of mixing apparatus and on the quantity of oil or non-volatile ingredients in the liquid phase. Dust up to 50 percent is well coated when two gallons of 20 percent of oil emulsion and 0.5 kg of dust is discharged per minute. Each particle of dust should be coated with oil equal to 20 to 35 percent of its weight. With decrease in size of particles, the velocity of air should be increased. However, the velocity of the air decreases with increase in the distance of the blower. In order to get better results, oils which do not volatalise rapidly and thereby reduce the size of the particles in air, should be used.

Following equation can be used to calculate minimum air velocity for maximum deposit of pesticides:

$$V_1 = 20,000 \, S/D^2$$

where, V_1 = minimum velocity in miles per hour

S = width of object in inches

D = diameter of particles in microns

Another equation used for this purpose, which is commonly employed in field operations is:

$$V_1 = 0.0000605 \times V_2^2$$

where, V_2 = velocity of air (mph)

The number of droplets, which fall on the target, depends on the diameter, the area to be covered and the velocity of air. Subsequently, the droplets suffer evaporation, the extent of which depends upon proportion of volatile and non-volatile liquids and solids, temperature, time of exposure of the droplets, velocity of air etc.

Droplets with diameter of 60 microns or less, have been found to show a loss due to evaporation of 75 percent or more. In case of the droplets of 100 microns or more diameter, the loss is around 80 percent. There is a tendency of large drops to work towards the outside of air stream and drift downward as velocity of air decreases. There are several layers of foliage to be covered from top to bottom. In order to achieve better coverage and penetration in the foliage, a low deposition efficiency on the first layer be maintained. In this way, deposition on the far side of the foliage can be obtained.

Velocity of air has to be efficient for atomization of spray so that good distribution and penetration are achieved from top to bottom of plants and underside of leaves.

The spray equipment should be checked and calibrated so that correct amounts of pesticides are applied. To do a good and safe job of spraying, it is necessary to check the operation and rate of application of the sprayer. As parts get worn out, the spraying becomes erroneous. Strainers and nozzle screens also become partially clogged and thereby the quantity of the spray is reduced. It is, therefore, necessary to check the operation and rate of application as per recommendation of the pump manufacturer for spray rates

Suppose, the width of swath = 100 inches

Length of swath = 1 mile

Area = 1 acre

Then, the number of acres covered per hour is given by:

$$= \frac{\text{Width of swath in inches} \times \text{mph}}{\text{Acres per hour}}$$

$$\text{Number of gallons acre}^{-1} = \frac{100 \text{ Gallons per hour}}{\text{Acres per hour}}$$

To test a nozzle, the sprayed liquid is collected in a graduated beaker or cylinder and its volume is measured. This should be comparable with the liquid that the nozzle should have delivered during a given interval of time.

Droplet Spectrum for Various Needs

Different droplet spectra are needed for the diverse situations in plant protection. Most of the flying insects require very fine and micro-range size spectrum i.e. 10-30 microns. Against locust and forest insects, a spectrum of 20-60 μm is considered suitable. Ultra low volume crop sprays provide a droplet spectrum of 70-150 microns. Low and medium volume sprayings provide a range of 150-220 microns. In case of high volume crop spraying and highly toxic chemicals, droplet spectrum of 240-300 μm may be used. In herbicide spraying a wide spectrum of droplets (400-1000 μm) is desirable to avoid drift hazards. Some droplet size spectra for different pest control situations are given in Table 9.3.

TABLE 9.3. APPROXIMATE DROPLET SIZE OF SPRAYS UNDER DIFFERENT USE CONDITIONS

Pest or use situation	Droplet size (Microns)
Flying mosquitoes and other insects	10-30
Locust hoppers	20-60
Forest insects	20-60
ULV crop spraying	70-150
Low volume spraying	150-220
High volume crop spraying	240-360
Highly toxic chemicals	250-400
Herbicide spraying	400-1000

The droplet spectra are generally classified as coarse spray, fine spray, mist, fog, smoke and vapour (Table 9.1) and are obtained by use of various types of application equipments, such as hydraulic sprayers, mist blower, micron sprayers, aerosol generators, fog machines, smoke generators, vapour generators etc. Selection of equipment has to be made depending on the technique of application and spray pattern desired.

EQUIPMENT AND MACHINERY

Success in control of insect pests and diseases is as much in proper selection and operation of equipment as with the use of correct toxicants in correct doses. Suitable equipment for application is needed particularly in view of the realization that rather small amounts of modern pesticides have to be applied over relatively large areas of land and/or crops. Selection of equipment poses a special problem these days since with the development of new chemicals, newer methods and techniques of their application have also been developed. Some old types of appliances are becoming obsolete. In recent years, the application of concentrated insecticides with low volume spray has gained popularity over high volume spray. The equipment requirements are also becoming unique as a single equipment is expected to be used for spraying and dusting or spraying, dusting and smoke generating and so on.

The plant protection machines can be broadly classified as follows:

1. Sprayers
2. Duster-cum-sprayer
3. Fog or smoke generator
4. Duster
5. Seed dressing machines
6. Fumigator
7. Flame throwers
8. Bird scarers
9. Miscellaneous

Sprayers

These are used to distribute small droplets of solution, emulsion or suspension, as evenly as possible, so that an adequate coverage of the surface to be protected is obtained with a minimum amount of pesticide. Atomization of a liquid in a sprayer is accomplished by several methods, the most common being by forcing a liquid by hydraulic pressure (e.g. rocking type or foot sprayers, ordinary power sprayers) or air pressure (compressed air sprayers, etc.) through a nozzle, or by the use of high velocity air streams striking either a jet or liquid or coarsely atomized liquid. In more recent type of sprayers (e.g. micron, microsol etc.), atomization of the liquid by the latter method is secured. Since in both the manual and the power operated sprayers, the atomization of a liquid is obtained through a nozzle, proper understanding of its function and construction is desirable.

The nozzle output depends on the orifice size, tank pressure and speed of operation. In a manually operated hydraulic spray, the rate of application of fluid (per unit area), depends on the nozzle aperture, the swath width and speed of the operator. The following mathematical equation is generally used for a good approximation:

Application rate $(1 \, m^{-2}) \times 10^4$ = Nozzle output $(1 \, min^{-1})$/Swath width (m) × speed $(m \, min^{-1})$

Nozzles

Various types of nozzles are used for dispersing pesticide sprays. Basically, the nozzles used in plant protection have three main functions viz. (*i*) to break up the spray liquid into drops, (*ii*) to spread the droplets on to the target, and (*iii*) to measure the dispersed spray liquid.

Nozzles are generally classified and named after the energy used to form the droplets, e.g. the hydraulic nozzles, gaseous pneumatic nozzles, centrifugal nozzles and thermal nozzles.

(i) *Hydraulic Nozzles:* These are mostly used in spray where large quantity of water is used. The hydraulic pressure determines the rate of discharge through atomization of droplets. These nozzles produce coarse droplets. Different types of nozzles are used for various requirements or purposes.

(a) *Impact Nozzles or Flood Jet Nozzles (Fig. 9.1):* These operate at low pressure (0.5 to 1 kg cm^{-2}) giving a coarse spray with no drift, as used for spraying herbicides and fertilizers. These are also supplied with wide angles, which are convenient for blanket spraying.

(b) *Flat Fan Nozzles (Fig. 9.2):* These are available with different discharge rates (500 to 3000 ml min^{-1}) and different angles ($60°$ to $110°$). The one most commonly used delivers 1500ml min^{-1} at 3 kg cm^{-2} and will have an angle of 80 degrees. The pressure gives moderately coarse spray with little drift. These are, therefore, especially suitable for booms because they will give a better overlap and distribution. The nozzles are also suitable for applications over water surface (e.g. anti-malarial operations) because the spray does not disturb the surface of water.

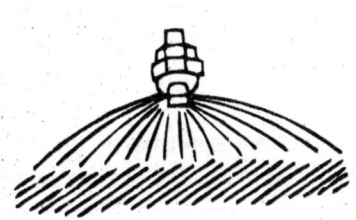

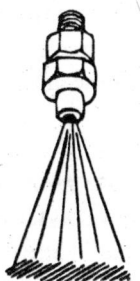

Fig. 9.1: Flood Jet Nozzle   **Fig. 9.2:** Flat Fan Nozzle

(c) *Cone Nozzle:* Generally, hollow cone nozzles rather than solid cone nozzles are used in agriculture (Fig. 9.3 and 9.4). These are cheaper and provide a large quantity of smaller droplets, which makes them quite suitable for spraying pesticides. Also, these penetrate well inside the vegetation. These are used most widely because of their low cost and reasonable quality, if used properly and carefully.

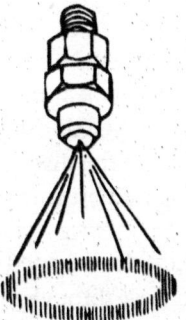

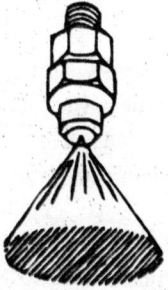

Fig. 9.3: Hollow Cone Nozzle   **Fig. 9.4:** Solid Cone Nozzle

(ii) *Gaseous/Pneumatic Nozzles:* These nozzles are commonly used with the mist-blowers. The atomization occurs by the impact of air blast meeting the spray liquid. To achieve good atomization, the volume of air discharged should be at least 1000 times more than the volume of liquid discharged. Transport of the particles is made by the air blast. A calibrated orifice in the liquid feed line accomplishes metering of the system. These nozzles provide fine droplets suitable for low volume spraying. The size of the droplets increases, as the flow of the liquid increases. The mist blower is always operated at full speed and the flow of air being maximum always, the ratio of the air blast divided by the flow of liquid is a variable factor.

(iii) *Centrifugal or Rotary Nozzles:* The spray liquid is fed into a rotating disc. The centrifugal force spreads the liquid and disintegrates it into droplets. This nozzle is used for spinning discs and other ULV sprayers. As the diameter of the droplets is directly proportional to the speed of the disc, fine droplets are obtained with this nozzle by rotating the disc at high speed. The size of the disc governs the tangential speed (speed of the points on the edge of the disc), which increases with the diameter of the disc. Since these nozzles do not throw the droplets to the target, a fan can be used for this purpose or the action of the crosswind will determine the direction and swath of the droplets.

(iv) *Thermal or Hot Tubule Nozzles:* A fog composed of very fine droplets can be produced by condensing a pesticide, which has been injected into a stream of hot gas to shear the liquid into droplets which are immediately vaporized. These vapours condense in very fine droplets, which float in the air. This air charging technique can be used successfully in green houses or in closed rooms for the control of flying insects.

These nozzles can be used alone or in combination depending on the requirement. Most of the hand sprayers use only one nozzle mounted at one end of a hollow metal tube called rig. The rig contains a suitable valve for regulating the flow of the liquid (Fig. 9.5). The other end of the rig can be connected to the sprayer tank with the help of a pressure tubing or hose.

Types of Sprayers
(i) Pneumatic Sprayers:

(a) *Hand Compression Sprayers:* Spray liquid is filled in tank and the air pressure is raised with the help of air pump. After the solution is discharged through the nozzle, the air is to be released for refilling the tank. Pressure drops as the discharge takes place. Occasional air pumping is necessary to build up the required pressure.

(b) *Pressure Retaining Sprayer:* Air is filled first and liquid is pumped with the help of charge pump. The lance has a regulator for uniform discharge through the nozzle. When discharge of liquid is completed, only liquid is to be fed with the help of charge pump. Proper discharge and droplets are maintained throughout the operation.

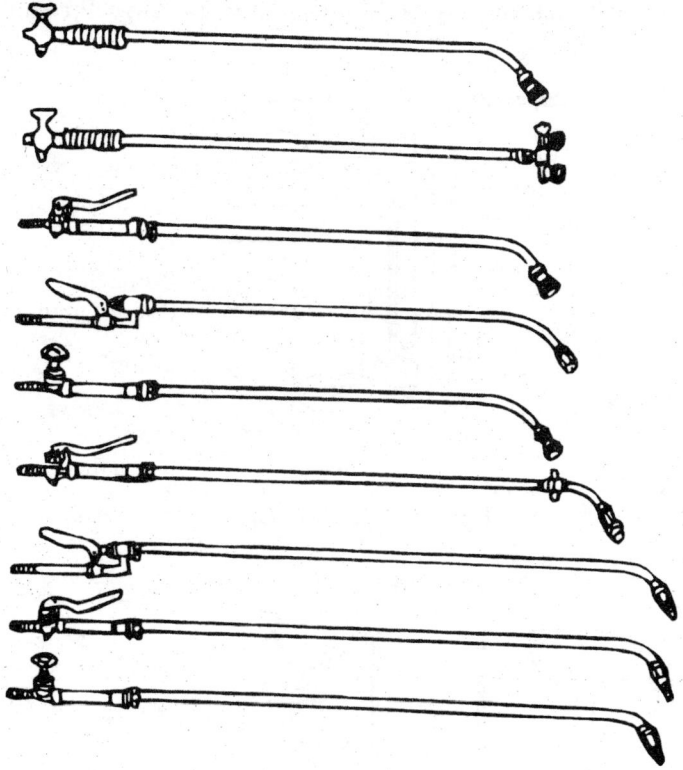

Fig. 9.5: Some Rigs for Agricultural Sprayers

(ii) Hydraulic Sprayers: These use fluid pressure for operation. Some of the important designs are described below.

(a) Continuous Knapsack Sprayer: It has a plunger or diaphragm pump mounted either on the outside or inside of the container and immersed in the liquid (Fig. 9.6). The pump is operated by hand lever, either under arm type or over arm type. Occasionally solid piston type pump may be used instead of plunger pump with a leather cup. The tank is not pressurized and continuous pumping is necessary to have uniform discharge through nozzles. The lance is provided with shut off regulator.

(b) Foot Sprayer: It is similar to the hydraulic pump, but the foot of an operator operates it (Fig. 9.7). It has suction and one or two delivery tubes, lances and nozzles. The equipment requires 2 to 3 operators. Continuous pumping is necessary to have uniform discharge of droplets. Pressure chamber is very strong to maintain high pressure.

(c) Rocker Sprayer: It is similar to foot sprayer but is operated by a long hand lever. It may be either single or double acting. Two operators are necessary to attend this work.

Continuous pumping is necessary to have uniform discharge. Pressure chamber is also provided.

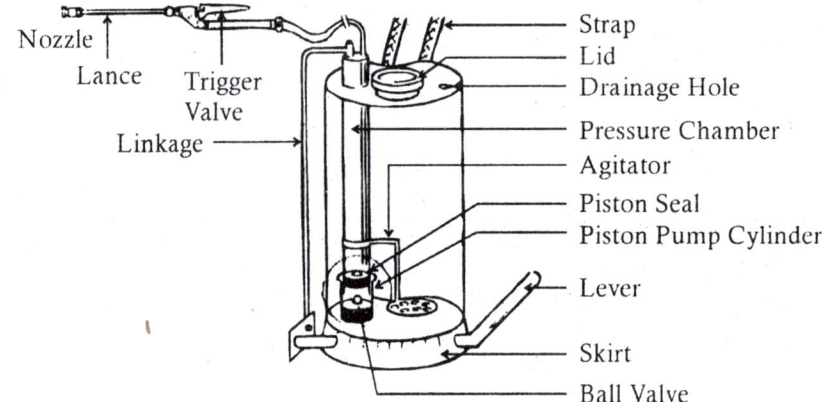

Fig. 9.6: Knapsack Sprayer

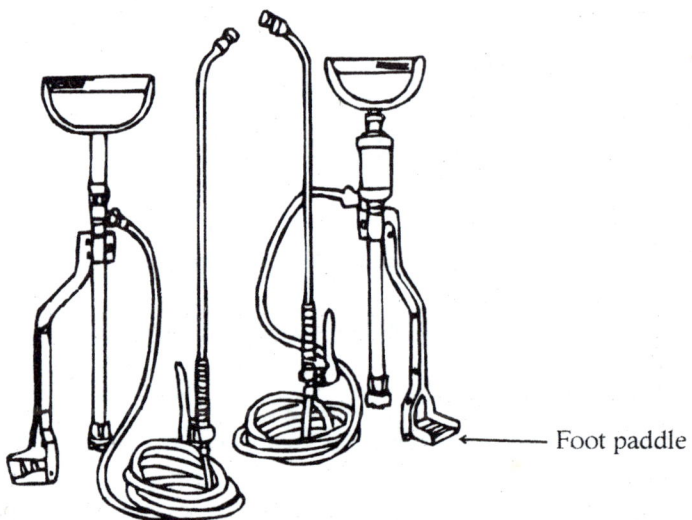

Fig. 9.7: Foot Sprayers

(d) *Stir up Pump (Bucket Sprayer):* It is hand or foot operated hydraulic pump, the inlet of which can directly be dipped in a bucket of the liquid formulation. Continuous pumping is necessary for getting continuous discharge. Small pressure chamber can be also provided for continuous and constant pressure.

(iii) *Power Operated Sprayers:*

(a) *Mist Blower:* A two stroke petrol engine drives the blower, part of which pressurizes the chemical tank. The inflow blows the air through the sprayer nozzle while solution flowing from the liquid tank gets atomized.

(b) ***ULV Sprayer:*** Spinning disc or any centrifugal nozzle produces very small droplets, which are carried on to the target with the help of wind or air blast.

(c) ***Traction Sprayer:*** The wheel of the sprayer dragged by bullocks or tractor drives the pump. It may be pneumatic or hydraulic. The wheel must have a diameter as big as possible so as to overcome the draught. These are used for field crops.

(d) ***Power Sprayer:*** Power sprayers generally include hydraulic pump. A 4-stroke petrol engine supplies the power. It has one or two delivery hoses. The pump is generally piston type in which the pass line is provided to regulate the pressure. It is useful for spraying trees and orchards.

(e) ***Tractor Mounted Sprayer:*** The pump is driven by PTO shaft of the tractor. the sprayer may be mounted to power lift or kept on trailer. Liquid tanks are mounted on the tractor and long spray booms are adopted for field crops. Fig. 9.8 shows some of the rigs (lance) and booms used for field spraying using power operated sprayers.

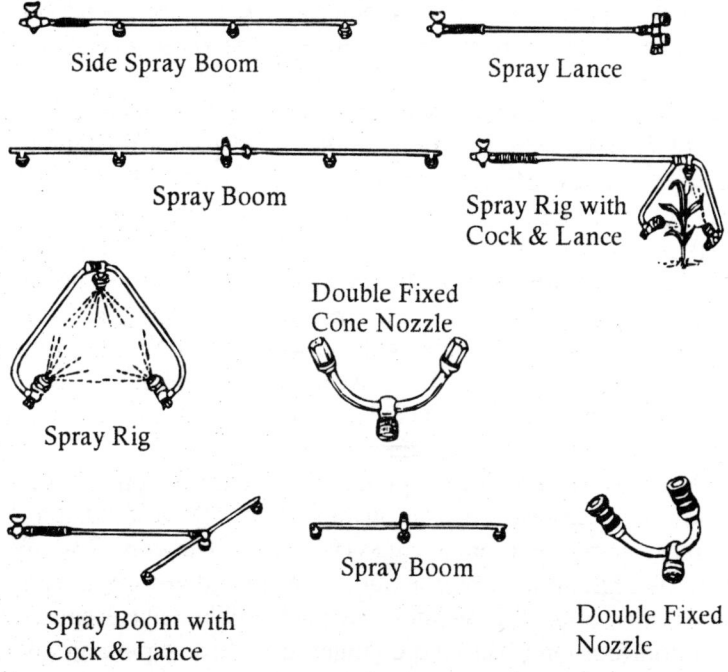

Fig. 9.8: Some Booms and Rigs for Power Sprayers

Duster-Cum-Sprayer

It combines the features of both sprayer and duster and is called spray-duster. By providing hopper for the dust and a pump with suction and delivery hoses, the duster can work as a mist sprayer. The blast created by the fan is used for blowing the liquid as well as the dust simultaneously with the advantage of wetting the dust, which settles down on foliage and is not easily blown off or washed away. Working as sprayer alone, the spray dusters work on low pressure (50-60 lbs inch^{-2}) but have high emission rate of 6 to 10 gallons per minute.

Knapsack Duster cum Sprayer

This type of dusters are light, generally powered by engines of one or two HP and can be carried on the back of the operator. The frame is provided with shock proof cushions which fit comfortably on the back by means of shoulder straps. The discharge is about one to two pounds of dust per minute. These dusters can also be converted into mist type sprayers or foggers by replacing its hopper by a tank carrying 1 to 1.5 gallons of the spray solution. The liquid is carried through a tube connected with the tank, by gravity or by air pressure drawn from the fan and is discharged at the opening of the delivery pipe from which it is blown outside by the air blast supplied by the fan. Such type of sprayers and dusters can be operated by one man, and are useful for operating in areas where other types of machines cannot be easily carried. A spray duster is specially suited for low volume spraying and can treat about 10 acres a day.

The conventional hydraulic sprayers are inefficient in delivering the pesticide spray on the undersurface of leaves, with less drift. This led to attempts to use an electrostatic charge on droplets.

Spinning Disk Sprayer is used for low volume and ultra low volume applications. These sprayers consist of a reservoir for the spray liquid that dispenses the liquid into the lancer through a hose by gravity flow. The lancer consists of the spinning disc, which functions as the nozzle, a battery compartment to house 4-8 batteries and an electrical switch. The reservoir fitting is optional as a shoulder-mounted reservoir or a one-litre bottle attached with tip of the lancer is also available. The 12 volt DC motor powered by two batteries energizes the spinning disk to the speed ranging up to 10,000 rpm. The liquid flowing by the gravity through the restrictor to the spinning disc is held above the crop downwind of the operator so that air turbulence distributes the spray within the crop canopy. The liquid dispensed at the centre of the spinning disk reaches the circumference. The spray liquid is delivered out of the spinning disc as ligaments, which disintegrate into final particles.

In many of the earlier versions, disc is plain with uniform circumference. In the modern spinning disc sprayers, the grooves are made on the surface of the disc radiating from the centre to the circumference ending in toothed edges (Bals 1970, Clayton 1992). While a smooth edged disc produces a fine spray, a combination of teeth and grooves gives better control of droplet size for a wider range of flow rates (Matthews, 1996). The droplet size and droplet spectrum depend on rotation speed and configuration of the spinning disc, the dynamic viscosity and surface tension of the spray liquid and the convection air current prevailing in field.

Control droplet application (CAD) is the development based on the need for particular droplet size within the narrow limits (Bals 1975). The narrow droplet spectrum has been achieved by use of spinning discs. Increasing the rotation speed of the disc can decrease the drop size. These sprayers enable application of pesticide at the rate of 5-15 l ha^{-1}. These are capable of producing a droplet spectrum of 50-100 microns (Matthews 1998, 1999). In many cases electrostatic charge is added to spray from a hydraulic nozzle or spinning disc by induction, ionized field or by direct charging of the spray (Merchant, 1980; Matthews, 1989). Law (1980) developed an induction charging system for an air-shear nozzle for use in tractor mounted booms and a hand held portable lance line system (Giles *el al.*, 1995). The technique utilizes a high voltage generator powered by four 1.5 volt batteries to charge the nozzle upto 25 KV. Emerging under gravity, the charged liquid forms regular ligaments and produces a very narrow droplet spectrum. At any fixed voltage, increasing the flow rate increases the droplet size.

The limited number of pesticides that could be formulated and used at less than $3 \, l \, ha^{-1}$ and the required resistivity are some of the factors that prevented commercial development of the system. While electrostatic charging of the fine sprays can reduce spray drift, the distribution of the charged spray within a crop may be unsatisfactory. Charged droplets are deposited on the nearest surface. So when a crop canopy is thick, there is a little penetration to the lower foliage. Using air assistance with electrostatistically charged sprays improves penetration. Use of 18 μm droplets in combination with $2 \, m^3 s^{-1}$ of airflow and air assisted cone nozzles, improved the leaf coverage and deposition on the cover.

Aerial Dusting and Spraying

Dusting and spraying from the air on crops against pests and diseases has become popular these days. Small, fixed wing single engine aircrafts are fitted with spraying or dusting equipment. As a sprayer, such an aircraft has a tank with a capacity of 40 to 190 gallons, a delivery control valve fitted in cockpit and a spray boom with low volume nozzles or other types of spraying attachments. As a duster, it has dust chamber of 500 to 2000 pounds capacity with a wind propeller type of agitator, a delivery slot with control valves and vent for air currents past the delivery slot designed to concentrate the flow. The flow of spray liquid and dust into the delivery system is generally by gravity or sometimes by compressed air. Now a days, airplanes have been specially designed to suit spraying or dusting of crops.

Helicopters have also come into prominence for crop spraying these days and have been found suitable, especially for small fields. An airplane can cover 200 to 1200 acres per day, depending on the type of plane used and the prevailing weather conditions.

Fog or Smoke Generators

Such type of machines have been developed to disperse the toxicant material as extremely fine particles (1 to 50 microns in diameter), which remain air-borne for a long time. This type of equipment can be used successfully where the foliage is dense and fog or smoke takes time to settle and does not drift with the wind. A fog machine consists of a tank, with or without an engine, and thermal or mechanical arrangements to break up the spray solution into desired droplet size and create a force for carrying fog from the machine. The units are mostly skid type to be mounted on a truck although those mounted on wheels are also available.

Dusters

Dusters function by producing air blast that is passed on to the dust chamber causing the dust to blow out through a delivery tube. Mostly, home gardeners and pest control operators use these in structures. Dusters may be hand or power operated. Hand dusters may consist of a squeeze bulb, bellows, tube, shaker, sliding tube or a fan powered by a hand crank. They have the advantage of light weight and provide dust with good penetration in confined spaces. The limitations are high cost, poor foliage adherence, drift and difficulty in directing dust to target.

Plunger Duster
It is made up of air pump, dust chamber and a discharge assembly. The duster is held in one hand and pumped with the other. Air blast creates dust cloud and passes it through the delivery outlet.

Bellow Type Duster
It has a small container for dust that is fed to the air stream producing dense cloud. Operating bellows create the air stream.

Push Duster
A small duster which can be pushed between the rows of crops. A perforated drum contains the dust. The dust falls while the drum rotates on wheels.

Rotary or Fan Duster
It consists of a fan geared to a hand crank and hopper holding the dust. The high speed gear moves the blower and the dust cloud enters the atmosphere through a spreader nozzle. Wind from fans carries the dust on to the target.

Hand Shake Duster
A small duster with perforations sufficient to allow the dust to enter the atmosphere and be carried by the wind to cover the basal portion as well as top surface of the crops.

Wet Dusting Equipment
In such equipment, the dust while passing out joins the spray liquid to form a wet dust. This type of machine is used in semi arid zones where water is a problem. Mist blower with dusting attachment can be combined to obtain wet dusting. Hand rotary duster and hand sprayers can be combined to make wet dusting equipment for small farmers.

Power Operated Dusters
A power-dusting machine comprises of a hopper containing one or more agitators, and a regulator to control the discharge dust. A fan or a blower produces an air blast for drawing dust from the hopper and blowing it out. The flexible pipes fitted with nozzles or field boom help in distributing the dust evenly on the crop. The hopper is provided with a mechanical agitator, which prevents caking of the dust, and assures a constant and uniform feed to the fan. The dust regulating device can also adjust the dust flow at any desired rate between 2 to 20 pounds per minute.

The fans used are mostly centrifugal type with the fan case having one or more outlets. The power of the engine in field power dusters varies from 1 to 6 HP. The power dusters are mounted on trolleys for carrying the frames. The heavier types of dusters, which are without trolleys, can be mounted on tractors or trucks or animal drawn high clearance carts.

Seed Dressing Machines

These are available both as hand driven and also engine powered models. The hand driven models carry a drum fitted on a stand, which can be rotated or turned upside down by means of a handle. For thorough mixing of seeds with the chemical, the drum is provided with small iron plates fixed at right angles on the inner surface or fitted horizontally with a wire netting. Such type of contrivances ensure greater disturbance in the seeds, which get uniformly coated with the chemical. Thirty to forty revolutions of the drum will mix the seeds and the chemical satisfactorily. The capacity of the drum varies from 20 to 60 kg. A hand seed dressing machine can treat 200 to 800 kg of seed per day, depending on capacity of the drum.

The engine powdered seed dresser can handle larger quantities of seed grains. About 4 tonnes of seed can be treated per hour in this type of power seed dresser and it is suitable for large farms.

Fumigators

Rat Fumigation Pumps

These are plunger type of dusters fitted with an air pump, a dust chamber made either of glass or metal of one pound capacity, and a discharge tube. It is provided a foothold for conveniently holding the pump while dusting. The delivery tube is sometimes provided with a small metallic tube at the end for inserting it into the rat hole. It is used for blowing dust into the burrows to give out poisonous gas to kill rats and porcupines.

Soil Injectors

Small hand operated soil injectors with a capacity of about two pounds are used for fumigating soil to a depth of five to seven inches for controlling soil pests, particularly root knot worms. The equipment is designed to ensure an accurate application of the prescribed dosage in the fields. For treatment of large areas, power driven trailer fumigators having cultivator type tines attached to them, are used. The hand operated soil injector machine can treat about an acre of cover in one day.

Flame Throwers

These are compressed air sprayers containing kerosene oil for producing the flame. The machines are commonly used to eradicate obnoxious weeds or swarms of insects from cropped or non-cropped land. It consists of a fuel tank with a lance modified to carry a burner to heat the oil. The flame thrower is used for burning settled swarms of locust and scrub weeds.

Bird Scarers

The equipment is a mechanical device to produce sound repeats at regular intervals to scare away birds and wild animals, which destroy the crops. The explosion results from the combustion of acetylene gas generated from calcium carbide, stored in such a way that water drips from a small container over calcium carbide kept in a separate chamber. The flow of water can be regulated which also controls the rate of explosion. The mechanical bird scarer may be timed to give two reports in one minute or only one in five minutes according to requirement. One such unit when suspended on a pole or a tree branch can cover three to five acres. Bird scarers are light to carry and are available in several designs. These should not be used at one place for long because the birds and animals get familiar and acclimatized, are then no more scared.

Miscellaneous

Following miscellaneous equipments are used along with those already described.

Bait Containers

These are used for placing bait and keeping it out of reach of the children and the animals. Different types of bait containers are being used but cutting an empty kerosene tin or a similar container could make an easy and effective one.

Rat Traps

Several types of mechanical devices for trapping rats are in market. Cages made of wire or wooden box with an instantaneous trap door, are available. A perforated box with door closing device or the one with spring board type inlet are commonly used in houses.

In fields, different locally made trapping devices are used, which are made by arranging pegs, bamboo sticks, bow, earthen pots, strings, rubber pieces etc. in such a way that rats fall an easy prey to the trap. Some of the rat traps used in the paddy fields in India consist of a bow or earthen pot resting on wooden pegs. These too vary in kind of arrangements in the bow type. The one used on coconut trees can be hung while those kept in the tapioca or paddy fields can be pegged. Some traps used in Bihar, have a wooden nail which when released pierces into the body of the rat and kills it. These rat traps do not cost much but require precision in fixing. Only one rat at a time is caught or killed in such contrivances.

Recent Developments

There have been significant advances in the field of pesticide application with the development of electrostatic spraying that has accompanied a greater use of laser and computer to measure spray droplets. More complex electronic equipment is now available to monitor and control application of pesticides. The use of controlled droplet application with tractor mounted spinning disc sprayers is on the increase.

Electrostatistically Charged Sprayers

A system of charging water and oil based sprays on a spinning disc which results in significantly increased deposition on upper and lower leaf surfaces is now well accepted. The electro dynamic nozzle uses high voltage, usually 16-30 KV, to charge pesticide formulations with a resistivity in the semi-conducting range up to 10^8 Ohm m^{-1} as it is metered through a narrow gap. An electric field using counter electrode, connected to the earth through a trailing wire, surrounds the nozzle so that a divergent electric field causes the liquid to form ligaments. Droplet size is adjusted from 40 to 200 μm by voltage and flow rate control affecting the number and thickness of ligaments. The uniformity of ligaments ensures a narrow droplet spectrum.

A dynamic cloud of charged droplets influenced by the electric field between the nozzle and the nearest target propels the droplets faster than their terminal velocity, so that the risk of drift of small droplets is greatly reduced. Excellent coverage of plants can be achieved, but penetration through dense canopies is limited. A plastic annular electro dynamic nozzle fitted to a bottle provides a disposable 'nozzle' container, ensuring accurate flow rate and droplet size with special formulations. The system eliminates the need to measure and dilute concentrated chemicals.

Controlled Droplet Application

Multiple discs are mounted on an electrically driven horizontal shaft fitted on a mast to accommodate higher flow rates and faster tractor speed. The mast can be raised or lowered above the crop, depending on wind strength and position of the sprayer in the field, in order to achieve overlapping swaths. Spray deposition, aided by air turbulence, is mostly within 30 metre down wind. 'Micro drop' is an alternative to multiple discs on the herbicide sprayers. The 'Micro max' is a single, large spinning cup designed to produce 250 μm droplets with flow rate of one litre per minute at 2000 rpm. Smaller droplets can be obtained at higher disc speed (5000 rpm) if flow rates is to be maintained. These units have been mounted on high speed self propelled vehicles for rapid application over extensive area and also on a small wheel driven unit for small scale farm use.

Droplet Sizing

Image analyzing computers can now speed up measurement of spray droplets collected on artificial surfaces, and electrostatic samplers to collect small droplets in the field are available. Measurement of droplets in flight is now possible with two laser systems. One measures the energy pattern on a special detector resulting from the diffraction of light caused by droplets passing through a laser beam within one focal length of the ions. The best fit of the data to a mathematical distribution is used to calculate the volume spray in each droplet size range. A system of measuring the shadow of droplets in a laser beam is also used. Both systems can provide data rapidly, allowing more detailed investigation of different nozzles under various operating conditions.

Care and Maintenance

General Maintenance

Following precautions are commonly desirable for longer life of costly application equipment:

(*i*) Clean outer surface with brush or cotton wads by using kerosene oil or water.
(*ii*) Apply lubricating oil or grease to the moving or rubbing surfaces of parts.
(*iii*) Filter or strain the chemical solution/fuel oil mixture while pouring into the tanks.
(*iv*) Flush the equipments with clean water to wash the inner parts of containers, tubes and the nozzles to make them free from chemicals.

Care and Upkeep of Hand Sprayers and Dusters

(*i*) Dry and sieved dust should be used for dusters.

(*ii*) Grease the duster's gear box once a month.

(*iii*) Clean the duster after each operation by removing all dust from the hopper.

(*iv*) Oil the leather washers of cup and bucket of sprayers frequently.

(*v*) Spray tank discharge lines and nozzles should be flushed with clean water after the day's work.

(*vi*) Lances and nozzles should not be kept on ground.

Care and Upkeep of Power Sprayers and Dusters

(*i*) Lubricating oil levels should be checked and maintained in four stroke engines daily.

(*ii*) Mixture of engine oil and petrol should be used in correct proportions for two stroke engines, the dusts if any stirred and strained.

(*iii*) Clean the air and fuel filters with petrol frequently.

(*iv*) All the nuts and bolts should be tightened once a week.

(*v*) Check the pressure gauges and safety valves frequently.

(*vi*) Drain the fuel tank after the day's work.

(*vii*) Stop two stroke engine by closing the petrol cock.

(*viii*) Belts should always be kept tightened to avoid slip and slackness.

(*ix*) Keep proper inflated pressure in the tyre wheels of power sprayers.

(*x*) Rubber tyre equipment should be rested on steel when stationed.

(*xi*) Rubber hoses should not be bent at angles and dragged on the ground.

(*xii*) Equipments should be stored in a clean, dry, cool store room.

Care and Upkeep of Plant Protection Equipment when not in Use

(*i*) Plant protection equipment should be arranged properly in a storehouse.

(*ii*) Equipment of one category should be kept at one place and not in a mixed up fashion i.e. do not dump the equipment.

(*iii*) Attachments like discharge lines; lances and nozzles should be kept attached to the equipment.

(*iv*) The equipment should be cleaned with waste cotton every day and polished once a month.

(*v*) The parts and washers are to be oiled or greased well once a week.

(*vi*) The equipment should be tested for its normal performance once a week.

(*vii*) The equipment in stores should be classified and labelled to indicate its condition as:

(*a*) Working condition

(*b*) Needs servicing and repairs

(*c*) Needs parts and repairs

(*d*) Not serviceable

(*viii*) Rubber tyres should be inflated regularly or they should be jacked and propped.

Care and Upkeep of Plant Protection Equipment when Taken to Field

(*i*) Always carry tools required for attending to field troubles.

(*ii*) Carry some spares like washers, filters, gaskets, pins etc.

(*iii*) Carry small quantity of kerosene, petrol, engine oil, grease in containers and also waste cotton.

(*iv*) Carry the plant protection equipment properly and carefully.

(*v*) Do not drop the equipment or attachments to the ground.

(*vi*) Clean the equipment before and after work.

(*vii*) Flush the equipment with clean water, after the work is over.

(*viii*) Oil the moving parts and apply grease on gears and in grease cups.

(*ix*) Filter the chemical liquids and fuel oil mixtures.

Trouble Shooters

Common causes for trouble shooting in application equipment during field operation and their remedies are listed in Table 9.4.

TABLE 9.4. TROUBLE SHOOTING, CAUSES AND REMEDIES IN APPLICATION EQUIPMENT

Name of equipment	Field trouble (s)	Cause (s)	Remedy
1. Stirring pump	No water suction	Leakage in gland nut	Tighten the gland nut, use fresh packing, check valve and replace, if necessary, check joints and mechanism
2. Head compression	Air pump does not pump air	Leather bucket washer not flexible	Add oil drops, change bucket leather washer
	Liquid enters air pump	Air check valve is not working	Check and replace the air check valve and assembly
	Trigger does not function	Rubber seat not proper	Check or change the rubber seating and its assembly needed
	Nozzle and filter get jammed	Dust and solid particles in liquid	Clean the filter element, nozzle pin, replace if needed.
3. Foot sprayer	Pump does not suck liquid	Air leaking through gland packing	Gland packing and nut to be checked, if necessary, replace
	Pump does not work	Leather washer is not flexible	Apply oil to leather washer or change the washer
	Foot pedal does not return freely	Springs are defective	Service the springs with oil or change springs
4. Rocker sprayer	Pump does not work	Piston not tight	Check the PVC piston and adjust the nut
	No suction of fluid	Strainer or suction pipe blocked or leaking	Check and clean the strainer and suction pipe. Replace hose, if leaking
	Pressure does not build up	Leakage in pressure chamber and joints	Tighten the air chamber and joints

5. Hand rotary duster	Crank does not rotate	Fan choking up with dust	Clean the fan housing, grease the gear box
	No dusting	Regulators not working	Sieve the dust and charge dry dust. Regulators to be cleaned
	Dusting not uniform	Improper speed	Maintain proper speed
6. Four stroke petrol engine used for power sprayer	Engine does not start	Lack of petrol, lack of spark, lack of compression, defective valves	Check up the petrol system, check up the spark system, check valve beating, tappet adjustment, change the piston rings
7. Two stroke petrol engine used for mist blower and micron sprayer	Engine does not start	Air cleaner blocked, petrol not available, spark not available, lack of compression	Service the air cleaner, service the carburetor, service spark plug, oil drops in cylinder
	Engine starts but stops immediately	Petrol not available continuously	Clean the petrol tank and petrol pipe
		Spark not available	Earthing at the bottom may be checked
		Silencer blocked	Service the silencer to remove carbon
	Engine does not pick up speed	The throttle slide and needle jammed	Check and adjust the throttle slide and needle
	Engine does not pick up load	Insufficient petrol supply, loss of petrol compression	Service and check up carburetor, air cleaner, change the rings and piston, if needed

PRECAUTIONS WHILE USING PESTICIDES

(*i*) Read label on containers before opening. This may save the men, animals and crop from serious injuries.

(*ii*) Necessary calculations to estimate the amount of formulation and water required.

(*iii*) Whether the pesticide application equipment is in working order, is to be checked. If not, it should be got repaired and kept ready for use.

(*iv*) Appropriate protective clothing should be available to all operators. Clean clothes should be available so that the workers can change them after pesticide application.

(*v*) Plenty of water, soap and towels should be available at the site of operation.

(*vi*) If there are apiaries in the neighbourhood, the owners should be informed about the pesticide application programme so that appropriate precautions can be taken.

(*vii*) Ripe fruits, vegetables and other edible parts of the plant should be plucked up before pesticide application.

Protective Clothing and Aids

The following basic protective equipments are a must during pesticide application:

(*a*) High neck cotton coveralls without pockets with full sleeves, buttons at the cuffs and narrow bottoms, (*b*) a handkerchief to cover mouth and nose,

(*c*) cap,

(*d*) goggles or plain glasses,

(*e*) gloves,

(*f*) shoes.

Depending on the circumstances, respirators, cartridge filters, masks, and breathing equipment may also be required. Clean protective clothing should be worn and should be washed regularly.

General Instructions

(*i*) Worker should not work alone while handling pesticides.

(*ii*) Children or irresponsible persons should neither be allowed near the application site nor handle pesticides.

(*iii*) Instructions or precautions for pesticide use should be strictly observed.

(*iv*) Pesticides should never be left unattended in the field. Care should be taken against curious children or animals who may get attracted and be harmed.

(*v*) The contamination of skin, especially of face and genitals should be avoided. If body gets contaminated with a pesticide, immediately wash with soap and water.

(*vi*) Eyes or face should never be rubbed while working with pesticides. Inhalation of dusts or vapours of pesticides should be avoided.

(*vii*) Drinking, eating, smoking or chewing is not to be done while working with pesticides.

(*viii*) Eating materials, drinking water, beverages, tobacco or cooking utensils should not be kept in the work area.

(*ix*) If a sprayer has been earlier used for applying herbicides, it should be washed thoroughly with washing soda solution (about $2 \text{ g } l^{-1}$) before using it for other pesticides.

(*x*) It is better not to mix or use pesticides on windy days.

(*xi*) Keeping adequate supervision on persons who have not applied pesticides earlier is necessary. An inexperienced operator has a greater possibility of getting harmed.

(*xii*) Arrangement of sufficient rest periods should be provided for workers engaged in pesticide application. Operators should not work for more than eight hours a day.

(*xiii*) Persons handling pesticides for sometime tend to take pesticide application as a routine activity and ignore precautions. Such tendency should be curbed.

(*xiv*) In hot season, pesticides should be applied early in the morning or late in the evening.

(*xv*) While applying granular formulation, gloves should be worn. Application of granules should not be done if there are cuts or wounds on the hand of worker.

Mixing and Handling

Handling of pesticide concentrates involves the greatest potential of hazard exposure.

(*i*) Mixing and preparation of pesticide should be carried out in the open or in well ventilated places. Wind should be guarded while mixing the pesticides.

(*ii*) Containers should be opened carefully to avoid splashing of the liquids or puffing up of the powder formulations. Special care should be taken while opening factory sealed containers Bags or sacks should be opened with scissors, knife or blade. It is not safe to try to tear open the bags as the jerky movements are likely to result in spillage.

(*iii*) While pouring the liquid pesticide, the container is to be kept close to the vessel to be used for making dilution. Mouth siphoning of a pesticide from a container should be strictly avoided.

(*iv*) For mixing the pesticides, it is desirable to use a long handled stirrer and a deep vessel. Bare hands should never be used for mixing.

(*v*) Food or beverage containers on buckets to be used later for bathing and storing drinking water should not be used for the preparation of spray fluids. Funnel should be used while pouring of pesticide solution into the spray pump. The liquid should not be allowed to splash.

Application Care

During Application

(*i*) Ensure that the pesticide container is not leaking. In addition to loss of costly pesticide, leaking equipment can contaminate the operator and his environment.

(*ii*) Irrespective of season, pesticides should never be applied without wearing protective clothing.

(*iii*) Mouth and nose should be covered with a handkerchief or other suitable piece of cloth.

(*iv*) The application equipment should be calibrated to deliver the correct amount of pesticide over a given area.

(*v*) Pesticides should always be applied in the direction of flow of wind. Application should be started near the downward edge of the field and should proceed upwind, with back to the wind, so that the operator is always in the untreated area.

(*vi*) If the wind is likely to cause a substantial drift, pesticides should not be applied. Gentle air movement is helpful in dispersing the pesticides, but wind faster than 12 km per hour will cause significant drift.

(*vii*) Care should be taken that human beings, animals, edible crops to be harvested, animal feed etc. are not contaminated, and that drift is not contaminating water sources, in the direction of flow of wind.

(*viii*) Any sprinkler, nozzle, hose or other part of the pesticide application equipment should not be blown, or sucked by mouth.

Post Application

(*i*) After tightly closing the container, the unused pesticide should be stored. Empty container should be disposed off appropriately in a suitable place.

(*ii*) Pesticide should never be left in the application equipment. If some excess pesticide is left in the equipment, it should be sprayed on barren land.

(*iii*) Empty pesticide applicator should be washed first with detergent and water, and then rinsed thrice with plenty of water.

(*iv*) Bathing with soap with plenty of clean water is essential. Clothes should be washed separately. Soiled clothes should not be taken home for getting them washed along with other clothes.

(*v*) Caps, hats, shoes, belts, goggles or spectacles worn during pesticide application should also be washed.

(*vi*) Proper post application care, as prescribed for different chemicals, should be taken about the treated substrates too.

REFERENCES

Bals, E. J. (1970). Rotary atomization. *Agricultural Aviation.* **12**: 85-90.

Bals, E. J. (1975) The importance of controlled droplet application (CAD) in pesticide applications. *Proceedings, VIIIth British Insecticide Fungicide Conference,* U.K. pp. 153-160.

Clayton, J. S. (1992). New developments in controlled droplet application (CAD) techniques for small-scale farmers in developing countries—opportunities for formulation and packaging *In: BCPC Conf.-Pests Diseases.* British Crop Protection Council. Surrey, U.K. pp. 333-342

Giles, D. K., Welsh, A., Stench, W. E. and Said, S. G. (1995). Pesticide inhalation exposure, air concentration and droplet size spectra from greenhouse fogging. *Transactions of the American Society of Agricultural Engineers,* **38**: 1321-1326.

Law, S. E. (1980). Droplet charging and electrostatic deposition of pesticide sprays--research and development in the USA. BCPC *Monograph,* 24: 85-94. British Crop Protection Council, Surrey, U.K.

Marchant, J. A. (1980). Electrostatic spraying—some basic principles. *In:* Proc. *BCPC Conf. Weeds.* British Crop Protection Council, Surrey, U.K. pp. 987-997.

Matthews, G. A. (1989). Electrostatic spraying of pesticides: A review. *Crop Prot.,* **8**: 315.

Matthews, G. A. (1996). Pedestrian sprayers: Equipments for ultra-low and very-low volume spraying. *EPPO Bulletin,* **26**: 103-110.

Mathews, G. A. (1998). Application techniques for agrochemicals. *In: Chemistry and Technology of Agrochemicals Formulations.* (ed. D. A. Knowles), Kluwer Academic Publishers, London, pp. 302-336.

Mathews, G. A. (1999). The application of pesticides. *In: Pest and Disease Control Handbook,* British Crop Protection Council, Surrey, U.K., pp. 33-51.

CHAPTER 10

Bio-efficacy

BIOLOGICAL ACTIVITY

A chemical is said to possess biological activity if it interferes with the metabolic processes going on in the living system. By interference it means that the substance either inhibits or enhances the metabolic processes or reactions. Therefore, the biological activity should be estimated by measuring the degree of inhibition or acceleration of such processes. In case of pesticides, the mortality, antifeedancy, growth regulation etc. are the indices of biological activity or response. For example, the organophosphorus compounds inhibit the activity of acetyl choline esterase enzyme and, therefore, possess biological activity.

The activity is now justified on the basis that the pesticide molecule has to possess the correct chemical configuration in order to interact with the receptor. A large number of *in vitro* studies have to be carried out to investigate the structure activity relationship with size and shape of the molecules to achieve the best possible results.

When an organic compound is sprayed or applied on a target, it is first absorbed and then translocated to reach the site of action. It has to pass through various barriers inside the living system before reaching the receptor site. Also, such a movement is possible only over short intermolecular distances, which are very necessary for interaction of the chemical with the site of action or the receptor. The path travelled by the chemical is depicted below:

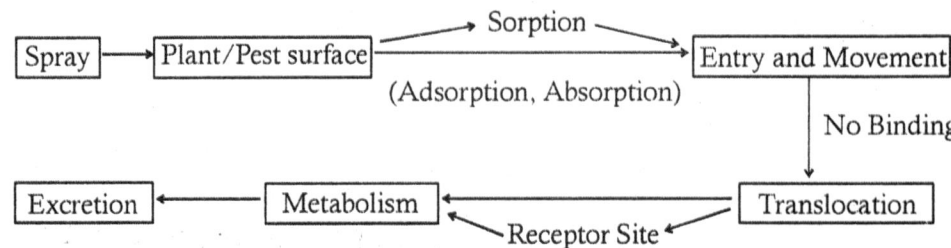

A similar path is observed in insects or other organisms. A number of factors including lipophilic adsorption, translocation, metabolism and reaction at the site of application/action, form the basis for correlating the activity of a chemical with the size or shape of the molecule. The pesticide should be able to penetrate through the corneum stratum of the mammalian skin, the lipophilic epicuticular layer of insect or the polysaccharide phosphoprotein material of the bacterial membrane. A chemical will be active if it successfully crosses all the barriers and is finally excreted as residual metabolic product. Sometimes, an agrochemical may be very active in the *in vitro* studies but prove to be inert or inactive *in vivo* or vice versa; underlining due caution before rejection or acceptance of a new molecule. The key parameters that determine the efficacy of a pesticide formulation are absorption, translocation, spreading, retention, penetration and permeation. Biological activity *in vivo* is also governed by related parameters such as stability, persistence, diffusion and many other physico-chemical considerations. The major principles are briefly discussed below. However, readers are advised to refer standard textbooks on physical chemistry for details.

Basic Considerations

A chemical used as pesticide should have necessary structural characteristics to be biologically active. For bio-dynamic action, it has to cross a number of barriers like cuticle of living insect or plants, etc. Most of the membranes have a similar structure containing a bimolecular layer of lipids, coated/covered on each side by a layer of protein. It has been observed that penetration of a pesticide molecule into the membrane depends on its lipid-water partition coefficient. Penetration can take place by sorption, diffusion, osmosis or some other physico-chemical process.

Diffusion

The pesticide molecule should have appropriate hydrophile-lipophile balance (HLB) to enable it to cross the membrane of a particular lipid/water partition coefficient. Movement of a pesticide from its solution continues by diffusion from higher concentration to lower concentration beyond the biological membrane. The tendency of a substance to diffuse may be expressed as its diffusion coefficient D which is defined by Fick's Law. According to this law, the rate of diffusion or flux of the solute J (the quantity per unit time, dw/dt) is proportional to the concentration gradient. Across the plane of area A, it is given by:

$$J = dw/dt = -DA\, dc/ds$$

where, D = diffusion coefficient

dc = change in concentration

ds = change in distance

The negative sign is introduced so that D will have positive value. It means a positive flow occurs with a negative concentration gradient.

According to Fick's Law the diffusion of pesticides obeys First Order Reaction rate. It means that the time taken by a given proportion of the reactant to diffuse is independent of initial concentration of the reactant and depends on the diffusion coefficient D. The time taken by half the initial concentration to diffuse is called half-life of diffusion.

If for a pesticide having initial concentration c, an amount x penetrates the membrane during time t, then rate of diffusion can be written as:

$$J = dx/dt = k(c - x)$$

or $\quad dx/(c - x) = k\,dt$

where, k is the rate constant.

On integration, the above equation can be rewritten as:

$$\ln c - \ln(c - x) = kt$$

or

$$k = \frac{1}{t} x \ln\left(\frac{c}{c - x}\right)$$

For diffusion, D can replace k, i.e.

$$D = \frac{1}{t} \ln\left(\frac{c}{c - x}\right) = \frac{2.303}{t} \log\left(\frac{c}{c - x}\right)$$

Diffusion being a first order reaction, the half-life for the reaction $(t_{1/2})$ can be easily calculated from the above equation, i.e.

$$D = \frac{2.303}{t_{1/2}} \log 2$$

$$t_{1/2} = \frac{2.303 \times 0.3010}{D} = \frac{0.693}{D}$$

Chemical Potential

Absorption and translocation of a chemical can also be explained in terms of chemical potential of the system. For ith component of concentration c, it is defined as rate of change in free energy of a system with variation of concentration of i at any constant temperature and pressure (keeping all other components and conditions almost unchanged). In diffusion the driving force for flow of solute is the negative gradient of its chemical potential. The chemical potential (μ_i) of a pesticide is given by

$$\mu_i = \mu_i^o + RT \ln c$$

where, $\qquad R$ = gas constant

$\qquad\qquad T$ = absolute temperature

$\qquad\qquad \mu_i^o$ = standard potential of i

$\qquad\qquad c$ = concentration of the chemical

If c varies with distance, there will be potential gradient as given below

$$\frac{d\mu}{ds} = \frac{RT}{c}\left(\frac{dc}{ds}\right)$$

where, s = distance or thickness of the membrane

On reaching a steady state

$$-\left(\frac{d\mu}{ds}\right) = -\frac{RT}{c}\left(\frac{dc}{ds}\right)$$

The negative sign shows that the chemical potential acts in the direction of decreasing concentration.

The free energy of a pure substance consists of its internal energy U and entropy S and it is, therefore, a constant quantity at a fixed temperature and pressure. The effective free energy of a solution and/or pesticide spray is less than that of the pure substance because there is always a decrease in free energy on formation of a solution. The chemical potentials of the components in a solution are their effective free energies. Let μ_A and μ_B be the chemical potentials of a binary solution of A and B. Then the free energy G of the solution is given by:

$$G = n_A \mu_A + n_B \mu_B$$

where, n_A and n_B are moles of A and B respectively.

If the solution obeys Rault's law, then the chemical potential can be expressed as under:

$$\mu_A = \mu_A^o + RT \ln x_A$$

$$\mu_B = \mu_A^o + RT \ln x_B$$

where, μ^o values are the standard chemical potentials of pure components.

Since solutions of pesticides are non-ideal, the mole fractions can be replaced with activities, so that

$$\mu_A = \mu_A^o + RT \ln a_A$$

$$\mu_B = \mu_B^o + RT \ln a_B$$

It is known that,

$$G_A = G_A^o + RT \ln a_A$$

$$G_B = G_B^o + RT \ln a_B$$

where,

G_A = standard free energy per mole of A

G_B = standard free energy per mole of B

a_A = activity of A in solution

a_B = activity of B in solution

The molar free energy G is also referred as chemical potential, μ, so that for component A,

$\mu_A = G_A$

The value of partial molar free energy G_A^o can be obtained from standard conditions. From the above equations one can derive that

$$\ln a_A = \frac{G_A - G_A^o}{RT}$$

This helps to calculate activity of A. The standard state for a pure solid or liquid is only one i.e. one mole of the pure substance at one atmospheric pressure so that the activity is unity ($a_A^o = 1$). But in case of a solution, the standard state for solute is the extrapolated state to very dilute solution where the concentration is equal to one mole. Partial molar free energy is an intensive property and not an extensive property because the value does not depend upon the amount of material but only on the composition at a given chemical potential.

According to the postulated standard state the value of μ^o for a given solution depends only on the temperature and is independent of pressure. The standard state of unit activity is usually selected as that of the pure liquid at the same temperature and atmospheric pressure

(760 mm of Hg). The chemical potential of a solute e.g. pesticide, is higher in the drop on plant leaves than that at receptor site. To represent a steady state, however, when equilibrium is established, it is presumed to be zero.

Standard Free Energy

Change in standard free energy at constant temperature T for any concentration c of a chemical is given by:

$$\Delta F = RT \ln c$$

Gibbs free energy is given by

$$G = G^o + RT \ln c$$

For transfer of a pesticide from the drop on the foliage by penetration causing a change from c_1 to c_2 (by movement inside the foliage), change in free energy ΔG is given by:

$$\Delta G = G_2 - G_1 = RT \ln c_2/c_1$$

or $$\Delta G = RT \ln c_2 - RT \ln c_1$$

ΔF and ΔG should both be negative if penetration were to take place. If ΔF is positive, there will be no penetration.

Free energy changes are more conveniently described in terms of enthalpy. At any given temperature,

$$\Delta G^o = \Delta H^o - T\Delta S^o$$

where, H = enthalpy

S = entropy

Since enthalpy changes do not vary much with pressure, superscript zero is not very important and often omitted. The thermodynamic functions related to free energy, which are important for pesticide formulation have been reported earlier in Table 2.23.

Mass Flow

The theory of mass flow may be applied to calculate the movement of chemicals such as herbicides from the point of application to the site of action, from higher chemical potential to the point of lower chemical potential, depending on the concentration gradient.

Hagen-Poiseulle equation has been applied to calculate the flow of solution through a tube:

$$\frac{dp}{ds} = \frac{8 k \eta V}{Ap}$$

where, η = viscosity of solution

A = cross-sectional area of the tube

p = hydrostatic pressure

k = proportionality constant

V = velocity of flow

For any given system at a constant temperature and pressure, the above equation can be written as

$$\frac{dp}{ds} = KV$$

The following expression has also been reported to calculate the velocity of translocation of the compound in plants.

Velocity = Volume trayunsfer/area $(ml\,cm^{-2}h^{-1})$

Velocity may also be given in litre per square meter per unit time, say hour $(1\,m^{-2}h^{-1})$

If the solution obeys Rault's Law, relative lowering in vapour pressure can be used to calculate the actual flow:

$$P_A = P_A^o x_A$$

where, P_A is partial pressure of A (the solvent), x_A is mole fraction of A and P_A^o is vapour pressure of pure solvent.

or
$$\frac{P_A}{P_A^o} = x_A$$

or
$$\frac{P_A}{P_A^o} = 1 - x_B$$

or
$$\frac{P_A^o - P_A}{P_A^o} = x_B = \text{(mole fraction of solute)}$$

i.e. relative lowering of vapour pressure is equal to the mole fraction of the solute in the solution.

Ionization Constant

Pesticides, which are weak organic acids or bases, have a greater tendency to pass through a membrane, and this depends on pH of medium and ionization constant of the toxicant. Suppose HA, a weak acid ionizes as under:

$$HA \rightarrow H^+ + A^-$$

For dilute solutions, ionization constant K_a is given as

$$K_a = \frac{[H^+][A^-]}{[HA - H^+]}$$

H^+ being infinitely small and the amount of water being very large, the concentration of water is taken to be constant.

$$K_a = \frac{[H^+][A^-]}{[HA]}$$

Symbols in brackets represent the concentrations of ions in moles per litre.

Taking $\log_{10}$ of both the sides, the above equation may be written as:

$$\log K_a = \log \frac{[H^+][A^-]}{[HA]}$$

$$= \log[H^+] + \log[A^-] - \log[HA]$$

$$-\log K_a = -\log[H^+] - \log[A^-] + \log[HA]$$

$$pK_a = pH - \log \frac{[A]}{[HA]}$$

where,

$$pK_a = -\log K_a$$

$$pK_a = pH - \log \frac{[\text{ionized acid}]}{[\text{unionized acid}]}$$

$$pH = pK_a + \log \frac{[\text{ionized acid}]}{[\text{unionized acid}]}$$

Since several pesticides are salts of weak acids or bases and soil acts as buffer, a little description of mathematical expression for the pH in buffer can be considered as below:

If a salt BA is almost completely ionized, the high concentration of A^- suppresses the ionization of weak acid, so that it may be assumed that the free A^- are entirely due to the salt. Therefore,

$$[A^-] = [\text{Salt}] \text{ and } [HA] = [\text{Acid}]$$

From the equation of K_a above,

$$[H^+] = K_a \frac{[\text{Acid}]}{[\text{Salt}]}$$

Taking log of both sides,

$$\log [H^+] = \log K_a + \log \frac{[\text{Acid}]}{[\text{Salt}]}$$

$$-\log [H^+] = -\log K_a - \log \frac{[\text{Acid}]}{[\text{Salt}]}$$

$$pH = pK_a + \log \frac{[\text{Salt}]}{[\text{Acid}]}$$

In case of a base:

$$BOH \rightarrow B^+ + OH^-$$

$$K_b = \frac{[B^+] + [OH^-]}{[BOH]}$$

$$\log K_b = \log [OH^-] + \log [B^+] - \log [BOH]$$

$$\log [OH^+] = \log K_b + \log \frac{[BOH]}{[B^+]}$$

In case of dissociation of salts of weak bases, the high concentration of B^+ ions produced by complete ionization of the salt, suppresses the ionization of the weak base so that it may be assumed that the free B^+ ions in solution are entirely due to the salt. Thus

$$[B^+] = \text{Salt, and } [BOH] = \text{Base}$$

$$\log K_b = \log[\text{OH}^-] + \frac{\log[\text{salt}]}{\log[\text{base}]}$$

$$\log K_b = \log[\text{OH}^-] + \log[\text{salt}] - \log[\text{base}]$$

$$\log[\text{OH}^-] = \log K_b + \log[\text{Base}] - \log[\text{Salt}]$$

$$-\log[\text{OH}^-] = -\log K_b - \log[\text{Base}] + \log[\text{Salt}]$$

$$\text{pOH} = pK_b + \log\frac{[\text{Salt}]}{[\text{Base}]}$$

Because ionization constant of water $K_w = [\text{H}^+][\text{OH}^-] = 10^{-14}$ and $-\log K_w = 14$

$$\text{pH} = 14 - \text{pOH}$$

$$\text{pH} = 14 - \left(pK_b + \log\frac{[\text{Salt}]}{[\text{Base}]}\right)$$

$$\text{pH} = -\log K_w - \left(pK_b + \log\frac{[\text{Salt}]}{[\text{Base}]}\right)$$

$$\text{pH} = -\log\frac{K_w}{K_b} + \log\frac{[\text{Base}]}{[\text{Salt}]}$$

The unionized form of the pesticide being lipophilic, it penetrates through the lipophilic biological membranes more readily. This means that highly polar compounds will not penetrate the membrane. However, once inside the living system pK and pH may play significant role.

The penetration of a pesticide through the foliage will, therefore, depend on the properties of both the plant leaf and the toxicant molecule. The chemical is absorbed first and then it penetrates easily through the membrane. However, the molecules of the pesticide have to pass through the cuticular layers, cell sap etc.

In case of mature leaves, the waxes are present in the cuticle and cuticular layers and not on the surface. The contact angle, which is a measure of hydrophobic nature of the foliage, generally decreases with the maturity of the plant leaves. It appears, therefore, that the absorption of the chemical is less in the matured leaf, probably due to the hydrophilicity of the surface.

According to Le Chatelier principle, if a system at equilibrium is disturbed, it shifts in such a way that the effect of the disturbance becomes the minimum. When a spray drop falls on the plant leaf, it is first absorbed and then movement of the chemical takes place from a higher chemical potential to the receptor site, which has a lower chemical potential.

Vapour Pressure

In case of volatile pesticides (with low vapour pressure), the kinetic energy of the molecule exceeds the forces with which the plant surface/leaves attract it. These molecules vaporize causing a loss of the active ingredient. To make them stay in soil or on the leaf for a longer time, their vapour pressure should be raised by substituting a suitable functional group in the molecule.

The heat required to evaporate a compound is called latent heat of vaporization, which is determined by Clausius Clapyron equation:

$$\log \frac{P_2}{P_1} = \frac{\Delta H(T_2 - T_1)}{2.303 R \times T_1 T_2}$$

or $\quad \log P_2 - \log P_1 = \dfrac{\Delta H(T_2 - T_1)}{4.576 \times T_1 T_2}$

where, ΔH is latent heat of vaporization that is denoted by L also.

If the logarithm of pressure is plotted against $1/T$, the latent heat of vaporisation is calculated from the slope of straight line.

For a pesticide dissolved in a pure solvent, the vapour pressure of either constituent in the binary system is depressed so that,

$$P_1 = P^0 \times N_1$$

where, N_1 is mole fraction of the component one and P^0 its vapour pressure in pure state.

In order, therefore, to increase the residual life of the volatile component, it should be mixed with a component, of lower molecular weight so that N_1, its mole fraction becomes larger.

Mixing chlorinated biphenyl with isopropyl 2, 4-D at 40 °C reduces evaporation losses from 8.5 to 4.17 percent.

In case of soil applied herbicides, a relatively longer residence time in soil may be desirable for a long lasting effect (weed control). If it breaks down or is destroyed to inactive products, it becomes ineffective. To have maximum biological activity, the chemical should be modified to have higher enthalpy (ΔH), which means that the compound will not be destroyed so soon. According to the Arrhenius equation, ΔH is related to reaction rates as under:

$$\log \frac{k_2}{k_1} = \frac{\Delta H(T_2 - T_1)}{2.303\, R\, T_2 T_1}$$

where, k_1 and k_2 are rate constants at temperatures T_1 and T_2 respectively.

In case of isopropyl N-phenyl carbamate (IPC) at 15° and 29°C, by putting chlorine in the benzene ring to get CIPC, the value of ΔH needed to decompose the herbicide increased from 7768 to 21237 calories. It is thus possible to introduce or attach another group in the compound, which will have higher bond energy so that the new compound can be made more stable.

Another important factor to be considered here is whether the compound will be active through the roots or leaves of the plant. It has already been discussed that the pesticide, which may be applied in the form of solution or as an emulsion, will penetrate from the higher chemical potential to the receptor site having lower chemical potential. This movement will be directly related to the Le Chatelier principle to minimize the effect of the external force. Due to these reasons the chemical, after penetration, gets translocated to different parts of the plant.

Hydrophile-Lipophile Balance (HLB)

Another important factor which influences the biological activity is the partition of the toxicant between the hydrophilic aqueous phase and the lipophilic phase. The lower the solubility of a compound in water, the lesser is the concentration required to act effectively, through a lipophilic barrier.

Surface Tension

The addition of surfactants (e.g. wetting agents) to the pesticide spray lowers its surface tension. Also, the contact angle of the drop with the treated surface decreases so that the drop spreads and wets it more efficiently. On plants, the layer of pesticide formulation/toxicant, often protects it against the attacking pests. On insects and weeds, it is a pre-penetration stage.

Quite often, a spray liquid containing the toxicant is diluted with another liquid in which it is insoluble. The liquid heavier than the other liquid (e.g. water) will sink to the bottom. However, a liquid, which is lighter than water, will spread over it giving a uniform thin film. This phenomenon is of great practical value in the control of aquatic pests such as mosquito larvae. Adding a solute to a solvent may lower the surface tension considerably but if the solute causes an increase in surface tension, the effect is small.

The oil containing the pesticide floats over the surface of water making an angle equal to zero degree. The various forces acting on the surface are γ_w, the surface tension of water, γ_o, the surface tension of oil and γ_{ow}, the interfacial tension between water and oil layer. When oil containing the toxicant is sprayed on water, for making a uniform layer or spreading, the spreading constant or factor, S ($S = \gamma_w - \gamma_o - \gamma_{ow}$), also called spreading coefficient, has to be positive. If it is negative, the spreading is indefinite. A positive value of this factor indicates a good uniform monolayer. The thickness and the area covered by the oil depend on the quantity of the oil sprayed on water. Since the oil contains a variety of constituents in addition to the pesticide, it shows a variety of results. It also exhibits interference colours when exposed to light.

The above fact is of importance in aquatic pest control such as malaria eradication programme using waste oil. The waste oil from vehicle maintenance, also called crank-case oil, is used for mosquito larvae control. In this case, oxidation of the oil takes place during combustion at high temperature in machine. It contains spreading agents. The oil forms a uniform layer over water in ponds. However, contaminants may interfere in its spreading. The spreading is further reduced if water has higher viscosity due to some dissolved materials.

Cohesion and Adhesion

The energy, which is applied to pull out a unit cross-sectional area in two parts, is called energy of cohesion. The energy required to pull out the oil spread on water is called energy of adhesion. The oil spreads on water making a zero degree angle. The energies of cohesion and adhesion may be calculated as under.

E_w, the energy of cohesion of water is given by $E_w = 2\gamma_w$

Similarly, E_o, the energy of cohesion of oil $= 2\gamma_o$

Then energy of adhesion $E_{ow} = \gamma_o + \gamma_w - \gamma_{ow}$

$$E_{ow} - E_o = E_{ow} - 2\gamma_o = \gamma_o + \gamma_w - \gamma_{ow} - 2\gamma_o$$

$$= \gamma_w - \gamma_o - \gamma_{ow}$$

The quantity $E_{ow} - E_o$ is also called Harkin's spreading coefficient. If energy of adhesion E_{ow} of oil and water is greater than energy of cohesion of oil E_o, it shows indefinite spreading, which means that $E_{ow} - E_o$ should be positive. It is the oil which spreads and not water and, therefore, E_w does not appear in this equation.

When a liquid is sprayed on a solid, the molecular forces similar to those acting on the spreading of oil on water are called into play. However, there may be some other forces acting on this system due to rough surface in the case of solid, say soil, which is immobile.

It has been observed that the effect of surfactants on activity of herbicidal toxicity is more pronounced when the film pressure at the liquid-solid interface is higher. The drop of liquid placed on solid surface spreads so as to increase the liquid-solid and liquid-air interfacial area. Simultaneously, the solid-air interfacial area decreases and the angle of contact, θ, between the drop and the solid is reduced to zero. The degree of spreading is governed by the surface tension of the liquid γ_{LA}, the surface tension of the solid γ_{SA} and the interfacial tension of the liquid and the solid γ_{SL}.

Let AB be the smooth surface of a solid and CC, a drop of a liquid lying on it. The tangent on the surface of the drop makes an angle θ with the solid which is called angle of contact. Suppose γ_{LA}, γ_{SA} and γ_{SL} are tensions at liquid-air, solid-air and solid-liquid interfaces respectively. Surface tension γ_{SA} tends to spread the drop i.e. to shift the contact point outwards. Spreading continues until the system reaches equilibrium. The degree of spreading is governed by the surface tension of the liquid, γ_{LA}, the surface tension of the solid, γ_{SA} and interfacial tension γ_{SL} (Fig. 10.1). The horizontal component of the tension, γ_{LA}, of the liquid (which is equal to γ_{LA} $\cos\theta$) acts in the opposite direction. At equilibrium, the resultant force is zero.

i.e. $\gamma_{SA} = \gamma_{SL} + \gamma_{LA}$

This equation is known as Young's equation.

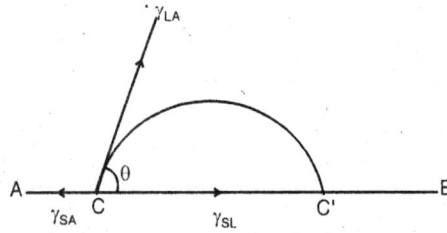

Fig. 10.1: Surface Forces Involved in Spreading of a Drop on Solid Surface

For spreading on a solid surface, the spreading coefficient S is given by

$$S = \gamma_{SA} - \gamma_{SL} - \gamma_{LA}$$

For good uniform spreading, S has to be positive. Also, for perfect adhesion or cohesion, the contact angle being zero, $\cos\theta$, is unity i.e. a solid is wetted completely by a liquid only if the contact angle, θ, is zero and that is made possible by a positive spreading coefficient. The spreading of the formulation or wetting of a surface by a spray will depend not only on the wind and spray direction but also on surface tension of the spray liquid at solid interface. To achieve this property by a formulation, an appropriate knowledge of chemistry and physico-chemical properties of surfactants is a pre-requisite.

It has been observed that the excess of energy of adhesion over the surface energy should be greater than the kinetic energy $(1/2\, mv^2)$ of the droplet on leaf so that the drop can adhere on it.

The surface energy of the drop E_o is $4\pi r^2\gamma_L$. If surface energy of the drops over the plant leaf or a solid is E_s, the energy barrier $[(E_o - E_s)/E_o]$ should be greater than the kinetic energy of the drop so that the drop is retained on the plant leaf and it does not suffer any reflection. On plotting $(E_o - E_s)/E_o$ versus contact angle, it has been observed that the reflection of the drop occurs when the contact angle is very high (around 179°). In case of a drop of 400 μm diameter, the contact angle is zero because the kinetic energy of the drop is equal to the surface energy, which is the condition for complete spreading. In fact the reflection takes place when the contact angle is more than 90°.

The absorption and penetration is a rate process and the toxicant has to pass through the energy barrier. In order, therefore, to ensure penetration, the energy barrier should be lowered. This is best achieved by the addition of surfactants.

Activation energy E_a, with and without the addition of surfactants, is calculated as per Arrhenius equation given below:

$$\frac{d\ln k}{dt} = \frac{\Delta E_a}{RT^2}$$

Also
$$\log\frac{k_2}{k_1} = \frac{\Delta E(T_2 - T_1)}{2.303 \times RT_1T_2}$$

This has been observed from the experimental results that E_a is lower in case of spray drops containing surfactant. In other words, the pesticide penetrates more readily and quickly in this case.

Arrhenius equation can be usefully employed to compare the function of different surfactants and the one, which lowers the surface tension of the spray solution more effectively, can be selected. Taking penetration process as a first order reaction, the following can be used to determine the rate constants.

$$\ln\frac{a}{a-x} = kt$$

where, a = initial concentration of the deposit, x = amount penetrated.

On plotting $\ln a/a-x$ vs. time t, the rate constant can be determined from the slope of the curve.

It has been observed that without the use of surfactants or other materials, which lower the surface tension, the penetration of ions and organic compounds in aqueous medium through the cuticle or the stomatal openings is negligibly low.

Adsorption

After application, the toxicant may be often held on the surface of a target rather than move to the site of action due to strong physical or chemical forces, resulting in adsorption. The properties of the solute, solvent, surface etc. influence such molecular forces and hence, the adsorption.

Amphipathic Adsorption: The non-polar molecules are adsorbed on the surface of water (or vice versa) and such an adsorption can be overcome by use of surfactants (refer Chapter 3). Adsorption of pesticides on solids (ex. soil) is often more important because of the larger surface presented.

In soil chemistry, which involves the behaviour of clays and organic matter for the cation exchange capacity in relation to the availability of cations like potassium, calcium etc., the adsorption by ions assumes special significance. The ionic adsorption of pesticides on clay soils is of great agronomic importance. The salts are soluble in solvents of high dielectric constants in the ionic form. The ions of dipoles possessing opposite charges attract each other. Surface of soil is heterogeneous and consists of a wide variety of minerals and organic matter. The surface contains both hydrophilic and hydrophobic regions and electrical charges. The pesticides are strongly adsorbed on the solid phase in a way that these are not biologically available for action. Whereas, the pesticides are strongly held up on dry soil, these are easily displaced on irrigation with water. Some of the pesticides in soil solution are readily degraded while others like dimethoate and picloram are not hydrolysed in soil solution and can easily translocate in plants through the root system.

The population of oppositely charged ions around a charged particle may be considered as adsorption. On removing the charges the soil particles are flocculated. The ions follow a general sequence in their tendencies for adsorption and coagulation i.e. K, Na, Li, Ba, Ca, and Mg which means that ions follow the order of their electropositivity and those with smaller hydrate radii are preferentially adsorbed.

Among herbicides, adsorption of triazines is less at higher pH in the alkaline medium and increases with the lowering of pH, being maximum when the pH is equal to the pK. This behaviour of clays appears to be due to the negative charge on clays and attraction between the protonated triazine molecules, which behave as weak acids (similar to an amino acid forming zwitter ions or double ion). The adsorption on clays in the buffered medium is lower than that in the unbuffered suspension.

Diquat and paraquat being positively charged cations, are very strongly adsorbed on clays. Adsorption of paraquat does not depend upon the pure acid clay to neutralize the cations because on a number of clays the chemical is non-available for interaction with target after application. The substitution of the position *ortho* to nitrogen in paraquat results in reduced absorption indicating that planarity of the molecule is essential for strong adsorption on clay. It is likely that some other forces are also involved in adsorption of the cations on clays.

Effect of Emulsions and Colloids: The phenomenon of adsorption and surface effects are of great practical importance in colloidal dispersions and in passage through biological membranes. Interface differs from the bulk phase by composition or molecular order. The droplets of a liquid and the fine particles of a solid possess a large volume and surface ratio per unit mass.

Free energy of a surface is equal to the work done in production of a uniform surface, and is given by the equation:

$$G = \gamma \times A$$

where, G = Gibb's free energy

γ = surface tension

A = area of the surface

Smaller the area and lower the surface tension, the lower will be the free energy. Qualitatively, adsorption is the difference between the concentrations of the bulk and surface phases. Only if a solute in solution lowers the surface tension of the solvent, can it migrate to the surface. If the surface of the solvent rises, the solute moves to the bulk phase and the surface of the liquid will be deficient in the solute. If a solid is in contact with a liquid or a gas, then the molecules of the solute or gas are adsorbed on the surface of the solid. It is, therefore, evident that adsorption depends upon the change in surface tension caused by concentration of the adsorbed material.

Consider a solution containing n moles of the solute, which move to the surface area A, causing a lowering of surface tension by $d\gamma$, then the free energy change $dG = A \times d\gamma$.

If the free energy increases due to an increase in concentration dc, then $dG = nRTd\ln c$.

At equilibrium, the sum of the free energy changes is equal to zero.

Then, $$Ad\gamma + (n)RTd\ln c = 0$$

$$n/A = \Gamma = -d\gamma/RTd\ln c$$

where, Γ = surface concentration of the solute per unit area of the interface.

Γ is generally dependent on nature of the surface phase. It is very difficult to devise measurement techniques for experimental verification of the equation. However, this Gibb's equation is used to calculate the amount of material adsorbed by a given interface.

For dilute solutions, surface tension is equal to the free energy per unit area of the surface. The enthalpy H is defined mathematically as:

$$G = H - TS$$

or $$H = G + TS$$

$$H = G - T\left(\frac{dG}{dT}\right)p$$

where, p = pressure

$$H \text{ (of the surface)} = G - T\left(\frac{dG}{dT}\right)p$$

$$= \gamma - T\left(\frac{d\gamma}{dT}\right)p$$

The inorganic compounds when dissolved in water increase the surface tension of the solvent and this results in a negative adsorption. Many organic compounds, which are soluble in water, however, lower the surface tension resulting in positive adsorption. This means that the water-soluble organic compound is surface active. The polar part of a molecule like hydroxyl, carboxylic, amino preferably solubilize in water and are directed towards the solvent whereas the hydrocarbon part forms a film on the surface. The surface active part of the hydrocarbon derivative in aqueous solution is near the surface of water and decreases the surface tension of water. Due to this reason the adsorption increases. The increase in adsorption can be calculated from the lowering of surface tension. Study of surface tension of aliphatic carboxylic acids has shown that surface activity increases with the length of hydrocarbon chain. Tranbe rule states that concentration of carboxylic acids to give a required surface tension is reduced by a factor of three for each added methylene ($-CH_2$) group. This is due to the reason that molecules are curled like spheres. The Gibbs adsorption equation can be applied to compute the surface tension of a solution. If A is the surface area in square centimeter containing n moles of the solute and if surface tension γ is related linearly from zero concentration to (rising up to) a concentration c, the following equation holds good:

$$\frac{n}{A} = \frac{\Delta\gamma}{RT}$$

or $$A\Delta\gamma = nRT$$

where, $\Delta\gamma$ = the tension gradient, also called surface pressure and denoted by ϕ

Thus, $$A\phi = nRT$$

Many detergents and soap materials e.g. sodium oleate, sodium dodecyl sulphate, and cetyl pyridinium bromide lower the surface tension of water in dilute solutions. The surface pressure ϕ is the lowering of surface tension and varies from surfactant to surfactant. It is a measure of concentration (theoretically) at which it forms a uniform monolayer on the solvent surface of water and any further addition of the surfactant to solvent will not have an effect on the surface tension.

For two immiscible liquids forming a o/w or w/o system, the surface between the two liquids is the interfacial film and the energy required to produce unit area of the surface is equal to interfacial surface tension. When octyl alcohol is shaken with water, two layers are formed. The interfacial tension between the two liquids, determined using a tensiometer, has been reported as 8.5 dynes cm^{-1}. The OH groups are oriented towards water, forming a film over it. If the concentration is decreased, it gradually forms a film at monomolecular level. The film forming capacity is due to the lower solubility of the surface active agent such as proteins and synthetic polymers. The thickness of the film can be evaluated. The usual method is to prepare a solution of the material in a suitable solvent like benzene, diethyl ether etc. and then spread over water. The volatile liquid evaporates leaving behind the insoluble film spread over water. The thickness of the monomolecular film may then be correlated with the dimensions of the molecule. The pressure, which is the force per unit length, is equal to the difference in the surface tension of the film on one side and that of the pure solvent on the other. Since the OH and COOH groups are arranged on the surface of water, the nonpolar parts are arranged vertically pointing upwards. The addition of methylene groups in the hydrocarbon chain increases the thickness of the film by 1.4 Å. The thickness of stearic acid molecule is around 21 Å and area about 21 Å^2. On adding more of the acid, there is a compression resulting in a multimolecular layer or film. Since the hydrocarbon parts are pointing upward, the area occupied by molecule is the cross-sectional area of the hydrocarbon chain. The molecular area of the glyceryl tristearate is 66 Å^2, that is about three times the area of a single molecule of stearic acid.

Adsorption on Solids: The solids possess surface tension and also surface energies that are greater than those on the liquid surfaces. The surface of solid does adsorb gas or liquid molecules (adsorbate) on it. The surface covered with the adsorbate molecules results in a decrease of surface energy. The rough and porous surface adsorbs more of the adsorbate to fill the cracks, pores and other irregularities. Since adsorption occurs on the surface, greater the area of the surface, greater is the extent of adsorption. The porous adsorbents of finely divided solid material may have a surface area approximately equal to 500 m^{2}g^{-1} and hence these have greater capacity for adsorption. The finely divided particles being almost spherical have volume equal to 4/3

πr^3 and the surface area is proportional to square of the mean radius. Silica gel, activated carbon, platinum black, finely divided nickel, animal charcoal are examples of the best adsorbents. Alumina is obtained as precipitate of aluminium hydroxide and heating at 500°C changes it into oxide that is available in a variety of surface treated products.

Condensation of gas molecules on the solid adsorbent is called physical or van der Waals adsorption. Another form of adsorption could be the chemical adsorption or chemisorption in which case the adsorbate is held very strongly with a valence bond accompanied by the evolution of heat (25-75k cal mole^{-1} as compared to 3-10 k cal mole^{-1} in the case of physical adsorption of gas molecules). The difference between the physical and chemisorption can be well illustrated by the adsorption of oxygen on charcoal at 0°C. After the equilibrium is established, the oxygen gas is extracted at a constant temperature under vacuum. However, some oxygen does remain attached to the charcoal and does not come out. This is the chemisorbed amount which forms monomolecular layer, containing oxygen gas fixed up by valence bond. Most of the oxygen is adsorbed due to physical adsorption. Oxygen gas can be extracted from monomolecular film by heating the charcoal which breaks the valence bond to set oxygen gas free.

The rate or amount of the gas adsorbed in a adsorption process is always proportional to the pressure of the gas at a given temperature. The relation between the amount of the gas adsorbed and a given temperature is called adorption isotherm. When the rate of gas desorbed at a given temperature and pressure becomes equal to the rate of adsorption, equilibrium is established. The molecules of the gas that strike the surface bounce off because there is no room for the additional molecules to form valence bond.

If ϕ is the fraction of the total number of sites occupied by a gas, the number of vacant sites
$$= 1 - \phi$$
Rate of adsorption $= K_a (1 - \phi) \times P$

The rate of sorption is proportional to number of sites occupied $= K_d \times \phi$
where, K_a and K_d are constants for adsorption and desorption.

However, at equilibrium, the two are equal i.e.

$$K_a(1 - \phi) \times P = K_d \times \phi$$
$$K_a P - K_a \phi P = K_d \phi$$
$$K_d \phi + K_a \phi P = K_a P$$

$$\phi = \frac{K_a P}{K_d + K_a P}$$

This is called Langmuir Isotherm.

For adsorption in dilute solutions, ϕ, the fraction of surface covered with solute will be proportional to the amount of solute (x) adsorbed by a given mass (m) of the adsorbent i.e. for monolayers

$$\phi = \frac{x}{m} = \frac{K_a P}{K_d + K_a P}$$

Dividing both the nominator and the denominator by K_d

$$\frac{x}{m} = \frac{K_a P / K_d}{1 + K_a P / K_d}$$

$$= \frac{KP}{1 + KP}$$

where, $\qquad K = K_a/K_d$

or $\qquad KP/(x/m) = 1 + KP$

The equation is similar to $Y = mx + C$. The value of constant can be calculated from slope of the curve obtained after plotting P against $P/(x/m)$.

Limitations of Langmuir Isotherm: The isotherm suffers from following drawbacks or limitations:

(*i*) It is applicable to monomolecular and monolayer adsorption.

(*ii*) It generally refers to chemisorption or chemical adsorption.

(*iii*) The adsorption is independent of the surface already covered. If the surface is non-uniform and of irregular shape, the value of constant K varies on account of different enthalpies for different sites. Even a pure solid presents different surfaces because crystal arrangement may terminate in a different pattern.

(*iv*) The attraction or repulsion between the neighbouring molecules may cause variation in the equilibrium constant, which will depend upon number of sites involved.

(*v*) It may not hold good for heterogeneous systems like soil. In soil there is heterogeneity since there are many types of mineral particles of different composition. The partly decayed organic matter adsorbs the pesticides if the minerals are coated with organic matter.

In spite of these limitations, the Langmuir isotherm gives pretty good results when applied to monolayer and/or chemisorption.

In case of physical adsorption that involves multimolecular or multilayer adsorption, Brunauer, Eemmet and Teller equation is applicable. In this case equilibrium is established when the rate of condensation of a gas is equal to the rate of desorption at a fixed or a given temperature and pressure. It is assumed that K has a constant value for first layer and a different value for the other layers. Also, the adsorption becomes rapid when the gas approaches condensation. (see Table. 2.23)

In a thermodynamic system at equilibrium, besides the force applied by a gas, there are some other forces in addition to the external forces. Of physico-chemical interest are the surface forces, the effect of which becomes apparent when the quantity of matter contained in the surface is larger as compared with the system taken as a whole. In case of two bulk phases, either two liquids or liquid and its vapour, separated by surface phase, the excess of a system is the excess of that constituent in the surface phase over the amount that would have been, if both bulk phases remained homogeneous up to the geometrical surface. The adsorption of unionized pesticides by organic matter in soil is similar to the partition between water and organic solvent.

If a pesticide is strongly adsorbed, then it becomes inactive unless it is desorbed. For example, paraquat is inactivated by soil when applied at normal rate because it is very strongly adsorbed by clay and due to this reason it is completely immobilized. The movement of pesticide through soil decreases with adsorption on soil and the organic matter content. The herbicides like monuron, propazine, atrazine and simazine have been reported to move through soil at a rate inversely proportional to the extent of adsorption. Because of these reasons, the biological availability of pesticides is greatly affected. The pesticides which are strongly adsorbed, remain on the surface of the soil until they are lost by evaporation or decomposition. This forms the basis of weed control by using pre-emergence herbicides. Water-soluble herbicides may undergo leaching from the surface of soil. It is, recommended that somewhat larger doses of soil pesticides be used on clay soil or organic soil than on light sandy soil. In addition, there could be losses by degradation or uptake of pesticides, by the organisms. When a pesticide is applied to soil, its uptake by plants depends on the concentration in soil solution and also on activity of the absorbing roots. The activity of the plant system depends on nature of the product function, time and capacity of the receiving organism to accumulate the toxic dose. The uptake of toxicant is, of course, a slow continuous process, which depends on the rate at which the plants can absorb the toxicant from the soil solution containing pesticide, especially by plant roots.

The surface of soil generally contains clay minerals and soil organic matter which are hydrophilic and are preoccupied by water. However, they can adsorb polar compounds. In dry soil, a pesticide is held strongly and, therefore, there is little chance of loss by evaporation. However, on irrigation, the pesticide is replaced by water. This is, therefore, evident that adsorption displacement does play a part. Water is held by the hydrophilic soil surface, which even when dry, contains around 2% water. At 100% relative humidity, pesticide may be lost in preference to water.

The accumulation of pesticides can be attributed to wick action due to which a pesticide would produce about 20-fold increase in concentration at the surface of soil as compared with the original concentration in one centimeter depth and 400 fold of that in two centimeter depth. Such an effect is important only when the day atmosphere is very dry. The chemicals, which resist degradation, transport as adsorbate on soil particles, can cause loss of residues due to surface action. The wind and the flowing water cause erosion of soil, which depends on wind speed, the particle size and bulk density of soil particles. The air borne dust has a diameter less than 0.1 mm but particles having diameter greater than 0.5 mm are not moved by wind. Effectiveness of soil applied pesticide decreases, if the topsoil suffers erosion. The effectiveness of removing pesticide from soil varies according to topography, the weather and properties of the chemical and the soil. It has been seen that rain, after the application, washes the herbicides down the slope and may cause sever phytotoxicity to the crops growing below. Adsorption of pesticides on soil should not damage the activity of the toxicant. Operator is interested in extending the time of contact of toxicant with the substrate so that its effect lasts longer. For example, in case of 2, 4-D the degree of esterification with polyvinyl alcohol is adjusted to make the product water-soluble and at the same time it gets strongly adsorbed on leaf surface. The ester slowly gets hydrolyzed on the leaf surface and the acid liberated would penetrate the cuticle.

Adsorption in Soils: When a pesticide solution or emulsion is sprayed in field, a large proportion of it is adsorbed on the soil depending on its mechanical composition. A soil high in clay content will adsorb more of the pesticide. The extent of adsorption can be measured by shaking a known weight of the soil with solutions containing different concentrations of the solute. Study of Langmuir isotherm, indicates that there is some reaction between a solid clay and the solute. A layer of the compound is formed on the surface of the clay and then no further adsorption takes place. As stated earlier, adsorption occurs on solids because of valence forces and also by the attraction of molecules on the surfaces. This depends on the nature of the solute and also on the nature and concentration of adsorbing material. The adsorption can be physical or chemical. The heat of adsorption is small in case of physical adsorption and this has a much higher value in case the solvents that react chemically with the adsorbent.

The amount of material adsorbed by the adsorbent depends upon the concentration of the solution. More and more of the solute is adsorbed on the surface of the solid but at the same time process of desorption also takes place. On increasing the concentration, this process continues till a new equilibrium is established between adsorption and desorption.

Micelles

The ions or molecules called amphiphiles exist as such in very dilute solutions but in the concentrated solution, they associate to form aggregates called micelles. These exist in equalibrium with the single molecule in solutions i.e.

$$nx = Y$$

where, $nx = n$ number of single molecules of x

Its aggregate number Y = micelle (or number of molecules forming a micelle).

The value of n may vary from 10 to 100 molecules in a micelle.

Such molecules show variable molecular weight in solution. For example, the molecular weight of sodium lauryl sulphate in extremely dilute solution is 144 because it ionizes to give sodium cation and lauryl sulphate (dodecyl sulphate) anion due to which the molecular weight decreases. As the concentration of the same solution is increased, the molecular weight starts rising indicating the beginning of association of the molecules. In other words, the individual particles start aggregation to form micelles.

The concentration of solution at which this association starts is called critical micelle concentration (CMC). The aggregation of ions and molecules is determined from the colligative properties like osmotic pressure, which registers a fall in its value. In order to determine size and shape of the micelle, hydrodynamic methods involving the process of sedimentation of particles are employed. At the CMC, the aggregates appear to be spherical in shape, surrounded by water from outside and the hydrocarbon chains are interlocked on the inside so that they are isolated from the solvent. Due to this reason, the micelles are protected from the repulsive forces of polar groups.

A very important property of micelles is that they make an insoluble organic compound soluble in water. In doing so, it appears that the CMC is lowered so that the insoluble organic molecule helps to make micelles stable by locking itself within the interior part of the micelle, which contains the hydrocarbon part of the surfactant molecules. It is due to this reason that micelles of associated colloids can normally solubilize water insoluble materials like dyes, drugs, pesticides etc. The non-polar lipophilic material is incorporated in the hydrophobic core of the micelle and undergoes orientation. The oppositely charged ions are taken into tightly bound counter ion sphere (Fig. 3.1). If the salt of the associated colloid with that ion is oil soluble, it dissolves in the core of micelles. This process of moving in and out of micelles goes on for more than 10^4 times per second in some cases. The dissolution of sparingly soluble solids can thus be increased. The solubility may be increased till the solid disappears (fully solubilized). Solubilization may actually decrease the total driving force in some extreme cases because the unsaturated solution has a lower activity than the undissolved material. This explains why many biocides lose their effectiveness when mixed with soaps.

The geometry and number of molecules in a micelle depends on the following factors:

    (*i*) Chemical nature of surfactant
   (*ii*) Length of hydrocarbon chain of the surfactant
  (*iii*) Size and solvation power of polar groups
  (*iv*) Nature of solvent
   (*v*) Temperature
  (*vi*) Ion strength of the medium and the electrostatic interactions
 (*vii*) Concentration.

Micelle Formation: The association of molecules of surfactants in a solution may be oriented to form bigger multimolecular particles (micelles). The surfactants are slightly soluble, both in water and the organic solvents (10^{-5} to 10^{-2} mol l^{-1}). The low solubility is due to their combined hydrophilic-lipophilic character or nature. However, the micelle formation occurs only after the concentration exceeds the minimum solubility limit, the CMC. In the case of water-soluble surfactants, the hydrophilic heads face the aqueous phase whereas the lipophilic tail portions of the molecules keep away from the aqueous phase. The reverse process of molecular orientation occurs when the surfactant is dissolved in an organic solvent.

On reaching the CMC, some physical properties of the solution are affected. The properties that are used to determine the exact value of CMC, include:

  (*a*) ***Refractive index***: Determine and plot the refractive indices of the solutions of varying concentrations of surfactants. A break in the curve gives CMC value.

  (*b*) ***Electrical conductivity***: Draw a curve to show the fall in electricl conductivity with a change in concentration. Abrupt sharp fall indicates the formation of micelles.

  (*c*) ***Surface tension***: The surface tension of the solution of surfactant falls rapidly when the concentration is increased. It becomes minimum at CMC and does not change with further increase in concentration of surfactant. The value of CMC can be obtained from the curve.

  (*d*) ***Light scattering***: Scattering of light that increases slowly with increase of surfactant concentration shows a steep rise at the CMC.

Sometimes, the surfactant being fairly soluble in the polar solvent results in poor aggregation. In such cases, addition of non-polar solvent facilitates formation of micelles. If there is excess aggregation, this can be corrected by adding polar organic solvent, and so on. This interaction of polar molecules in non-polar solvents has been explained by the reaction of soft and hard acids, with the soft and hard bases. Lewis acids that accept electrons are hard acids, but possess poor polarizability whereas soft acids, which have more polarizability, react with soft bases. The surfactants which possess a hard base (carboxylic group) react readily with cations like

Na^+, K^+ which are hard acids and form bigger aggregates. Addition of water lowers their interaction and can decrease the size of micelles. Solvents like water alcohol etc., which have a strong hydrogen bonding are hard Lewis acids and, therefore, they hinder the interaction of hard base carboxylic acid group with the cations. Due to this reason the formation of micelle is greatly affected resulting in a decrease in their size and number. This shows that aggregation of molecules or formation of micelle is higher in the case of a strong interaction of cations with the anions. This helps to manipulate the solubility of surfactants to obtain maximum stability for the system.

Solubilization of sparingly soluble salts is a slow process and it involves binding of molecules or ions by the micelles. This process continues till the solubilized particles leave the micelle as rapidly as they are taken up again so that equilibrium is established. When solid is dissolved, the total driving force for the transport in the unsaturated solution has a lower activity than the undissolved material. It has been established that no significant contribution is made by the micelle in transporting the material through the membrane although they transport the material up to the membrane and then pick it again on the other side. Some toxicant molecules, which may get entrapped in the micelle during its formation, may enter the root system with the micelle in water stream. The increase of temperature raises both the degree of aggregation in the organic solvent and also the value of CMC. This is due to an increase in activity of the molecules/ions with the rise of temperature. The activation energy of micellisation is much smaller both in the aqueous and organic system. However, when the solution above the CMC is agitated, there occurs a decrease in the surface tension, which is of course, a temporary process. As soon as the agitation is stopped, the monomers again combine to form micelles. The surface tension, therefore, increases again after stirring is stopped.

Others

Intermolecular Forces: The intermolecular forces exist between atoms and molecules and also between the electrically charged particles namely ions. Such forces are not the result of covalent bonds between them, and are called van der Waals forces. These forces are essentially electrical in origin.

Suppose q_1, q_2 are the electrical charges on two particles and r is the distance between them, Then, according to Coulombs law,

$$F = \frac{q_1 q_2}{kr^2}$$

where, k is the dielectric constant

When two oppositely charged particles approach each other, they lose energy. When we pull them apart, then work has to be done against the force of attraction. In case of similarly charged particles their energy increases if they are made to approach each other.

The Coulomb attractive forces stabilize the solid phase. A molecule is electrically neutral but on polarization, the charges get separated within the molecule, the magnitude of which is called dipole moment. Since the distance between the atoms is 10^{-8} cm and the electrical charge on the electron 10^{-10} e.s.u., the dipole moment is 10^{-18} e.s.u. cm^{-1}, which is also called one Debye. In the case of a molecule of water, the oxygen atom is more electronegative than hydrogen draws the bonding electrons closer to it and also there are unshared electrons on the oxygen that causes a permanent dipole in the water molecule. In addition to this, ions in the solution create additional dipoles, when two particles having permanent dipoles approach one another with proper orientation from end to end or side to side, then the forces of attraction are more than the forces of repulsion.

The Hydrogen Bond: The hydrogen bond occurs when a hydrogen atom is attached to an electron-attracting atom like oxygen, fluorine or nitrogen so that hydrogen is the positive end of an electric dipole and is then attracted by the negative end of another dipole. The hydrogen atom has the bonding pair of electrons in its neighbour and there is little repulsion of the second atom by the negative charge, as it stays opposite the electronegative atom to which hydrogen is attached with covalent bond. The hydrogen bond has an energy equal to 10 k cal mole^{-1} compared to 1 or 2 k cal per mole for van der Waals forces. The small size of the atoms of nitrogen, oxygen and fluorine and their strong tendency to attract negative charges makes the hydrogen bond to be very significant. In case of water, the oxygen-oxygen distance is between 2.4 to 2.8 Å. The longer distance between the oxygen and the hydrogen is due to hydrogen bond and the shorter distance due to covalent bond. The H—O—H bond angle is 105°, both in the liquid and vapour phase, but in ice the hydrogen bonds are slightly bent. The hydrogen bonding in water is shown below.

Similarly, carboxylic acids and secondary amines act, both as donors and acceptors of hydrogen. Intramolecular hydrogen bonds occur when the geometry of the molecule permits the donor group and acceptor group to approach each other. In case of maleic and salicylic acids, it is shown below:

Maleic acid   Salicylic acid

Interfacial Tension: At the interface of two partially miscible or completely immiscible liquids, there is existence of tension similar to surface tension. The interfacial energy is the work which is called into play to increase the area of the interface by one square centimeter. The methods required to measure the interfacial tension are the same as for surface tension. Soaps and detergents are surface active agents and lower the surface tension of aqueous solutions. Solutes which lower the surface tension of the solvent are called capillary active e.g. alcohol, organic acids, esters, amines and ethers.

One liquid may spread over the surface of another liquid forming a film on the interface. This is dependent on the spreading coefficient which is equal to the difference in the surface tension of two liquids minus the interfacial tension. In case this difference has a positive value, then the liquid of lower surface tension will spread spontaneously over the surface of the other liquid. If this value is negative then there will be no spreading because of the increase in the free energy of the surface.

According to Gibbs equation if the processes occur spontaneously, there is always a decrease in free energy of the system. In other words, it means that if there is a decrease in surface tension ($d\gamma/dc$), then one liquid will be adsorbed over the other liquid. This is similar to the spreading of oil on water forming a film due to a decrease in the surface tension.

Liquid-Solid Interface: When a pesticide formulation is sprayed on plants, it spreads on surface of the leaf. The spreading depends on the contact angle that the drop makes with the leaves and in case of spontaneous spreading of liquid; the contact angle may be zero. This process is similar to the rise of liquids in capillary tube where the contact angle of the liquid with the side of the capillary tube is zero. In case of those liquids, which make a contact angle of more than zero, the rise of liquid is less. For example, mercury makes a contact angle greater than 90 degrees and therefore, there occurs a depression in the capillary tube. The contact angles are, therefore, important in considering the spreading of the pesticide spray and wetting of the plant leaves.

Reaction Kinetics: Knowledge of rate of reaction is sometimes helpful in the prediction of bioefficacy of a formulation.

Rate of a Reaction: It is expressed as the change in concentration of either the reactants or the products per unit time.

e. g. $2HI \rightarrow H_2 + I_2$

Decrease in concentration of HI or formation of iodine divided by the time taken gives rate of the reaction. The velocity of the reaction depends upon the molar concentrations of the reactants, which decrease with the passage of time so that speed/rate of reaction decreases with the time depending upon the progress of the reaction. The rate of a reaction is, therefore, the instantaneous rate of change of concentration of any one of the reactants or the products. The rate of reaction at any point can be determined as follows:

For the reaction A + B = C + D

The concentrations of the reactants (x) or the products (a), at regular intervals, say every minute, are determined and a graph is plotted between the concentration x and the time period (t). In order to know the rate of a reaction at any time interval, a tangent is drawn to the curve against the time t. The slope of this tangent gives the rate of reaction. The distance along the ordinate gives changes in concentration x and distance along abscissa gives time interval t. The rate of reaction is given by $-\dfrac{dx}{dt}\, or\, \dfrac{da}{dt}$.

There are analytical methods available to monitor the chemical reactions. Samples may be withdrawn periodically and determinations done by the methods that may involve titration, refractive index, spectrophotometric, electrochemical and other measurements. The sample of the solution which is taken out for analysis should be immediately cooled/ frozen to stop the reaction.

It is, however, convenient to follow the reaction without removing the sample from the reaction mixture. Changes of pressure, volume, conductivity can be followed with the reaction mixture. Also, the nuclear magnetic resonance, or infra red spectrum show absorptions which can be assigned to some of the reactants.

Expressing Reaction Rates: The rate of reaction is usually expressed as dx/dt for the products and if the reaction is losing concentration then rate of reaction $= -dx/dt$.

Homogenous Reactions: The reaction between nitrogen and oxygen to from nitric oxide (all present in a single phase).

$$N_{2(g)} + O_{2(g)} \rightarrow 2NO_{(g)}$$
$$-d(O_2)/dt = k(N)^2 \cdot (O)^2$$

Heterogeneous Reactions: The reaction between copper oxide and hydrogen gas contains two solid phases and one gaseous phase.

$$CuO_{(s)} + H_{2(g)} \rightarrow Cu_{(s)} + H_2O_{(l)}$$
$$-d(CuO)/dt = k \cdot (CuO) \cdot (H)^2$$

Decomposition of hydroiodic acid

$$2HI \rightarrow H_{2(g)} + I_{2(s)}$$
$$-d(HI)/dt = k \cdot (HI)^2$$

The rate of a chemical reaction at any particular point is proportional to the product of molar concentrations of each reactant raised to the power equal to the molar concentrations of the species for a reaction

$$aA + bB + cC + \dots = K + M + \dots + \dots$$
$$\text{Rate} = -dx/dy = k[A]^a \cdot [B]^b \cdot [C]^c \dots$$

For a chemical reaction to occur, the species must collide with each other. Under ordinary conditions of temperature and pressure, magnitude of such collisions is nearly 10^{25} per minute. In actual practice most of the reactions take a much longer time and in fact many of the reactions are slow because only a small fraction of the total number of collisions (called effective collisions) are effective in bringing about the chemical reaction. Rate of reaction is thus, a function of effective collisions.

If z is the number of collisions per unit volume per second and f, the fraction of number of effective collisions, then the rate of reaction is given by

$$-dx/dt = fz$$

It may be noted that the fraction f may vary from zero to nearly unity for chemical reactions depending upon the nature of reactants and temperature.

Molecularity of a reaction is the number of reacting species that must come together or collide to bring about a chemical reaction. The reactions are mono, di, tri etc. molecular depending upon the number of colliding particles. The reaction may take place in a number of steps and the rate of reaction is determined from the step that is slower e.g. in reaction.

$$aA + bB \rightarrow Products$$

Rate of reaction $= -dx/dt = k(c_A)^a \cdot (c_B)^b \ldots$

where, $c_A, c_B \ldots$ are concentrations of the reactants in rate governing step.

Order of Reaction: It is the number of reacting particles whose concentration determines the rate of reaction. However, molecularity is the number of particles, which participate in the reaction, whether or not their concentration affects the rate of reaction. The order of reaction refers to the actual number of reactants affecting the rate of a reaction. Consider the hydrolysis of ethyl acetate.

$$CH_3COOC_2H_5 + HOH \rightarrow CH_3COOH + C_2H_5OH$$

The molecularity of the reaction is two whereas the order of reaction is one since water is present in such a large quantity that its concentration is practically constant.

Similarly, the hydrolysis of cane sugar to form glucose and fructose is the reaction of first order.

$$C_{12}H_{22}O_{11} + HOH \rightarrow C_6H_{12}O_6 + C_6H_{12}O_6$$

This is evident, therefore, that the order of a reaction is determined experimentally. The majority of chemical reactions are not portrayed in a single step because they occur in a sequence of individual steps. The rate constant for each step is equal to rate constant multiplied by the concentration of reactants. The fate of most of the pesticides viz. stability, release, diffusion, degradation etc. is governed either by zero order or first order kinetics, the two orders are being briefly described mathematically.

(*i*) **First Order Reaction:** Take a reaction $A \rightarrow B$, which is of first order.

Let concentration of A $= a$

After a time t, the concentration of product B $= x$

Concentration of A at that time $= a - x$

Then, rate of reaction

$$dx/dt = k(a - x)$$

Integrating this equation

$$\ln a/(a - x) = kt$$

or $\quad 2.203 \log a/(a - x) = kt$

The half life $t_{1/2}$ of the reaction is the time for half of reaction or disappearance of half the concentration of A, i.e.

$$\ln a/(a - a/2) = 2.303 \log 2 = k t_{1/2}$$
$$t_{1/2} = 0.693/k$$

This means that $t_{1/2}$, the half life of the reactant is independent of concentration.

If a curve is drawn by plotting from the experimental results $\log a/(a - x)$ versus time, t; the value of k can be calculated from the slope. A straight line shows that it is a first order reaction. The rate of reaction generally increases with a rise in temperature. In some cases, however, the rise in temperature may have an adverse effect on the reaction as in case of enzymatic processes (here the energy of activation rises first and then falls with further increase in temperature).

(*ii*) *Zero Order Reaction*: This type of reaction is independent of concentration e.g.

$A \rightarrow B$

The integrated rate equation is $kt = x$. The rate constant is determined by plotting x, the concentration of the product versus time, t. The rate constant k is obtained by measuring the slope of the plot.

Applied Aspects

Coverage, retention, losses, etc.

Coverage, retention, losses, etc. of pesticides on different targets govern the ultimate performance of a product.

The absorption and penetration of drops of spray solutions of pesticides in plants is additionally governed by parameters such as roughness of leaf surface, the air between drop and leaf, nature of groups in the epicuticular waxes, pH, humidity, light, temperature and nature of the surfactants viz. cationic, anionic, non-ionic etc.

The epicuticular wax contains a number of organic compounds/groups linked together. For example, alkanes, alcohols, ketones, aldehydes, acids and esters are usually present. Their presence renders the plant surface hydrophilic and the leaves become wet. The variation in wettability of plant leaves may be caused by the difference in the packing of the exposed groups. The leaf surfaces may have different chemical and morphological characteristics, yet the absorbtion of pesticides depends on the amount of toxicant deposited on the target, the extent of surface covered and its persistence. The contact angle of the drops on smooth hydrocarbon surfaces is about 105°.

When the toxicant is sprayed on a plant, some of the drops are deposited on the leaf which itself is making an angle with the branch/stem. The leaf is not parallel to the ground but is bending at an angle, say α. This means that the drops and the leaf are pulled down by gravitational force that is equal to $mg \sin \alpha$ where, m is mass of the drop and g is acceleration due to gravity. A force of adhesion on the leaf, in turn, opposes this, and as soon as the gravitational force exceeds the force of adhesion, the run off occurs. The following relationship holds for adhesion.

$$mg \sin\alpha = \gamma_{LA} \times w \, (\cos\theta_R - \cos\theta_A)$$

where,

γ_{LA} = surface tension of the liquid

w = width or diameter of drop

θ_R = receding contact angle

θ_A = advancing contact angle

The retention factor (F) of the applied spray in terms of the number of drops retained on a surface may be assessed as follows:

$$F = \frac{\theta_M [\gamma_{LA}(\cos\theta_R - \cos\theta_A)]^{1/2}}{P}$$

where,

θ_M = mean of contact angle i.e. $\cos\theta_M = (\cos\theta_A + \cos\theta_R)/2$

P = pressure within plant pores or capillaries

The area A, covered by spray drops on plant leaf is given by:

$$A = k/[\gamma_{LA}(1 - \cos\theta)]^{1/2}$$

where, k = constant depending on drop diameter and speed of the spray.

The leaf contains a large number of pores/capillary tubes through which the spray drop enters it. The following relationship gives the pressure, with which the toxicant enters the capillary:

$$P = 2 \times \gamma_{LA} \times \cos\theta / r$$

where, P = pressure

R = radius of capillary

θ = contact angle

The contact angle should be less than 90° to get a positive value of the pressure. In case the contact angle is greater than 90°, $\cos\theta$ has a negative value. This means that the liquid will not be able to penetrate into the capillary.

$\gamma_{LA}\cos\theta$ is called surface pressure on the leaf, such that

$$\gamma_{LA}\cos\theta = \gamma_{SA} - \gamma_{SL}$$

where, γ_{SL} = interfacial tension at the solid liquid interface.

γ_{SA} = surface tension at the solid air interface

It means that the activity of sprayed pesticide, is directly proportional to surface pressure.

Spreading and wetting have been distinguished so that the spreading involves the formation of liquid solid interface but wetting is taken only after the excess liquid is drained off.

Taking into consideration smooth and rough surfaces:

$$\cos\theta_S = R\cos\theta_r$$

where, R = ratio of real to apparent surface

θ_S = contact angle on smooth surface

θ_r = contact angle on rough surface

This means that if the θ_S is less than 90°, then the contact angle due to roughness decreases so that water is not repelled from the rough surface. The surface becomes more water repellent if the contact angle exceeds 90°.

In nutshell, the surface tension plays the most important part in all liquid formulations. Because pure solutions (i.e. single solute dissolved in pure liquid) are seldom sprayed on crops due to obvious reason of phytotoxicity and other application constraints, it becomes a matter of great importance to choose a proper blend of solvents, surfactants and other adjuvants. The art of formulation becomes more complicated when dealing with a systemic fungicide or herbicide that requires penetration and movement in the living tissues (xylem or phloem).

Penetration

In order to exert its characteristic effects on a target *in vivo*, a drug or toxicant must arrive at the biochemical site(s) of action in sufficient quantity for efficient interaction. If a method is available for measuring the interaction of a toxicant (pesticide) with the target itself on a molecular basis, then the potentially attainable toxicity (intrinsic toxicity) can be ascertained. From the time of its contact with an organism, the toeicant is subjected to numerous processes. Therefore, molecular design has to take into account the effect of structure on these intermediary processes. Pharmacodynamics involves the study of absorption, distribution, biotransformation and excretion of a drug as well as its mechanism of action at the target. The first four properties comprise the pharmacokinetic phase of drug action, and determine the amount of drug that is biologically available for interaction with the target site.

In Plants

The initial penetration of any pest control chemical into an organism is of major importance for its bioefficacy. The chemicals are absorbed by the roots or leaves or seeds and then translocated into the plants before they show any bioactivity. Following are some basic criteria that play a major role in penetration and effective pest control using a pesticide formulation.

Permeability of the Skin or Cuticle: It is a rate limiting stage in the exchange of chemicals. The measurement of permeation of the membrane involves changes in the concentration in the environment after the chemical penetrates the cuticle.

Suppose a membrane has a thickness T and A is its area exposed to the penetrant. The flux, F is given by:

$$F = \frac{P \times A}{T \times \Delta c}$$

where, c = concentration of the penetrant

Δc = difference in the concentration of the penetrant inside and outside of the membrane

P = permeability constant, similar to diffusion coefficient, D.

The two are related as $P = D \times c$

where, c = the concentration inside the membrane.

In this equation Δc can be replaced by ΔP, the pressure difference, when one measures the permeability of a gas or vapours.

According to ultra filtration theory, the cell membranes have pores of certain sizes and the rate of permeation decreases as the pore size increases. However, some solutes, which have high solubility in organic solvents, penetrate the membrane much faster than would be predicted on the basis of molecular size. It is, therefore, evident that membranes have lipophilic pores that allow rapid permeation of even larger lipophilic molecules.

The mass flow from the point of application may be expressed as follows:

$$\text{Velocity cm h}^{-1} = \frac{\text{Mass transferred (g h}^{-1})}{\text{Concentration (g ml}^{-1})}$$

The Poiseuille formula has been modified to calculate pressure gradient as under:

$$P = (8R_1 \eta l)/r^2$$

where,
P = pressure in dynes cm^{-2}
R_1 = velocity in cm sec^{-1}
η = viscosity in poises
l = length of tube in cm
r = radius of conducting element in cm

Taking into consideration that the sieve plate openings are converging at the entrance and diverging at the exit, a new formula to explain the laminar flow, has been suggested:

$$P = (4R_1 \eta l)/r^2$$

The pressure calculated from this equation would be one half of that calculated from the Poiseuille equation. All plants do not have the same potential growth and, therefore, variations can occur within the same species.

The cell wall in plants is pierced by plasmodesmi and has a high permeability to minerals and organic substances. These interact with the cell wall by molecular adsorption, ion exchange etc., or combine with the constituents of the cell wall with stronger bonds. The ions that bind to the cell wall are easily desorbed into the free space within the cell wherefrom they are absorbed by the cytoplasm. The other barriers include the surface membranes. The biological membranes surrounding the living cell are intricate and highly specialized. These separate the contents of the cell from the environment.

The system of active transmission through the biological membranes is the first step in the absorption of and interaction with the molecules of the surface structure of the cytoplasm. In this process, the links in the membrane transport may be the adenosine triphosphate (ATP), activated by Mg^{2+}, K^+ and Na^+ cations.

There are three mechanisms of permeation:

1. Diffusion among the flexible and thermally agitated units of molecular structure
2. Diffusion through pores
3. Flow through the pores under the pressure or concentration difference.

The penetration of pesticides into the cells of living organisms takes place by diffusion through membranes along the gradient of concentration. Many of the pesticides are highly active organic compounds, which are lipophilic and applied as emulsifiable concentrate. They readily dissolve in the lipids of cell membranes and then diffuse into the cells. The mineral pesticides enter the cell as ions or undissociated molecules.

There are platelets of wax embedded in the amorphous cutin and the molecules of penetrants are oriented perpendicular to the plane of lamina. Closely packed paraffin mole-cules offer resistance to permeation in the direction of molecular orientation. It is assumed that the embedded wax forms a pathway for the diffusion of nonpolar/lipophilic molecules.

A saturated, crystalline wax column would not permit any diffusion through its inner surface. Transfer could occur only around the outer surface of the block, if its holes are imperfectly filled. This is called "waxy pathway". Flooding with alkaline emulsion would be a possible technique for accelerating entry of leaf cuticle with wax platelet structure. In normal condition, vine leaves lose water 10-30 times as fast as the fruits on them. On exposure to petroleum vapour, the rate of dehydration increases 10-40 times in case of leaves and 2.5 times in case of fruits.

The cuticles of shaded ivary and tomato leaves offer one third of the resistance to penetration than the normal leaves. When solute constituents of natural oils and waxes of the leaf cuticle are removed by soaking in chloroform, the permeation in all cases increases by a factor of 40-50 in case of ivy leaf cuticle and 10 in case of tomato leaf cuticle. The wax removal facilitates the penetration of polar compounds.

Swelling of Cuticle: Cuticle swelling has been considered to affect permeability by loosening the postulated wax plugs in pores. Many organic solvents affect the cuticle. Alcohols, though readily absorbed into the paraffinic cutin probably due to hydrogen bonding within the structure acting as cross link, restrain swelling. The composite cuticle formed by detachment from the cells, has physical properties differing between the outside and the inside. The differences are evident in wettability and swelling. The shiny outer surface of most separated cuticles accepts water only with a finite contact angle. The removal of soluble wax, however, does not always result in a decrease in the contact angle.

Entry Through Stomata: The stomata are specified openings in the leaf surface that permit continuity of the air filled cavity below them with the outside air. These are flanked by two guard cells, which are very well adapted to delay the entry of liquid. Mobile non-polar liquids enter into the leaves through the stomatal cavity. Using fluorescent tracers it has been shown that penetration can be achieved with aqueous sprays if they contain enough wetting agents.

The stomatal penetration, even if proved to occur, cannot be exploited because (*i*) it would receive sprays only when it is open, which depends on circadian cycle modified by light intensity, humidity and temperature, (*ii*) It would require aqueous sprays, and uneconomic rates of surfactants, and (*iii*) many crop plants and weeds have stomata only on the lower surface of the leaf. However, penetration through leaf can be correlated with the stomatal density in response to uptake of only herbicides or systemic insecticides or fungicides.

Entry Through Trichomes and Grooves: Many leaves are hairy. The botanists use the term 'trichomes' for all single or multi celled protrusions on the surface. These are specially adapted to water intake but in most crop plants, their significance for penetration of pesticides is incidental. Structure of the hair is such that it repels water but collects small mist droplets in the hair canopy outside the leaf surface. It has been reported that the cuticle around the hairs has more than average residual density of applied pesticide after spraying.

The veins of leaves form ribs on the abaxial surface and corresponding grooves on the adaxial surface. The grooves permit faster spread of a liquid than the plane areas and provide preferential drainage channels in a wet surface. Pesticide residues, therefore, tend to deposit more over the veins than interveinal areas.

Diffusion Through Pores: If a solution of viscosity η passes through a pore of radius r so that ΔP is the pressure difference between the top and the lower end of length l cm, then V, the mean velocity is given by:

$$V = \frac{r^2}{8\eta} \times \frac{\Delta P}{l}$$

Flux, $\qquad F = V \cdot A \cdot c_m$

$$= \frac{r^2}{8\eta} \times \frac{\Delta P}{l} \times A \times c_m$$

where, $\qquad A$ = area of the cross-section

$\qquad\qquad c_m$ = mean concentration of the solute or active ingredient

In case of diffusion through pores, flux is given by:

$$F = -D \times A \times \frac{\Delta c}{l}$$

where, D = coefficient of diffusion

$$\frac{\Delta c}{l} = \text{ concentration gradient}$$

In case diffusion and mass flow are taking place together, the above flux become equal i.e.

$$A \times c_m \times \frac{r^2}{8\eta} \frac{\Delta P}{l} = D \times A \times \frac{\Delta c}{l}$$

If c_o be the concentration of initial spray and because concentration of the active ingredient is zero in the receiver, $\Delta c = c_o$ and $c_m = c_o/2$

Therefore, $\Delta P = 16\,\eta D/r^2$

The value of ΔP can be calculated from this relation.

Molecular Size: The size of the penetrant molecule plays an important role in permeation of compound. The diffusion coefficient of molecules of small size is smaller as compared with the size of solvent molecules, which are larger. However, the values of diffusion coefficient D of smaller molecules show more variation, as is apparent from the values of the order of 10^9 for polyisobutene as compared to those for propane 4.81, isobutane 1.45, n-butane 3.24, isopentane 1.32, n-pentane 2.64 cm s^{-1} respectively. The branching in aliphatic chain also effects penetration e.g. in a series of isomeric butyl phosphate insecticides, the mobility decreased in order of normal, secondary, *iso*, tertiary (60 : 40 : 20 : 1).

Concentration: Concentration gradient plays important role in penetration of pest control formulations. Penetrants, which swell the cuticle, penetrate at the rates which increase rapidly with the concentration. The flux F reported above, is given by

$$F = -D \times A \times dc/dl$$

It is seen that D increases with an increase in c, the concentration. A great variation in diffusion coefficient with a change in concentration has been reported.

In Pest

The most common means of contact between an insecticide and a living organism is either through accidental or deliberate application to the integument or in the food ingestion. In mammals and insects, it is followed by absorption through walls of the alimentary canal. Having passed these barriers in insects, the insecticide enters the insect haemolymph, in which it may be carried to all parts of the organism in solution or bound to protein or dissolved in lipid particles, depending on its physical properties. During the penetration and distribution process, some of the toxicant will be taken up by inert tissues, such as adipose tissue, and so become biologically unavailable. During penetration and distribution, non-enzymatic chemical conversion and enzymatic biotransformations, lead to more toxic or less toxic molecules. Finally the distribution of the insecticides or their metabolic products will depend on their ability to cross semipermeable membranes separating the various biophases within the organism. The various parameters (disassed above) influcing penetration of chemicals into plants may also hold good to explain situation based penetration of chemicals in the insect pest system.

An insecticide applied to insect cuticle, first comes in contact with the epicuticular wax, which overlies more polar layers consisting largely of chitin and tanned protein. The best contact insecticides are generally relatively non-polar compounds, though an inverse relationship between lipophilicity and rate of cuticular penetration may also occur. The partition coefficient of the insecticide between its carrier and epicuticular wax determines the actual concentration, so that the nature of the carrier markedly affects the results. Not only its great resistance to water loss by evaporation characterizes the cuticle ware but also it possesses great power to absorb water. In considering the rate of penetration of any material through the insect cuticle, two factors *viz.* (*a*) the nature of the cuticle as a membrane and (*b*) the nature of the material (particularly its polarity) may be of relevance. For optimum performance, the rate of penetration must be directly proportional to the concentration of the material administered and exponentially to the penetration time.

Absorption Through Gut: The toxicants may be absorbed from the mouth, stomach, or gut on oral ingestion by insects. The mucosal lining of the mouth behaves as a lipoid barrier through which non-ionized compounds can diffuse rapidly into the blood stream.

If a weak electrolyte cannot pass through aqueous pores, the distribution achieved will depend on pKa and pH gradient across the membrane. The degree of ionization of an organic electrolyte at any pH is given by the following equation:

(for acids) $\quad \log \dfrac{c_m}{c_i} = pKa - pH$

(for bases) $\quad \log \dfrac{c_i}{c_m} = pKa - pH$

where, c_m and c_i are concentrations of the molecular and ionized forms respectively.

The relative total concentrations (ionized + un-ionized) of the electrolyte on either side of the membrane govern the lethal action of the molecule.

Studies on penetration of ^{14}C labelled carbaryl, parathion, and dieldrin in isolated foregut of *Apis mellifera* indicated a rapid initial uptake of insecticides from the lumen of the foregut into the foregut tissue. The toxicant in the tissue was proportional to the partition coefficient of these compounds. Total penetration from the lumen through the foregut and into the serosal fluid, however, was not proportional to the lipophilic character of the insecticide, the relative order of penetration being; parathion followed by carbaryl and dieldrin. The penetration of dimethoate analogues apparently followed first order kinetics.

Uptake and Release by Blood: When the compounds are not ionizable, an optimal balance between their polarity and lipophilicity is necessary for facile penetration through the gut wall. Compounds with lipophilic character in excess of the optimum, tend to penetrate the gut tissues and accumulate there, passing only slowly into the serosal liquid beyond. It is well known that the foreign compounds (xenobiotics) entering the blood may become bound to plasma proteins and then transported. Albumin is the most commonly involved protein but other plasma proteins may participate if the albumin becomes saturated.

Factors Influencing Rate of Penetration

Chemical Nature of Insecticides: The rate of penetration depends on polarity of the compound and is generally directly related to its lipophilicity since the outer layers of the cuticle are highly apolar. The mode of penetration depends not only on the constitution or chemical nature of the compound but also on the method by which the compound is introduced to the surface of the cuticle. However, the penetration can be increased by topical application. By doing so, the solute is directly introduced into the lipoidal part of the cuticle, so that it does not have to penetrate the lipid; rather it is mixed with the lipid. Thereafter, the molecule has to penetrate relatively polar material, such as procuticle (consisting largely of relatively polar chitin and protein molecules), and so, highly apolar compounds tend to stay with the epicuticular waxes.

Cuticular Components: The rate of penetration of solute through an excised cuticle, containing water on both of its sides is proportional to the water solubility of the solute. Once the pesticide is picked up by the cuticular lipid, it must be gradually dispersed from the point of contact until it reaches an equilibrium dictated by the partitioning properties of the compound in components of body tissue. For polar pesticides, the presence of lipids should increase the rate of penetration of the pesticide already there and for apolar lipophilic compounds the opposite is expected.

Three major factors appear to play important role in deciding the actual amount of insecticide penetrating the insect body. These are lipid solubility (or polarity), affinity for the cuticular components other than lipids and solubility in haemolymph. Absorption into the cuticular components could play an intermediate role as a factor in aiding the initial pick up. At the same time the pesticide in the cuticle would act as both a reservoir and the mediator between the lipid and the aqueous layers of the cuticle.

Carriers and Diluents:

Oils: The addition of oil to insecticidal preparations often increases toxicity. The oils used as insecticide carriers are relatively apolar and lack reactive groups. Lipophilic insecticides readily dissolve in such solvents. The mineral oils remove lipids from the cuticle. This also paves way for polar compounds to penetrate, provided they are not too highly dissociated. In contrast to the effect of solvents or polar carrier oils of low boiling point, too much externally applied apolar oil has detrimental effect on penetration, particularly of very lipophilic insecticides.

Surfactants: Surfactants form a "bridge" between lipophilic and water soluble substances and thus enhance penetration of insecticides. They not only help in penetration of the lipid layers of the epicuticle but also the protein layers of the endocuticle. The properties that render a surfactant most effective are:

- (*i*) Enough liposolubility to penetrate and emulsify the epicuticular wax
- (*ii*) Sufficient solubility in water
- (*iii*) Ability to penetrate the outer cementing layers of the epicuticle
- (*iv*) Reduce surface tension of the spray droplet.

Dusts: Controlling insects with dusts such as soot, ashes, or road dust is an ancient and effective practice. Now activated charcoal, oxides and carbonates of magnesium and calcium, siliceous minerals and finely powdered alumina are used. The mode of action of dusts involves withdrawal of body water and lipids from the insect, i.e. desiccation through abrasion of the cuticular surface. The efficiency of dust increases with increasing dryness of the air.

Particle size of dusts is very important factor that decides the adherence to surface. Finer the dust more is the adherence to the surface. The optimum size of particles is about 10 microns because the force, which tends to dislodge them depends on gravity. To dislodge 100 micron particle only 4000 times of gravitational force is required whereas to dislodge a particle of size 10 microns almost 500,000 times of gravitational force is required.

To summarize, the mode of penetration of pesticides into pests is still open, and for a given methodology, the situation is undoubtedly influenced by inter-specific differences in cuticular structure as well as the properties of the toxicant and formulants.

Translocation

In Plant

Entry into Free Space within the Tisse: Water and solutes after penetrating the leaf cuticle pass into free space by diffusion from the external environment. The cuticle controls the loss of water from the plant and aqueous solutions of many organic molecules can be transported across it, although a majority of chemicals enter by the root tips. Diffusion of an aqueous spray from leaves into the free space will occur only while a liquid film remains on the leaf surface. In case the speed of drying of the spray is a limiting factor, the uptake may be enhanced by addition of a humectant e.g. glycerol.

Translocation of chemicals in plants may be considered in three stages:

Movement in Xylem: Xylem vessels serve as water pathways allowing movement from the environment by free diffusion. The bulk of movement of water and soluble minerals from roots to leaves is via the non-living xylem and this process does not involve expenditure of metabolic energy.

Movement in Phloem: Movement within the living parts of the cell requires metabolic energy. The chemicals, which have arrived at the leaves through the xylem, are distributed to the growing tissues of the plant via the phloem. The movement of chemicals as discussed above occurs passively in the transpiration stream but requires evaporation at the surface and so can be reversed by immersion of the plant leaf in water, which checks evaporation at the immersed surface. This is also dependent on the metabolic activity and can be prevented by treatments that inhibit the metabolism or immobilize the phloem.

Apoplast, Symplast and Free Space: The contents of phloem are eventually continuous with the intracellular protoplasm, forming symplast. Similarly, the contents of xylem vessel are continuous with the intercellular pectinacious fluid which is called apoplast. It has proved helpful in studying the movement of systemic pesticides. Between these two systems, there is always the fluid permeable plasma lemma.

The uptake of ions into roots takes place in two phases. The first phase is a passive one. The chemical diffuses into the parts of the tissue which communicate reversibly and relatively freely with the external source. This phase is usually relatively rapid and is followed by a slower, more extended phase, in which the accumulated amount increases at an approximately constant rate. The extent of the first phase can be characterized by the amount of solute accumulated in the initial period of rapid uptake. For many experiments, the apparent volume has been calculated from the concentration of the original solution and has been referred to by such terms as 'outer space', 'apparent free space', 'diffusion free space' or 'effective volume' (Crafts and Crisp, 1971).

Systemic Insecticides: Plant systemic insecticides have been described as compounds which are absorbed by the plants and then stored or translocated to other parts of the plant. Many compounds can penetrate to a varying extent into plant tissues thus occupying an intermediate position between non-systemic and true systemic plant insecticides. These may be exemplified as methyl parathion, malathion, diazinon, nicotine, dieldrin, DDT, lindane etc.

The epidermis and plasma membrane are the natural barriers through which an insecticide has to penetrate in order to get into the living cell. This movement requires forces to break hydrogen bonding between the entering molecule and water and also to overcome van der Waals forces. Another mode of transport through the membrane involves enzymatically controlled mechanism of active transport.

A systemic insecticide should have sufficient water solubility to enable it move (either symplast or apoplast) in the plant sap. Translocation from the leaves occurs in a slow downward movement in the phloem and a faster upward movement (towards the apex) from the xylem. There is more absorption at higher temperatures through younger leaves. Entry of systemic insecticides into seed occurs through openings in the seed coat i.e. micropyle and the hilium and is controlled by humidity. A majority of systemic organophosphorus insecticides are also transported in the xylem. If the compound is to move essentially in aqueous plant sap, it must have sufficient aqueous solubility or be converted to such a compound after metabolism in plant tissue. On the other hand, if a compound penetrates via the foliage, it is transported across the waxy leaf cuticle due to appreciable lipid solubility. To function as a systemic foliar pesticide, the compound must, therefore, have a reasonable lipophilic-hydrophilic balance. If the compound is too lipophilic, it will remain held up in the cuticular waxes and if it is too hydrophilic, it will not penetrate the cuticle. Thus in the series of O, S-dimethyl N-alkyl phosphoroamidates, a parabolic relationship was established between systemic movement and the logarithm of the octanol-water partition coefficient (P), or relative hydrophilic constant (π). This is defined as the difference between the logarithm of the partition coefficient of the parent compound (P_0) and the derivative (P) so that $\pi = \log P/P_0$. This implies the existence of an optimum lipophilic-hydrophilic value for maximum systemic translocation in plants. In this series, π was calculated to be 1.19. The N-propyl derivative, with π value of 1.31, nearest to the optimum value, showed the greatest systemic movement in cotton leaf petiole. Some of the organophosphorus systemic insecticides are also translocated unchanged (e.g. mevinphos) while in other cases the active metabolites translocate in the plants (e.g. schradan).

Symplastic Systemic Insecticide: Systemic insecticides pass through the plasmalemma into the living symplast (i.e. living protoplast of the plant), diffuse from one cell to the other through plasmodesmata and move along the animilate stream in the phloem tissue to other areas of the plant.

The successful performance of the endometatoxic reaction involving systemic insecticides provides a method of converting lipophilic compounds into hydrophilic compounds *in vivo*. It is, therefore, a useful tool for transport across the lipophilic plant cuticle and its translocation in the aqueous system that is evident in the conversion of sulphide to sulphoxide and to sulphone in demeton, disulfotan, phorate and aldicarb. The alkyl thio group has many useful properties for incorporation into a variety of systemic molecules of phosphate, phosphonate and carbamate insecticides. The N-methyl group of bidrin and phosphamidon may also act as delay factor to prevent *in vivo* oxidation.

The symplastic insecticides are obtained by developing an insecticide with a COOH group that would be metabolized *in vivo* to yield the parent compound as a product of reaction after penetrating the plasma membrane. The insecticides that may be acid activated are trichlorofon, dimethoate and methomyl. A carboxylic acid moiety when attached to a hydroxyl group of trichlorofon apparently enhances long distance symplastic translocation. Translocation studies performed with other insecticides namely, malathion, naled, azodrin, metacid, zectram, indicate that they are apoplastic systemic insecticides. Amino triazole, acetyltriazole and 4-N dimethyl tolyloxyacetic acid behave as symplastic systemic molecules.

Apoplastic Systemic Insecticide: Malathion is translocated almost exclusively by the transpiration stream within the apoplast which behaves like a cavity of the non-living xylem tissue. Most of the organophosphorus type insecticides and the carbamates like baygon, carbaryl and zectran are apoplastic.

Systemic Fungicides: The diffusion of systemic fungicides through cuticle revealed that chemicals applied to abaxial surface are transferred more rapidly than those applied on adaxial surface.

Addition of humectant to delay drying may increase the time of exposure of a pesticide to get absorbed into the plant system. Humectants have not been used commercially. Glycerol and polyethylene glycols have been reported to increase the uptake of streptomycin. The apoplast can also be used for injection of the systemics into the tree trunk e.g. injection of carbendazim and benomyl against Dutch elm disease and thiabendazole and carbendazim against *Venturia pirina* on peas.

Cycloheximide, 2, 4, 6-trichlorophenoxyacetic acid and the oxathiin derivatives like carboxin and oxycarboxin also show apoplastic movement. Apoplastic transport seldom seems to be specific to a particular plant. Chemicals like benzimidazole derivatives namely thiabendazole and benomyl or its metabolite carbendazim have been observed to move well from treated leaf to next of the plant.

Most of the systemic fungicides including several antibiotics like streptomycin show xylem movement. The first generation of commercial systemic fungicides such as benomyl is mainly transported through the xylem, but the pyrimidine fungicides (e.g. ethirimol and dimethirimol) appear to move unchanged in the phloem. Sometimes downward translocation may be valuable for control of soil fungi as in the case of pyroxychlor.

Herbicides: Movement in the phloem has been observed for chemicals like phenoxyacid herbicides. Compounds like dalapon and maleic hydrazide move more freely from roots to xylem enabling them to be widely distributed in plants. Certain chemicals move from leaves to roots via phloem and may be released in soil. Hormone herbicides viz 2, 3, 6-trichloro-and 2, 3, 4, 6-tetrachlorobenzoic acids get exuded into soil and can produce malformation in adjacent plants.

In Pest

Following their absorption, drugs and insecticides get distributed in body tissues by circulation. Since most of the insecticides have only limited aqueous solubility, their transport in blood occurs not in solution in the plasma water but primarily through binding to plasma proteins, particularly albumin. Indeed, almost all types of chemicals undergo measurable protein binding through ionic, hydrophobic, or van der Waals forces or dipole-dipole interactions such as those responsible for hydrogen bonding. Protein binding is a reversible process so that the free and bound forms of a material are in equilibrium at all times, their ratio being dependent on the association constant of the reaction. Since a drug forms a high molecular weight complex, it is unavailable for membrane transport. The protein binding can have a significant effect on distribution of a given material in the tissue by limiting its availability for enzymatic modification, molecular filtration, or tubular secretion. Thus, protein binding often regulates the action of a drug or insecticide and any modification in the binding equilibrium will affect its biological effecacy.

The insecticide, before translocation and distribution, has to penetrate or pass through many biological membranes or barriers. Some barriers such as insect cuticle, consist of several layers of cells of about 100 Å thickness. These structures contain minute water filled pores (usually about 4 Å in diameter) which allow a rapid penetration of small, water soluble compounds, while the lipid membranes are readily penetrated by lipophilic substances. Thus, the molecules, that are highly hydrophilic, do not rapidly pass through such membranes unless they are small enough to move through the pores. Since the undissociated form of ionizable toxicants are highly lipid soluble, the degree of ionization is important and their rate of transfer through membranes, is, therefore, strongly pH dependent. Most foreign compounds move through the lipid membrane by passive diffusion, down a concentration gradient. The diffusion of hydrophilic molecules and low molecular weight ions through aqueous pores also follows a concentration gradient and is called filtration. In other two processes of facilitated diffusion and active transport, the compounds presumably pass through the membranes in the form of complexes with endogenous substances. Both the processes occur down a concentration gradient, and the active transport, also requires additionally, the metabolic energy.

The permeability of various biological membranes affects the transfer of insecticide and depends on a series of dynamic equilibria between the tissues and the circulatory system. The movement of insecticides in the insect body is by no means clearly understood. A further complication in distribution studies is that the afore mentioned equilibria observed *in vivo* may be profoundly affected by factors such as age, sex and nutritional status of the animal and the presence of xenobiotics, other than the one administered (for the purpose of control), which may cause induction or inhibition of the drug metabolizing enzymes.

In case of pyrethroid penetration, the initial phase corresponds to a rapid transfer of insecticide into areas of the cuticle, inaccessible to the more polar rinsing solvents. In the absence of metabolism, the rate of inward diffusion decreases as the external concentration decreases, so that equilibrium is eventually established between external toxicant, toxicant in the inner layers of the integument, and the toxicant in the tissues (Brooks, 1966). The second phase of penetration appears to correspond to approach the internal plateau level of toxicant. If a metabolic process is present, a steady state will first be achieved in which metabolic rate balances penetration rate. As metabolism of internal toxicant continues, the external toxicant eventually becomes insufficient to maintain the steady state and the internal level, therefore, begins to decline from its maximum value. The knockdown by the pyrethroids has been observed to be due to some effect on the central nervous system that is related to their rate of penetration.

For a number of compounds, the linear third phase of the penetration corresponds to the period of constant internal concentration, and this portion is, therefore, a measure of both penetration and metabolic rates. The data from various sources show that organophosphorus compounds penetrate the cuticle quite rapidly whereas chlorinated insecticides and carbamates penetrate slowly. Compounds like kerosene and thanite act as "quasisynergists" that act by increasing the cuticle permeability without necessarily altering the rate of metabolism. Assuming that any increased toxicity observed when synergist and toxicant are injected together should reflect only the inhibition of detoxification, while that observed on topical application should also include any effects on penetration, then following equation should hold good:

$$\text{Synergism due to increased penetration} = \frac{\text{Synergism by topical application}}{\text{Synergism by injection or oil spray}}$$

The pharmaco-kinetics of insecticide behaviour in both mammals and insects has been studied in a semi quantitative manner. For quantitative assessment of role of suitable surfactants and prediction of appropriateness of a formulation, a method is needed to describe the integrated effect of penetration, distribution, metabolism, final interaction with the target site (or toxicity *in vivo*) in relation to the molecular structure of toxicant.

The general importance of toxicant partitioning between hydrophobic and aqueous phases of an insect is evident from that the haemolymph is the major transporting phase in all cases. For optimal toxicity, a compound must first partition readily from integument to haemolymph and subsequently from haemolymph to nervous tissue. Ideally, the partitioning between haemolymph and other tissues, especially those such as gut and body fat, which are known to be involved in detoxification, should be in favour of the haemolymph.

Fumigants: There are many sufficiently toxic compounds, which are volatile at ordinary temperature to fall within the definition of fumigants. Fumigants can control insects within open structures, or inside commodities and in cracks or crevices into which penetration of other insecticides is difficult or scarce. To study sorption and penetration of fumigants, Graham's Law of diffusion has to be known. Graham's law of diffusion states that the velocity of diffusion of a gas is inversely proportional to the square root of its density and also densities are proportional to their molecular weights. The rate of diffusion is also directly related to temperature, so that a given gas will diffuse more quickly in warmer than in cold environment. The distribution and penetration can be aided and hastened by the use of blowers and fans. Sorption is the most important factor affecting the action of fumigants. It accounts for the total uptake of gas resulting from the attraction and retention of the molecules by any solid material present in the system. It removes some of the molecules of the gas from the space so that they are no longer able to diffuse freely throughout the system or to penetrate further into the interstices of the material. In fumigation practices, collision with air molecules tends to slow down gaseous diffusion through the material, and sorption takes place gradually.

The amount of fumigant must be sufficient, both to satisfy the total sorption during treatment and also to leave enough free toxicant to kill the flying, hiding pests. Sorption (absorption and adsorption) under a given set of conditions determines the dosage to be applied.

Among the fumigants, chloropicrin is absorbed the most by food commodities. It also persists for a longer time. Methyl bromide has high penetrating ability, but is sorbed very weakly. It is especially important in fumigation to establish the duration of gassing or exposure. Many harmful organisms can live for quite a long time in a poisoned atmosphere, with closed spiracles at the expense of the oxygen contained in their tracheal system and will perish only after all of this oxygen is used up. It has been established that small addition of carbon dioxide to fumigants may stimulate the respiratory movement and opening of the spiracles in insects and consequently, may increase the toxic action of the poison.

Losses

Soil, water and plant systems are responsible for the chemical attack on pesticides by hydrolysis, oxidation etc. Soil microorganisms are responsible for the microbial and enzymatic processes leading to ether cleavage, hydrolysis of amides, dealkylation and other similar processes. Simultaneously, these can induce reduction of nitro to amino group. Photochemical processes occur in soil due to solarization. This includes energy transfer processes resulting in the decomposition of pesticide molecules by oxidation, hydrolysis, ring fission and replacement of halogens by other groups. There is ozone layer in the atmosphere that absorbs the solar ultraviolet rays of wavelength lower than 290 nm but even then solar energy can decompose several kilograms of pesticides per hectare per day.

Pesticides differ much in stability and persistence in the environment. Stability and persistence depend on concentration, temperature, radiation and other factors. The rates of chemical reactions are proportional to concentration of the molecular species taking part in the degradation process. The degradation continues till the reactants are consumed. Several chemical reactions occur simultaneously. However, for all practical purposes the rates of degradation depend on the concentration of pesticide molecules and follow a first order reaction kinetics. The rate of the reaction is given as:

$$- dc/dt = kc$$

where, k = rate constant,

                     c = concentration

                     t = time

This equation is also applicable to cases where microbiological degradation of both pesticide and organic matter occur. The microbial degradation of the soil applied pesticides is related to organic matter content of the soil. The value of the rate constant k fluctuates because under field conditions, the breakdown of pesticide cannot be controlled.

The rate of reaction can be expressed as decrease in the concentration of pesticide (reactant) per unit time. Since the concentration of chemical keeps on decreasing with passage of time, the rate of reaction is not constant and must also decrease with passage of time. Rate of a reaction is, therefore, the instantaneous rate of change of a concentration of pesticide at a given moment of time. However, the conditions affecting the breakdown of pesticides in field are variable and, therefore, the persistence has only been empirically related. The receptiveness of leaf to evaporation in the early stage tends to reduce the overall rate of loss so that the deposit becomes more uniform. The redistributed pesticide then continues to evaporate from the solution. The rate of evaporation increases rapidly with increasing temperature and wind speed. It is very sensitive to the changes in relative humidity. Evaporation of pesticide from field soils is not a simple process because both water and pesticide evaporate simultaneously. Water is distributed in the fine pores and, therefore, the rate of disappearance depends on diffusion process. Also, the diffusion coefficient is not a constant quantity because the surface soil is drier than the lower layers. The process of adsorption plays a very important role in transport of dissolved pesticides. The phenomenon can be compared to chromatography in which the chemicals are separated by their affinities for the soil resulting in downward leaching of the pesticides.

Evaporation of Pesticide Deposits: After pesticide is sprayed, it is important to calculate its residual life. The rate of evaporation of pure chemical from a continuous surface is dependent on the exposed surface area and is independent of the thickness of the deposit. Hence, the loss percentage ($-dw/dt$) is a constant quantity where w represents the amount of deposit per unit area.

In case the spots of deposit are so near to each other as to form a uniform layer, then the rate of evaporation should be based on total area of deposit. However, if the surface is neither smooth nor inert (e.g. plant leaf), a part of the deposit may dissolve in the oils/waxes associated with the leaf cuticle. It, therefore, follows that the deposit evaporates uniformly till it separates into smaller spots or pools after which, if evaporation occurs, the rate is proportional to the radii of the pools. This is one of the most important factors governing retention of a lipophilic pesticide in the oil of cuticle or in the cutin layer itself. The cuticle of a plant has an area density of 1 to 5×10^{-4} g cm^{-2}, and that of the surface wax from 0 to 3×10^{-4} g cm^{-2}.

The losses may also occur due to the evaportion of pesticides. According to Clausius-Clapeyron equation

$$d(\ln P) = \frac{\Delta H}{RT^2} dT$$

where, P = vapour pressure of solution at temperature T

ΔH = latent heat of vaporisation

R = gas constant

$$dT = \text{change in temperature, say } T_1 \text{ to } T_2.$$

So that $\quad\quad \log_{10} P = \dfrac{\Delta H}{2.303 \times R} \times \dfrac{T_2 - T_1}{T_1 T_2}$

$\Delta H / R$ is given by the slope of the curve obtained by plotting log P against $1 / T$. If the latent heat ΔH is known, the above equation can be used to calculate vapour pressures at different temperatures.

The Rault's Law can also be employed to lower the vapour pressure as according to this law the vapour pressure is directly proportional to concentration of the solvent and relative lowering in vapour pressure is equal to the mole fraction of solute in solution as mentioned below.

$$P_A = P_A^o x_A$$

where, P_A is partial pressure of A (the solvent), x, the mole fraction of A and P_A^o, the vapour pressure of pure solvent.

or $\quad\quad\quad \dfrac{P_A}{P_A^o} = x_A$

or $\quad\quad\quad \dfrac{P_A}{P_A^o} = 1 - x_B$

where, x_B is mole fraction of solute.

or $\quad\quad\quad \dfrac{P_A^o - P_A}{P_A^o} = x_B = \text{mole fraction of solute}$

Thus, the relative lowering of vapour pressure is equal to the mole fraction of the solute in the solution.

Adsorption of pesticides in soil depends on both its organic matter and clay content. It is also affected by partitioning in organic matter and water in soil pores (soil solution), the partition coefficient (K_d) being equal to the uptake per unit mass of soil organic matter divided by that in the soil solution at equalibrium. Rate of adsorption, K_a is equivalent to the distribution coefficient K_d divided by the organic matter content. This value is almost similar for many soils so that it helps us to compare their relative affinities for soil surface. The extent of adsorption by a soil can be interpreted after determination of organic matter content.

Adsorption of unionized pesticide may be similar to partition between water and organic solvent. K_d values for triazines range from 2 to 20. The systemic pesticides, which belong to this category, are strongly adsorbed and due to this reason become less active. In order, therefore, to get better results, the systemic pesticides should be more useful if applied locally (like ethirimol, which gives better results if it is applied as seed treatment). Dry surface of soil adsorbs pesticide vapours strongly so that there is less loss by evaporation. Since the soils are generally moist, the vapours are not absorbed strongly and evaportion losses do occur. Adsorption at the air water interface has a pronounced effect on the transport of pesticides by molecular diffusion, if the soil water is static. However, effect is much lower if the water forming the interface with air is flowing. It is also evident that the air-water surface maintained between solid particles at field capacity is expected to hold up the adsorbed pesticides against flow. The material, however, is liberated on drying. The relative extent of adsorption in soils for a given unionised pesticide depends on following factors:

(*i*) Organic matter content of soil

(*ii*) Mechanical analysis of soil

It appears that the organic matter from different sources has similar adsorbent properties. Microbial degradation of pesticide will therefore be greatly influenced by the adsorption on soil surfaces. The equilibrium may reach slowly in case of soil with well-developed structure.

Effect of Radiation on Reaction Rates: Radiation has a strong influence on degradation of pesticide molecules sprayed in field. In such cases the energy is absorbed in the form of photons of different frequencies. The energy E of a photon is equal to hv where h is Planck's constant and v, the frequency of light. As soon as the absorbed radiation energy reaches a level so that it crosses the energy barrier and is absorbed by the molecule , degradation may take place. Such a reaction is called a photochemical reaction where the following relationship holds good:

$$E_a \text{ (forward)} = E \text{ (threshold)} - E_r \text{ (reactant)}$$

$$E_a \text{ (backward)} = E \text{ (threshold)} - E_p \text{ (products)}$$

$$\Delta E = E_p - E_r$$

or $$\Delta E = E_a \text{ (forward)} - E_a \text{ (backward)}$$

The persistence of a pesticide in environment is usually expressed in terms of its half life, which is the time taken for half the amount originally present to disappear. This half life is constant and independent of the initial concentration and inversely proportional to rate constant. It is, however, not possible to understand precise implications of the degradation of pesticides under variable conditions in the environment. In case the order of reaction is greater than unity, then the proportion degrading per unit time decreases so that disappearance of last traces of the pesticide is very slow. The degradation of pesticide molecule depends on pH, temperature and concentration and also composition of the medium, say soil. It is generally not possible to predict exactly the rate of degradation under so many variable conditions of interaction of solid surface within the heterogeneous soil medium. The dusts and wettable powders are not very stable because of the surface catalysed reactions that occur due to the acidic or basic sites on the carriers.

Improving/Reducing Persistence

Sometimes there may be a case where the pesticide being unstable rapidly disappears. In such cases it is necessary to change the centre of instability while activity is retained. A very interesting example is that of synthetic pyrethroids, which are very safe to mammals and potent against many insect species. The natural pyrethroids were very unstable in sunlight. The development of synthetic pyrethroids has helped to induce persistence in such unstable materials. Permethrin is a very good example in which the centres of instability have been removed although the structural features necessary for activity have been retained. Stability of a pesticide can be increased by suitable formulation, but in actual practice this part holds good only for shelf life of a formulation. In field application use of antioxidant and enzyme in the sprays is more popular.

Pyrethrin-I

Permethrin

The addition of antioxidants/stabilizers to several pesticides, particularly the modern bio-botanicals such as pyrethrum, neem materials etc. is assuming increased importance (Kumar and Parmar 1999).

Improving Persistence:

Use of Progenitor: A progenitor is a modified toxicant molecule that may be non-toxic *per se* but releases the toxic ingredient inside the living system. Modified molecules of active ingredients or addition of suitable formulants may lead to this effect. The advantages through their use include:

(*i*) Effective penetration and transport to the target.
(*ii*) Lower cost
(*iii*) Stability during storage
(*iv*) Easy/efficient application to the target
(*v*) Increased persistence at site of action
(*vi*) Reduced degradation losses

For a chemical mixed with soil, it is desirable to prevent the rapid escape of the chemical before it produces the desired result. The use of fumigant progenitor sodium monomethyl dithiocarbamate in aqueous solution extends the period of retention in the soil by releasing the active compound slowly. A progenitor in the form of dichlorophon is available that changes into dichlorovos, an insecticide used in public health. It is more stable in acidic medium.

The chemical containing thioether linkage can be converted to sulphone, as in organo phosphates like demeton, disulphoton and phorate. Thus these molecules also become more hydrophilic and hence more potent inhibitors of acetylcholine esterase. Similarly, to save an insecticide from leaching, decreasing polarity can be advantageous.

$$C_2H_5O \diagdown \underset{\underset{C_2H_5O}{\diagup}}{P} \overset{\overset{S}{\parallel}}{} - S - CH_2CH_2 - S - C_2H_5 \qquad C_2H_5O \diagdown \underset{\underset{C_2H_5O}{\diagup}}{P} \overset{\overset{S}{\parallel}}{} - S - CH_2CH_2 - \underset{\underset{O}{\parallel}}{\overset{\overset{O}{\parallel}}{S}} - C_2H_5$$

Disulphoton   Disulphoton sulphone

Reducing Persistence: Persistence of a pesticide can be problematic too. For example, the organochlorine insecticides like DDT have affected many non-target organisms due to their persistence. Ways and means are to be found out to decrease the persistence of such pesticides by developing more biologically degradable products, by the use of apporiate formulants such as photosensitizers, through enrichment of microbial species capable of degrading them and so on.

Another method of decreasing persistence of pesticides is through soil surface application instead of deep ploughing in soil. Pesticide is lost from the surface by evaporation, photochemical degradation etc. Repeated cultivation also helps to bring the pesticides to the surface so that the photochemical processes can easily degrade them. A decrease in the level of DDT by 25% and aldrin by 38% has been reported by repeated cultivation. The degradation of pesticides can also be increased by:

(*i*)   Adding energy rich materials like sugar or compost
(*ii*) Maintaining optimum moisture level of the soil
(*iii*) Maintaining aerobic/anaerobic conditions in the soil
(*iv*) Use of degradation promoters.

Transformation of DDT to DDE is quicker in anaerobic conditions compared to aerobic conditions. Addition of reagents like calcium polysulphide to soils increases degradation of triazine herbicides.

Selectivity

The pesticides should control pests selectively and be harmless to mammals and other non-target organisms. However, at times it is difficult to pinpoint safety. For example, redenticides can be harmful to almost all other mammals. The selectivity of pesticides can be established based on:

(*i*)   The rate of penetration and movement to the site of action
(*ii*) Nature of metabolism in different organisms
(*iii*) Selectivity of the site of action (in case of weedicide)

In methyl parathion the $P = S$ group is a weak inhibitor of acetyl choline esterase and it is the $P = O$ group that is more active. Methyl parathion on oxidation forms methyl paraoxon which is very harmful to mammals and not recommended for use as an insecticide. Similar oxidation reactions also occur in other organophosphorus compounds namely dimefox, schradan etc. The organophosphorus compounds which are systemic in action break down to form the products that are not injurious to mammals. Malathion also acts in a similar way by oxidation to form malaoxon at the site of action in the target but being a diester it is more prone to hydrolysis and degradation.

$$CH_3O \diagdown \overset{\overset{S}{\|}}{P} - O - \langle \bigcirc \rangle - NO_2$$
$$CH_3O \diagup$$

Methyl Parathion

$$CH_3O \diagdown \overset{\overset{O}{\|}}{P} - O - \langle \bigcirc \rangle - NO_2$$
$$CH_3O \diagup$$

Methyl Paraoxon

$$CH_3O \diagdown \overset{\overset{S}{\|}}{P} - S - CH - COOC_2H_5$$
$$CH_3O \diagup \qquad\qquad\quad |$$
$$\qquad\qquad\qquad\qquad CH_2 - COOC_2H_5$$

Malathion

$$CH_3O \diagdown \overset{\overset{O}{\|}}{P} - S - CH - COOC_2H_5$$
$$CH_3O \diagup \qquad\qquad\quad |$$
$$\qquad\qquad\qquad\qquad CH_2 - COOC_2H_5$$

Malaoxon

N-acetyl derivative of mexacarbate, which is very active against the harmful insects, is safe to mammals. N-arylsulphenyl and N-thioalkyl derivatives of aldicarb, carbofuran and propoxur are less toxic to mammals but are very effective against mosquito larvae.

Note: It is possible that a formulator may hit upon a wonderful formulation of a toxicant just after a few trials and with few errors. It is not possible to study all the foretold interactions for even a single pesticide before going for its formulation. However, a little forethought given to above aspects before developing a formulation goes a long way in selection of proper formulants and an appropriate formulation for a particular purpose.

REFERENCES

Brooks, G. T. (1966). Progress in metabolic studies of cyclodiene insecticides and its relevance to structure-activity correlation. *World Rev. Pest Control*, **5**: 62.

Crafts, S. H. and Crisp, C. E. (1971). *Phloem Transport in Plants*. Freeman, Sanfransisco, U.S.A.

Kumar, J. amd Parmar, B. S. (1999). Stabilization of azadirachtin A in neem formulations: Effect of some solid carriers, neem oil and stabilizers. *J. Agric. Food Chem.*, **47**: 1735-1739.

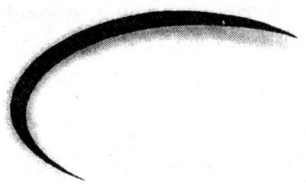

Glossary

INTERNATIONAL CODES, TERMS AND DEFINITIONS FOR TECHNICAL AND FORMULATED PESTICIDES

Code	Term	Definition
AB	Grain bait	Special form of bait
AE	Aerosol dispenser	A container-held formulation which is dispersed generally by a propellant as fine droplets or particles upon the actuation of a valve
AL	Any other liquid	A liquid not yet designated by a specific code, to be applied undiluted
AP	Any other powder	A powder not yet designated by a specific code, to be applied undiluted
BB	Block bait	Special form of bait
BR	Briquette	Solid block designed for controlled release of active ingredient into water
CB	Bait concentrate	A solid or liquid intended for dilution before use as a bait
CF	Capsule suspension for seed	A stable suspension of capsules in a fluid to be applied to seed, either directly or after dilution
CG	Encapsulated granule	A granule with a protective or granule release-controlling coating
CL	Contact liquid or gel	Rodenticidal or insecticidal formulation in the form of a liquid/gel for direct application, or after dilution in the case of gels
CP	Contact powder	Rodenticidal or insecticidal formulation in powder form for direct application. Formerly known as tracking powder (TP)

CS	Capsule suspension	A stable suspension of capsules in a fluid, normally intended for dilution with water before use
DC	Dispersible concentrate	A liquid homogeneous formulation to be applied as a solid dispersion after dilution in water (Note: there are some formulations which have characteristics intermediate between DC and EC)
DP	Dustable powder	A free-flowing powder suitable for dusting
DS	Powder for dry seed treatment	A powder for application in the dry state directly to the seed
DT	Tablet for direct application	Formulation in the form of tablets to be applied individually and directly in the field, and/or bodies of water, without preparation of a spraying solution or dispersion
EC	Emulsifiable concentrate	A liquid, homogeneous formulation to be applied as an emulsion after dilution in water
ED	Electrochargeable liquid	Special liquid formulation for electrostatic (electrodynamic) spraying
EG	Emulsifiable granule	A granular formulation to be applied as an oil-in-water emulsion of the active ingredient after disintegration in water, which may contain water insoluble formulants
EO	Emulsion, water in oil	A fluid, heterogeneous formulation consisting of a solution of pesticide in water dispersed as fine globules in a continuous organic liquid phase
ES	Emulsion for seed treatment	A stable emulsion for application to the seed either directly or after dilution
EW	Emulsion, oil in water	A fluid, heterogeneous formulation consisting of a solution of pesticide in an organic liquid dispersed as fine globules in a continuous water phase
FD	Smoke tin	Special form of smoke generator
FG	Fine granule	A granule in the particle size range from 300 to 2500 µm
FK	Smoke candle	Special form of smoke generator
FP	Smoke cartridge	Special form of smoke generator
FR	Smoke rodlet	Special form of smoke generator
FS	Flowable concentrate for seed treatment	A stable suspension for application to the seed either directly or after dilution
FT	Smoke tablet	Special form of smoke generator
FU	Smoke generator	A combustible formulation, generally solid, which upon ignition releases the active ingredient(s) in the form of smoke

Special forms of smoke generators:

Smoke candle	(FK)
Smoke cartridge	(FP)
Smoke pellet	(FW)
Smoke rodlet	(FR)

Smoke tablet	(FT)	
Smoke tin	(FD)	
FW	Smoke pellet	Special form of smoke generator
GA	Gas	A gas packed in pressure bottle or pressure tank
GB	Granular bait	Special form of bait
GE	Gas generating product	A formulation which generates a gas by chemical reaction
GF	Gel for seed treatment	A homogeneous gelatinous formulation to be applied directly to the seed
GG	Macrogranule	A granule in the particle size range from 2000 to 6000 mm
GL	Emulsifiable gel	A gelatinized formulation to be applied as an emulsion in water
GP	Flo-dust	Very fine dustable powder for pneumatic application in greenhouses
GR	Granule	A free-flowing solid formulation of a defined granule size range ready for use

Special forms of granules:

Encapsulated granule	(CG)	A granule with a protective or release-controlling coating
Fine granule	(FG)	Particle size range from 300 to 2500 μm
Macrogranule	(GG)	Particle size range from 2000 to 6000 μm
Microgranule	(MG)	Particle size range from 100 to 600 μm
Grease	(GS)	Very viscous formulation based on oil or fat
Water soluble gel	(GW)	A gelatinized formulation to be applied as an aqueous solution
HN	Hot fogging concentrate	A formulation suitable for application by hot fogging equipment, either directly or after dilution
KK	Combi-pack solid/liquid	A solid and a liquid formulation, separately contained within one outer pack, intended for simultaneous application in a tank mix
KL	Combi-pack liquid/liquid	Two liquid formulations, separately contained within one outer pack, intended for simultaneous application in a tank mix
KN	Cold fogging concentrate	A formulation suitable for application by cold fogging equipment, either directly or after dilution
KP	Combi-pack solid/solid	Two solid formulations, separately contained within one outer pack, intended for simultaneous application in a tank mix
LA	Lacquer	Solvent-based, film-forming composition
LS	Solution for seed treatment	A clear to opalescent liquid to be applied to the seed either directly or as a solution of the active ingredient after dilution in water. The liquid may contain water insoluble formulants

ME	Micro-emulsion	A clear to opalescent, oil and water containing liquid, to be applied directly or after dilution in water, when it may form a diluted micro-emulsion or a conventional emulsion
MG	Microgranule	A granule in the particle size range from 100 to 600 μm
OF	Oil miscible flowable	A stable suspension of active ingredient(s) in a fluid concentrate (oil miscible intended for dilution in an organic liquid before use as suspension)
OL	Oil miscible liquid	A liquid, homogeneous formulation to be applied as a homogeneous liquid after dilution in an organic liquid
OP	Oil dispersible powder	A powder formulation to be applied as a suspension after dispersion in an organic liquid
PA	Paste	Water-based, film-forming composition
PB	Plate bait	Special form of bait
PC	Gel or paste concentrate	A solid formulation to be applied as a gel or paste after dilution with water
PO	Pour-on	Solution for pouring on the skin of animals in a high volume (normally more than 100 ml per animal)
PR	Plant rodlet	A small rodlet, usually a few centimetres in length and a few millimetres in diameter, containing an active ingredient
PS	Seed coated with a pesticide	Self-defining
RB	Bait (ready for use)	A formulation designed to attract and be eaten by the target pests

Special forms of baits:

Block bait	(BB)
Grain bait	(AB)
Granular bait	(GB)
Plate bait	(PB)
Scrap bait	(SB)

SA	Spot-on	Solution for spot application on the skin of animals in a low volume (normally less than 100 ml per animal).
SB	Scrap bait	Special form of bait
SC	Suspension concentrate (= flowable concentrate)	A stable suspension of active ingredient(s) in a fluid, which may contain other dissolved active ingredients(s), intended for dilution with water before use
SE	Suspo-emulsion	A fluid, heterogeneous formulation consisting of a stable dispersion of active ingredients in the form of solid particles and fine globules in a continuous water phase
SG	Water soluble granule	A formulation consisting of granules to be applied as a true solution of the active ingredient after dissolution in water, but which may contain insoluble inert ingredients

SL	Soluble concentrate	A clear to opalescent liquid to be applied as a solution of the active ingredient after dilution in water. The liquid may contain water insoluble formulants
SO	Spreading oil	Formulation designed to form a surface layer on application to water
SP	Water soluble powder	A powder formulation to be applied as a true solution of the active ingredient after dissolution in water, but which may contain insoluble inert ingredients
SS	Water soluble powder for seed teatment	A powder to be dissolved in water before application to the seed
ST	Water soluble tablet	Formulation in form of tablets to be used individually, to form a solution of the active ingredient after disintegration in water. The formulation may contain water insoluble formulants
SU	Ultra-low volume (ULV) suspension	A suspension ready for use through ULV equipment
TB	Tablet	Pre-formed solids of uniform shape and dimensions, usually circular, with either flat or convex faces, the distance between faces being less than the diameter

Special forms of tablets:

Tablets for direct application	(DT)
Tablets for dissolution in water	(ST)
Tablets for dispersion in water	(WT)

TC	Technical material	A material resulting from a manufacturing process comprising the active ingredient, together with associated impurities. This may contain small amounts of necessary additives
TK	Technical concentrate	A material resulting from a manufacturing process comprising the active ingredient, together with associated impurities. This may contain small amounts of necessary additives and appropriate diluents. For use only in the preparation of formulations
TP	Tracking powder	Discontinued term. Refer to CP
UL	Ultra-low volume (ULV)	A homogeneous liquid ready for use through ULV liquid equipment
VP	Vapour releasing product	A formulation containing one or more volatile active ingredients, the vapours of which are released into the air. Evaporation rate is normally controlled by using suitable formulations and/or dispensers
WG	Water dispersible granule	A formulation consisting of granules to be applied after disintegration and dispersion in water
WP	Wettable powder	A powder formulation to be applied as a suspension after dispersion in water
WS	Water dispersible powder for slurry seed treatment	A powder to be dispersed at high concentration in water before application as a slurry to the seed
WT	Water dispersible tablet	Formulation in the form of tablets to be used individually,

to form a dispersion of the active ingredient after disintegration in water

XX   Others   Temporary categorization of all other formulations not listed above

---

"Catalogue of Pesticide Formulation Types and International Coding System". Technical Monograph No. 2, GIFAP, Brussels. Revised February 1989.

OTHER TERMS RELATING TO PESTICIDES (IUPAC RECOMMENDATIONS 1996)

Abiotic degradation
Degradation of a pesticide via purely physical or chemical mechanisms. Examples include hydrolysis and photolysis.

Absorption
Transfer of a component from one phase to another (Gold et. al, 1987). Movement of a pesticide from the environment (e.g. water, ingested food, leaf surface) across a biological membrane into an organism.

Acceptable daily intake (ADI)
Estimate of the amount of a pesticide in food and drinking water, which can be, ingested daily over a lifetime by humans without appreciable health risk. It is usually expressed in milligrams per kilogram of body weight. See also 'Tolerable daily intake'. (Duffus, 1993)

Action level (regulatory)
1. For food commodities, an administrative maximum residue limit (MRL) used by regulatory authorities to initiate action where no legally defined MRL has been established. 2. For the environment, concentration of a pesticide in air, soil or water at which emergency measures or preventative actions are to be taken. (Duffus, 1993)

Action limits (analytical quality control)
Limits for measurements on reference material or spiked samples which indicate when an analytical procedure is not performing adequately and requires immediate action before data can be reported

Active ingredient (ai)
Pesticide present in a formulation as described by the common name. The part of a pesticide formulation from which the biological effect is obtained. (FAO, 1990)

Accuracy (of measurement)
Closeness of agreement between the result of a measurement and the (conventional) true value of the measurement. Note 1. Use of the term precision for accuracy should be avoided. Note 2. True value is an ideal concept and, in general, cannot be known exactly. (ISO, 1984)

Acute toxicity
Ability of a substance to cause adverse effects within a short period following dosing or exposure. (Duffus, 1993)

Adjuvant
Formulant designed to enhance the activity or other properties of a pesticide mixture

Adsorption
Enrichment of one or more components in an interfacial layer. (Gold et. al., 1987)

Adverse effect
Change in morphology, physiology, growth, development or lifespan of an organism which results in impairment of functional capacity or which increases susceptibility to the harmful effects of other environmental influences. (IPCS, 1989)

Aerobic
Conditions under which molecular oxygen serves as the terminal

electron acceptor in respiration or in metabolic oxygenation. See also redox potential (Gold *et. al.,* 1987)

Aerosol
System of fine solid or liquid particles (< 30 µm in dia.) dispersed in a gas. Aerosol cans using an inert compressed propellant are a common means of dispensing insecticides for domestic use. See also nebulisation

AFID
Alkali flame-ionisation detector or detection for gas chromatography (cf NPD and TID)

Aged residue
Pesticide and degradates present in an environmental system after application and following a period long enough to allow transport, adsorption, metabolism, and dissipation processes to alter the distribution and chemical nature of some of the applied pesticide

Aggregate sample
Sample made up of set proportions of other samples, typically an average by weight. See also composite sample

Aglycon
Non-sugar part of a glycoside or glucuronide conjugate derived from the pesticide. See also exocon

Agrochemical
Agricultural chemical used in crop and food production including pesticide, feed additive, veterinary drug and related compounds

Aliquot
Known fractional portion of a homogeneous material (Horwitz, 1990). The term is usually applied to volumetric sub-sampling of fluids

Anaerobic
Condition under which reductive conditions prevail. See also redox potential

Analytical portion
See test portion

Analytical sample
See test sample

Analytical range
Measurement range of a test method where the performance has been validated and quality standards such as action limits have been developed

Analytical standard (pesticide)
Pesticide reference material of high and defined purity (generally > 95%) for preparation of calibration standards

Bait
A food or other substance used to attract a pest to a pesticide or trap where it can be destroyed

Batch
Quantity of material which is known or assumed to be produced under uniform conditions (Horwitz, 1990)

Benthos
Non-planktonic animals (not being suspended in water) associated with freshwater substrata (upper layer of the sediment in rivers and ponds) at the sediment-water interface (Wetzel, 1983)

Bioaccumulation
Progressive increase in the amount of a substance in an organism or part of an organism which occurs because the rate of intake exceeds the organism's ability to remove the pesticide from the body (Duffus, 1993)

Bioactivation
Transformation of a pesticide within an organism into a more biochemically active metabolite

Bioconcentration
Process leading to a higher concentration of a pesticide in an organism than in environmental media to which it is exposed (Duffus, 1993)

Bioconcentration factor (BCF)
Ratio between the concentration of pesticide in an organism or

	tissue and the concentration in the environmental matrix (usually water) at apparent equilibrium during the uptake phase (Rand and Petrocelli, 1985)
Bioavailability	Extent to which a pesticide residue can be taken up into an organism from its food and environment, and the rate at which this occurs (Duffus, 1993)
Biodegradation	Conversion or breakdown of the chemical structure of a pesticide catalysed by enzymes *in vitro* or *in vivo*, resulting in loss of biological activity. For hazard assessment, categories of chemical degradation include
Primary *Environmentally acceptable* *Ultimate*	loss of specific activity loss of any undesirable activity (including any toxic metabolites) mineralisation to small molecules such as water and carbon dioxide (Duffus, 1993)
Biological indicator	Species or group of species which is representative and typical for a specific status of an ecosystem, which appears frequently enough to serve for monitoring and whose population shows a sensitive response to changes, e.g. the appearance of a pesticide in the ecosystem (US-EPA, 1992)
Biological assessment of exposure	Assessment of exposure of a living organism to pesticides using biological specimens (blood, urine etc.) taken in the environment (workplace, field etc.) with analysis either directly by chemical determination of parent or metabolite, or indirectly by measurement of a relevant biochemical parameter (e.g. plasma cholinesterase activity for organophosphorus compounds) (Duffus, 1993)
Biomagnification	Bioaccumulation of a pesticide through an ecological food chain by transfer of residues from the diet into body tissues. The tissue concentration increases at each trophic level in the food web when there is efficient uptake and slow elimination (Rand and Petrocelli, 1983)
Biomarker	Indicator (molecular, biochemical, cellular or organism) signalling an event or condition in a biological system or sample and giving a measure of exposure to, effect of, or susceptibility to, a xenobiotic (Duffus, 1993)
Biomass	The total living mass in a defined segment of an ecosystem expressed as the living weight per unit area or mass. Soil microbial biomass is often used as an indication of potential microbial activity level in soil
Biometer flask	Experimental apparatus commonly used in laboratory studies of pesticide degradation in soil. Contains separate compartments for aerobic incubation of soil and for media to trap carbon dioxide and volatile products
Biopesticide	Pesticide of biological origin including microorganisms e.g. *Bacillus thuringiensis* and natural products e.g. rotenone, pyrethrins
Biotransformation	Conversion of the chemical structure of a pesticide catalysed by enzymes *in vitro* or *in vivo*. See also biodegradation
Biotransformation pathway	Sequence of the changes occurring in the structure of a pesticide when it is introduced into a specific biological test system

Blank material (sample) — Laboratory simulated test material known to be free of the pesticide being analysed. A portion of blank material is used to test the method, apparatus and reagents for interferences or contamination. See also control sample (Thompson and Wood, 1995)

Bound residue — Chemical species in soil, plant or animal tissue originating from a pesticide, (generally radio labelled) that are unextracted by a standard method, such as Soxhlet solvent extraction, which does not significantly change the chemical nature of the residues. These unextractable residues are considered to exclude small fragments recycled through metabolic pathways into natural products (Roberts, 1984)

Breakdown — See degradation

Buffer zone — Distance for environmental protection between the edge of an area where pesticide application is permitted and a sensitive non-target area e.g. water course

Carrier — Solid formulant added to a technical material as an absorbent or diluent (FAO, 1990)

Carryover (analytical) — Unintended contamination of a sample undergoing analysis with material from a previous sample

Carryover (field) — Persistence of pesticide residues in soil after use in one crop such that uptake is observed in a succeeding, possibly more sensitive, crop

Catabolism — Oxidative biodegradation of a pesticide to provide chemically available energy and generate metabolic intermediates (IUPAC, 1992)

Catchment — Land and water confined within a single drainage basin

Certified reference material — Reference material, accompanied by a certificate, whose pesticide concentrations are certified by procedures which establish their traceability and for which each certified concentration is accompanied by an uncertainty at a stated level of confidence. Storage conditions and period for which the certification remains valid may also be included for unstable materials (Thompson and Woods, 1995)

Chronic effect — Consequence of exposure to a pesticide which arises slowly and has a long-lasting, often irreversible, course (Duffus, 1993)

Chronic exposure — Continued exposures occurring over an extended period of time, or a significant fraction of the lifetime of the exposed individuals or test species (Duffus, 1993)

Chronic toxicity — Capacity for a pesticide to produce injury following chronic exposure or to produce effects which persist whether or not they occur immediately upon exposure or are delayed (Duffus, 1993)

Co-metabolism — Microbial metabolism of a pesticide where the derived energy is not used to support microbial growth. Cf. catabolism

Common moiety — Molecular sub-unit which is common to the structures of several pesticides or metabolites

Community — Assembly of populations of different species of living organisms (quite often interdependent on and interacting with each other) within a specified location in space and time. See also ecosystem (US-EPA, 1992)

Compartment

Part of an organism or ecosystem considered as an independent system for purposes of assessment of uptake, distribution and dissipation of a pesticide (US-EPA, 1992)

Compliance (GLP)

See GLP compliance statement

Compliance (residue)

Meeting of official maximum residue limit (MRL) standards by residue levels in food consignments sampled and tested by approved methods

Composite sample

Combined increment samples, or combined replicate samples, or combined samples from replicate trials. Preferred term to bulk sample which is ambiguous (Horwitz, 1990). See also aggregate sample, primary sample

Concentration-effect relationship

Association between the exposure concentration of the pesticide and the magnitude of the resultant continuously graded change either in an individual organism or in a population (Duffus, 1993)

Conjugation

Biosynthetic reaction in which a pesticide or its metabolite is linked to an endogenous compound. See also endocon, exocon, phase II metabolism

Contaminant

1. Minor impurity in a substance
2. Extraneous material added to a sample prior to or during chemical or biological analysis
3. Unintended pesticide residue in an agricultural commodity or environmental compartment (e.g. ground water). See also pollutant

Control sample (field)

Sample from a field test plot to which no pesticide was applied (a zero rate sample) or which received chemical treatments identical to the test plots except for the test chemical

Critical concentration

Lowest concentration of a pesticide in an environmental compartment at which adverse effects on organisms are likely to be observed (95% probability)

Critical load

Amount of a pesticide leading to a critical concentration when received by an environmental compartment (US-EPA, 1992)

Cumulative effect

Overall adverse change which occurs when repeated doses of a pesticide have biological consequences which are additive (Duffus, 1993)

Cut-off value

Numerical value set by regulatory authorities representing the limit of acceptability for a property or behaviour of a compound for the final step in tiered assessment schemes. See also trigger value

Degradation

Process by which a pesticide is broken down to simpler structures through biological or abiotic mechanisms. Synonyms include breakdown and decomposition. See also biodegradation, mineralisation (OECD)

Desorption

Depletion of one or more components in an interfacial layer (Gold, 1993)

Detoxification

Processes of chemical modification which make a pesticide less toxic (Duffus, 1993)

Diluent

Liquid or solid material used to dilute a concentrated pesticide formulation prior to application. Most commonly water for spray application

Dislodgeable residue
Portion of a pesticide residue on treated vegetation that is readily removable and may be used as an index for risk to farm workers. Generally measured by the residue removed when leaf discs are shaken briefly in water

Dissipation
Loss of pesticide residues from an environmental compartment due to degradation and transfer to another environmental compartment

Dissipation time 50% (DT_{50})
Time required for one-half the initial quantity or concentration of a pesticide to dissipate from a system. No assumption as to the rate equation is made. See also half-life

Dispersible granule
See water-dispersible granule

Dose-effect relationship
Graded relationship between the dose of the pesticide to which the organism is exposed and the magnitude of a defined biological effect, either in an individual organism or in a population. See also concentration-effect relationship (Duffus, 1993)

Dose-response relationship
Association between dose and the incidence of a defined biological effect in an exposed population (Duffus, 1993)

Drift control agent
Formulant that controls the distribution of spray droplet sizes and prevents production of excessive fines

Dry weight basis
Pesticide residue concentration reported as if the residue were wholly contained in the dry matter of the sample, i.e. analytical results are corrected for the water content of the test sample. Residues in soils and feeds, and maximum residue limits (MRLs) for feedstuffs are expressed on a dry weight basis

Dustable powder (DP)
Free flowing powder suitable for dusting (GIFAP, 1995)

EC_{50}
See median effective concentration

ECD
Electron capture detector, used in gas chromatography

Ecosystem
Assembly of populations of different species (often interdependent on and interacting with each other) interacting with their surroundings within a specified physical location and forming a functional entity. See also community (Rand and Petrocelli, 1985)

Ecotoxicologically (environmentally) relevant concentration (ERC)
Concentration of a pesticide (active ingredient, formulations, and relevant metabolites) that is likely to affect a determinable ecological characteristic of an exposed system. It is related to the toxicity characteristics, generally the no observable effect concentration, to the most sensitive species or groups of species (US-EPA, 1992)

ELISA
See immunoassay

Emulsifiable concentrate (EC)
Liquid homogeneous formulation of a pesticide with emulsifiers in an organic solvent which forms a dispersion when added to water as a diluent (GIFAP, 1995)

Emulsifier
Surfactant used to facilitate the preparation of a colloidal dispersion of one liquid in another liquid with which it is not miscible (Gold, 1987)

Endocon
That portion of a conjugated metabolite which is derived from natural products of the metabolising organism such as sugars and organic acids. See also exocon, phase II metabolism

Endpoint   Measurable ecological or toxicological characteristic or parameter of the test system (usually an organism) that is chosen as the most relevant assessment criterion (e.g. death in an acute test or tumour incidence in a chronic study)

Enforcement method   See regulatory method

Enhanced degradation   Increased rate of degradation of a pesticide in soil or other environmental matrix by a population of microorganisms that has adapted to metabolise it through previous exposure to it or a similar chemical. Synonyms include accelerated degradation and enhanced biodegradation

Enterohepatic circulation   Cyclical process in which a pesticide residue is absorbed and transported to the liver, metabolised (often including conjugation), transported to the intestine by the bile, reabsorbed (often after deconjugation), and transported to the liver for further metabolism (Duffus, 1993)

Environmental impact assessment   Assessment of the potential releases of a pesticide to the environment and their potential effects upon the environment and its components including man. See risk assessment

Environmental risk   Probability that an adverse effect on humans or the environment will be observed for a given exposure to a pesticide based on the frequency of occurrence and the sensitivity of the system. See risk assessment

Estimated daily intake (EDI) Prediction of the daily intake of a pesticide residue, based on the most realistic estimation of residues in food items and the best available food consumption data for a specific population. Residue levels are estimated taking into account known uses of the pesticide, the proportion of commodity treated and the quantity of contaminated commodities. The EDI is expressed in milligrams of residues per person (WHO, 1989)

Estimated environmental concentration (EEC)   Predicted concentration of a pesticide within an environmental compartment based on estimates of quantities released, discharge patterns and inherent disposition of the pesticide (fate and distribution) as well as the nature of the specific receiving ecosystems (US-EPA, 1992)

Estimated maximum daily intake (EMDI)   Prediction of the maximum daily intake of a pesticide residue, based on the assumptions of average daily food consumption per person and maximum residues in the edible portion of a commodity, corrected for the reduction or increase in residues resulting from preparation, cooking, or commercial processing. The EMDI is expressed in milligrams of residues per person (WHO, 1989)

Exocon   Portion of a conjugated metabolite which is derived from the parent pesticide. See also aglycon

Exposure   Concentration or amount of a pesticide that reaches the target population, organism, tissue or cell, usually expressed in numerical terms of concentration, duration and frequency. Also the process by which a substance becomes available for absorption by the target population, organism, tissue or cell, by any route (Duffus, 1993)

Exposure assessment   Process of measuring or estimating concentration, duration and frequency of exposures to pesticide present in environment or, if estimating hypothetical exposures, that might arise from the release of the pesticide into the environment. See also risk assessment (Duffus, 1993)

Extractability Degree to which a pesticide residue may be removed from a matrix (e.g. soil) through use of appropriate extraction techniques. See also bound residue

Extraneous residue limit (ERL) Maximum concentration of a pesticide residue, arising from environmental sources (including former agricultural uses), other than from the use of a pesticide directly or indirectly on the commodity, that is recommended to be permitted in or on a feed or food commodity (FAO)

Fat basis Residues and maximum residue limits (MRLs) of fat-soluble pesticides in animal commodities may be expressed in terms of their concentration in the fat rather than the whole product

FID Flame ionisation detector for gas chromatography or GLC

Field drainage Removal of excess water from soil and transport to surface waters in order to improve soil productivity and trafficability

Flowable See suspension concentrate

Food chain-primary producers Autotrophic organisms (e.g. algae, higher plants) which convert inorganic compounds during the process of photosynthesis or chemosynthesis into organic compounds (cell material) of higher energy content. These organisms represent the first trophic level of the food chain

Food chain-secondary producers Heterotrophic organisms (e.g. animals) using organic substances as a carbon and energy source

Food chain-primary consumers Heterotrophic organisms (e.g. filter feeding invertebrates such as daphnia species) using organic substances directly from primary producers (e.g. algae) as a carbon and energy source

Food chain-secondary consumers Heterotrophic organisms (e.g. predator animals) feeding on primary consumers

Food chain-primary decomposers Heterotrophic organisms (e.g. bacteria) using dead organic matter from all trophic levels as a carbon and energy source

Food chain-secondary decomposers Heterotrophic organisms (e.g. certain soil fungi, collembola, worms) using already partially decomposed organic matter as a carbon and energy source

Formulant Any added material in a pesticide formulation other than the biologically active ingredient(s). This may include carrier or other substances which enhance the biological activity or physio-chemical properties of the formulation. See also adjuvant, diluent, inert, sticker, surfactant, vehicle (CIPAC, 1980)

Formulation See pesticide formulation

Fortified sample See spiked sample

FPD Flame photometric detector for gas chromatography

Fresh weight basis Pesticide residues are reported on the laboratory sample as it is received, with no allowance for the moisture content. Maximum

	residue limits (MRLs) and pesticide residues in food commodities are expressed in this way
Freundlich isotherm	Empirical relationship describing the adsorption of a solute from a liquid or gaseous phase to a solid in which the quantity of material adsorbed per unit mass of adsorbent is expressed as a function of the equilibrium concentration of the sorbate. See also Kd
Frozen storage stability	See storage stability test
Fumigation	Use of a pesticide in gas or vapour form
FTIR	Fourier transform infrared spectroscopy
GC-EC	Gas chromatography with electron capture detector
GC-MS	Gas chromatography-mass spectrometry
GC-MSD	Gas chromatography with mass-selective detection (usually low resolution mass spectrometry using selected ions)
GLC	Gas liquid chromatography
GLP	See Good Laboratory Practice
GLP archive	Location (room, filing cabinet etc.) where study plans, raw data, final study reports, laboratory inspection reports, study audit reports, retention samples or specimens are stored. GLP records and material must be retained for the period specified by the appropriate authorities
GLP certificate	Certificate of test facility compliance with a national GLP programme
GLP chain of custody	Set of procedures and traceable records that demonstrate an unbroken control over, or custody of, a document, or raw data, or a sample from its collection through to its final deposition
GLP compliance statement	Declaration by study director that the study was conducted in compliance with the principles of GLP and relevant national statutes. Any aspects of non-compliance should be described in this statement
GLP inspection	Check of a test facility, a study or parts of a study by an internal or external authority to ensure compliance with GLP guidelines. Internal inspections are carried out by the quality assurance unit. See also GLP study audit
GLP principal investigator	Person nominated in the study plan who has the delegated responsibility to supervise certain phases of a study where the study director cannot exercise direct control
GLP protocol	See GLP study plan
GLP quality assurance programme	Internal control system containing written procedures to ascertain that studies are in compliance with GLP
GLP quality assurance statement	Statement prepared by the quality assurance unit specifying the dates inspections were made and any findings which were reported to management and to the study director. This statement is part of the final report of a study
GLP quality assurance unit (QAU)	Sub-section of the test facility, separate from actual testing, responsible for internal audits of the facility and its Study Reports to ensure compliance with GLP. The QAU is also generally responsible for the administration and training in all aspects of the quality assurance system

GLP standard operating procedure (SOP)	Written procedure, authorised by management which describes how to perform a certain routine test or activity normally not specified in detail in study plans or test guidelines, e.g. arrival, identification and storage of samples, standards or reagents; operation, maintenance, and calibration of apparatus; preparation of reagents; quality assurance procedures
GLP study	Experiment or set of experiments conducted under GLP
GLP study audit	Review by the quality assurance unit of an interim or final report, including raw data from a study, confirming that the study was carried out in accordance with the study plan and standard operating procedures and that it has been accurately and completely reported in compliance with GLP
GLP study director	The person responsible for the overall conduct of a study i.e. ensuring that all phases of the study are conducted under GLP according to the study plan
GLP study plan	The document which determines the entire scope of a study conducted under GLP. A written study plan must be completed and approved by the Study Director before a study starts. It contains information such as the title of study; name or code of test and reference substances; name and address of sponsor, test facility, study director and principal investigators; dates for start and end of study; methods including relevant standard operating procedures (SOPs); list of material to be archived
Glucuronides	Components resulting from the conjugation of a pesticide or its metabolite with glucuronic acid
Glycosides	Mixed acetals (ketals) resulting from the conjugation of a pesticide or its metabolite with a saccharide or saccharide derivative. In plants and insects the saccharide endocon is commonly an aldohexose
Good agricultural practice (GAP)	Nationally authorised safe uses of pesticides under actual conditions necessary for effective and reliable pest control. It encompasses a range of levels of pesticide applications up to the highest authorised use, applied in a manner that leaves a residue which is the smallest amount practicable. Authorised safe uses include nationally registered or recommended uses, that take into account public and occupational health and environmental safety considerations. Actual conditions include any stage in the production, storage, transport, distribution and processing of food commodities and animal feed (IPCS, 1989)
Good experimental field practice	The formalised process for designing and recording the practices used in the performance of field investigations with pesticides, and which assure the reliability and integrity of the data. See GLP
Good laboratory practice (GLP)	The formalised process and conditions under which laboratory studies on pesticides are planned, performed, monitored, recorded, reported and audited. Studies performed under GLP are based on the national regulations of a country and are designed to assure the reliability and integrity of the studies and associated data. The US-EPA GLP definition also covers field experiments (see Good experimental field practice) (OECD, 1992)

GPC	Gel permeation chromatography (cf SEC)
Granule	Solid formulation comprising particles of defined size $> 80 \mu m$ diameter, for application without further dilution, usually to soil
Ground water	Water present in the saturated subsurface zone of the soil profile, where all open spaces/pores in the sediment and rock are filled with water
Guideline level	Maximum concentration of a pesticide residue in or on a feed or food commodity, resulting from a use reflecting good agricultural practice, but where an acceptable daily intake has not been estimated
Guideline value	Maximum recommended pesticide residue in an environmental medium that ensures aesthetically pleasing air, water or food and does not constitute a significant risk to the user (Duffus, 1993)
Half-life ($t_{0.5}$/or $t_{1/2}$)	Time taken for the concentration of a pesticide in a compartment to decline by one half. Usually an estimate based on observed dissipation over several half-lives that can be described by first order kinetics with rate constant k, $t_{0.5} = 0.693/k$. See also dissipation time 50%
Hazard	Set of inherent properties of a pesticide which gives potential for adverse effects to man or the environment under conditions of its production, use or disposal, and depending on the degree of exposure (Duffus, 1993)
Hazard assessment	Determination of factors controlling the likely effects of a hazard such as mechanism of toxicity, dose-effect relationships and worst case exposure levels. This is the prelude to risk assessment (US-EPA, 1992)
Hazardous distance for the most sensitive effect (HDSE)	Statistically determined safety margin corresponding to a distance from treated areas at which protection of the terrestrial environment can be adequately achieved as measured by the most sensitive non-target species. See also buffer zone, margin of safety
Health advisory level (HAL)	Estimate of upper concentration limit for a pesticide in drinking water that can be consumed for a lifetime without adverse effects. HALs generally do not have formal legal significance but have been used, particularly in the USA, for preliminary risk assessment
HPLC	High performance liquid chromatography
HPTLC	High performance thin layer chromatography
HRGC	High resolution gas chromatography (GLC with narrow bore capillary columns)
Hydrolysis	Reaction in which a chemical bond is cleaved and a new bond formed with the oxygen atom of a molecule of water
Identification	Process of unambiguously determining the chemical identity of a pesticide or metabolite in experimental or analytical situations
Immobilization	Process leading to restricted mobility of a pesticide in plant or soil due to strong binding
	Incorporation of terminal pesticide degradates into complex organic forms in microbial or plant tissue
Immunoassay	Ligand-binding assay based on antibodies capable of specific binding to the pesticide analyte. Most commonly used in a competitive

	binding format where analyte molecules compete with a specific antigen complex labelled for detection using a radioisotope (radio immunoassay - RIA) or enzyme (enzyme-linked immunoassay - ELISA)
Impurity	By-product of the manufacture or storage of a pesticide. Impurities require definition, evaluation and regulation (if toxicologically significant)
Increment sample	An individual portion (unit) of material collected by a single operation of a sampling device from bulk materials or large units (Horwitz, 1990)
Incurred residue	Residue in a commodity resulting from specific use of a pesticide, consumption by an animal or environmental contamination in the field, as opposed to residues from laboratory fortification of samples
Inert ingredient	Formulant which by itself does not add materially to effectiveness for the purpose for which the preparation is intended e.g. solvent, emulsifier, diluent, carrier
In-life phase	Phase of a study following treatment in which the test system is alive/growing
Integrated pest management (IPM)	Pest management system that, in the context of the associated environment and the population dynamics of the pest species, utilises all suitable techniques and methods in as compatible a manner as possible to maintain the pest populations at levels below those causing economically unacceptable damage or loss (FAO, 1987)
In-vitro	'In-glass', referring to use of a cell line, microorganisms or biochemically active fraction derived from an organism (e.g. hepatocytes) in laboratory studies of biological activity, or metabolism or toxicity
In-vivo	Use of the whole living organism in studies of biological activity, metabolism or toxicity
Kd	See soil partition coefficient
Koc	See soil organic partition coefficient
Laboratory sample	Sample or subsample(s) sent to or received by the laboratory
Lag phase	Period which may precede commencement of rapid degradation of a pesticide by a microbial population. It is the period needed either for induction of microbial enzymes or for growth of the microbial population to adequate size. See also enhanced degradation (US-EPA, 1992)
LC$_{50}$	See median lethal concentration
LD$_{50}$	See median lethal dose
Leaching	Process by which a pesticide moves downward through the soil profile in the aqueous phase
Leachate	Aqueous phase percolating through a soil profile or a soil column
Limit of detection (LOD)	Lowest concentration of a pesticide residue in a defined matrix where positive identification can be achieved using a specified method
Limit of quantitation (LOQ)	Lowest concentration of a pesticide residue in a defined matrix

where positive identification and quantitative measurement can be achieved using a specified method. The term limit of quantitation is preferred to limit of determination to differentiate it from LOD. LOQ has been defined as 3 times the LOD (Keith, 1991) or as 50% above the lowest fortification level used to validate the method (US-EPA, 1986)

Limit of reporting

Practical limit of residue quantitation at or above the LOQ. The limit of quantitation for a defined matrix and method may vary between laboratories or within the one laboratory from time to time because of different equipment, techniques and reagents

Lipophilicity

Affinity for fat as described by partitioning behaviour between water and an immiscible organic solvent, favouring the latter, and which correlates with bioconcentration. See also octanol-water partition coefficient (Duffus, 1993)

Lot

Quantity of material which is assumed to be a single population for sampling purposes. See also batch

Lysimeter

Device for measuring leaching losses from a column or block of soil. The simplest lysimeters may be devices for sampling a portion of the water leaching through a natural sediment or soil (e.g. suction lysimeter), whereas more elaborate lysimeters may involve the confinement of an entire segment of soil from which all leachate is collected (e.g. monolithic lysimeter)

Macropore

Soil pore larger than 1 mm in diameter including interparticle void, earthworm or rodent burrow, drying crack, and decayed root channel. See preferential flow

Margin of safety (MOS)

Ratio of the highest estimated or actual level of exposure to a pesticide and the toxic threshold level (usually the NOEC or NOEL). See also uncertainty factor (US-EPA, 1992)

Market basket survey

Pesticide residue monitoring on a wide range of food items collected from consumer points of sale and in proportions approximating consumption patterns in the local population. Samples are prepared for analysis according to Codex guidelines i.e. minimal preparation. See also total diet study

Matrix

The material or component sampled for pesticide residue studies

Maximum residue limit (MRL)

Maximum concentration of a residue that is legally permitted or recognised as acceptable in, or on, a food, agricultural commodity or animal feedstuff as set by Codex or a national regulatory authority. The term tolerance used in some countries is, in most instances, synonymous with MRL. Normally expressed as mg kg^{-1} fresh weight (FAO, 1986)

Maximum tolerated dose (MTD)

Highest dose of a pesticide in chronic toxicity testing that is expected, on the basis of sub-chronic studies, to produce only limited toxicity when administered for the duration of the test period (Duffus, 1993)

Median effective concentration (EC$_{50}$)

Statistically derived concentration of a pesticide in an environmental medium expected to produce a certain effect in 50% of the test organisms in a given population under defined conditions (Duffus, 1993)

Median lethal concentration (LC_{50})	Statistically derived concentration of a pesticide in an environmental medium expected to kill 50% of test organisms in a given population under defined conditions (Duffus, 1993)
Median lethal dose (LD_{50})	Statistically derived dose of a pesticide expected to kill 50% of test organisms in a given population under a defined set of conditions. Normally expressed as mg of test material per kg of body weight of the organism (Duffus, 1993)
Mesocosm	See model ecosystem
Metabolism	Sum total of all physical and chemical processes that take place within an organism; in a narrower sense, the physical and chemical changes that occur for a pesticide within an organism. It includes uptake and distribution within the body, changes (biodegradation), and elimination of pesticides and their metabolites
Metabolites	Any intermediate or product resulting from metabolism (Duffus, 1993)
Microcosm	See model ecosystem
Mineralisation	Conversion of an element from an organic form to an inorganic form. Mineralisation of pesticides most commonly refers to the microbial degradation to carbon dioxide as a terminal metabolite. See also immobilisation
Model	Experimental or mathematical simulation of chemical behaviour in a specific environment (ASTM, 1984)
Model calibration	Testing of a model with known input and output information for adjustment or estimation of factors for which data are not available (ASTM, 1984)
Model, computer	Assembly of numerical techniques (algorithms), bookkeeping, and control language (i.e. the computer programme) comprising a mathematical model and which carries out acceptance of input data and instructions through to delivery of output (ASTM, 1984)
Model, conceptual	Qualitative depiction of a specific environment that describes the linkages between the different compartments. A conceptual model is required before a quantitative simulation model can be developed (Cohen *et al.*, 1995)
Model ecosystem	Man-made study system containing associated organism and abiotic components that is large enough to be representative of a natural ecosystem, yet small enough to be experimentally manipulated. There is some subjective differentiation between larger, outdoor model ecosystems (mesocosms) and smaller, generally indoor model ecosystems (microcosms)
Model validation	Comparison of model results with numerical data independently derived from experiments or observations of the environment (ASTM, 1984)
Model verification	Examination of the numerical technique in the computer code to ascertain that it truly represents the conceptual model and that there are no inherent numerical problems with obtaining a solution (ASTM, 1984)
Multiresidue method	Analytical method which measures a number of pesticide residues simultaneously

Nebulisation	Formation of an aerosol of very small liquid particles (fog) or solid particles (smoke) from a pesticide formulation, generally for fumigation of an enclosed space such as a glass-house
NMR	Nuclear magnetic resonance spectroscopy
Non-target organism	Organism affected by a pesticide although not an intended object of its use
No-observable effect concentration/level (NOEC/NOEL)	Highest concentration or amount of pesticide in the test system that causes no observable biological effect to the target organism (US-EPA, 1992) See also Ecotoxicologically relevant concentration, PNEC
NPD	Nitrogen-phosphorus detector or detection for gas chromatography (cf AFID and TID)
OC	Organochlorine pesticide. Generic term for pesticides containing chlorine but commonly used to refer to older persistent materials including aldrin, BHC, chlordane, DDT, dieldrin, heptachlor, lindane and toxaphene
Octanol/water partition coefficient (Pow)	Partition coefficient for a pesticide in the two-phase system octan-1-ol/water. The Pow is a distribution coefficient reflecting the relative lipophilicity of a pesticide and its potential for bioconcentration. For convenience, the value is often expressed in logarithmic (base 10) form (log Pow)
OP	Organophosphorus pesticide. Generic term for pesticides containing phosphorus but commonly used to refer to insecticides consisting of cholinesterase inhibiting esters of phosphate or thiophosphate
Partition coefficient	Ratio of the concentrations of a substance in solution in two phases which are in equilibrium. See Koc, Pow
Parts per billion (ppb)	Ratio of amounts expressed as parts pesticide per 10^9 sample. Strictly the quantities should be the same i.e. weight to weight (solids) or volume to volume (liquids or gases) e.g. 1ppb = 1 $\mu g\ kg^{-1}$. A common usage is for weight to volume but to avoid confusion it is recommended that the SI units are used rather than ppb; e.g. $\mu g\ l^{-1}$ (Mills *et al.*, 1993)
Parts per million (ppm)	Ratio of amounts expressed as parts pesticide per 10^6 sample e.g. 1 ppm = 1 mg kg^{-1}. As with ppb it is recommended that SI units are used rather than ppm, particularly for weights to volume i.e. mg l^{-1}
PED	Plasma emission detector
Pellet	Solid formulation of pesticide, larger than granule, often used for molluscicide formulations
Persistence	Residence time of a chemical species (pesticide and/or metabolites) subjected to degradation or physical removal in a soil, crop, animal or other defined environmental compartment
Pest	Organism that attacks food and other materials essential to mankind, or otherwise affects human beings adversely (Duffus, 1993)
Pesticide	Substance or mixture of substances intended for preventing, destroying or controlling any pest, including vectors of human or animal disease, unwanted species of plants or animals causing harm or otherwise interfering with the production, processing, storage,

transport, or marketing of food, agricultural commodities, wood, wood products or animal feedstuffs, or which may be administered to animals for the control of insects, mites/spider mites or other pests in or on their bodies. The term includes substances intended for use as a plant growth regulator, defoliant, desiccant, or agent for thinning fruit or preventing the premature fall of fruit, and substances applied to crops either before or after harvest to protect the commodity from deterioration during storage or transport. See also agrochemical, plant protection agent (FAO, 1986)

Pesticide chemical name Scientific name of a pesticide following the recommendations of IUPAC for naming of chemical compounds or other accepted naming convention (e.g. Chemical Abstracts)

Pesticide common name Simple name assigned to a pesticide active ingredient by the International Organisation for Standardisation (ISO) to be used as a generic or non-proprietary name (FAO, 1986)

Pesticide formulation Pesticide product offered for sale. It generally comprises active ingredient(s), adjuvant(s) and other formulants combined to render the product useful and effective for the purpose claimed (FAO, 1986)

Pesticide residue Substance(s) which remains in or on a feed or food commodity, soil, air or water following use of a pesticide. For regulatory purposes it includes the parent compound and any specified derivatives such as degradation and conversion products, metabolites and impurities considered to be of toxicological significance (FAO, 1986)

Pesticide residue definition The pesticide, its metabolites, derivatives and related compounds to which the maximum residue limit (MRL) applies, as specified by Codex or a national regulatory authority

Pesticide residue enforcement Pesticide residue monitoring programme where the intention is regulatory action against non-complying consignments

Pesticide residue monitoring Sampling and analyses of pesticide residues in biological and environmental samples taken according to pre-arranged schedules

Pesticide trade name Proprietary name assigned to a pesticide or its formulations by the company manufacturing or selling it

Phase I metabolism Initial biotransformation of a pesticide. These are mainly oxidative, reductive and hydrolytic processes

Phase II metabolism Biotransformation where the pesticide or phase I metabolite is conjugated with a naturally occurring compound (e.g. sugars, glutathione)

Phloem Part of the plant's vascular system adapted to the transport of photosynthetic products from leaves to the rest of the plant

Photolysis Chemical reaction caused by light in which a bond is cleaved (Calvert, 1990)

Plant growth regulator (PGR) Naturally occurring or synthetic substance which influences plant development or reproduction but has no nutritive value

Plant protection agent Pesticide product intended for use in agriculture to protect crops

Pollutant Undesirable substance introduced into a solid, liquid or gaseous environmental medium totally or partially by human activities. See also contaminant (Duffus, 1993)

Population
Assemblage of individual organisms of defined ages and growth stages belonging to one species within a specified location in space and time (US-EPA, 1992)

Post-emergence
Period after a crop or pest has appeared. Herbicide usage can be referred to as post-emergence (weeds) or post-emergence (crop)

Precision
Closeness of agreement between independent test results obtained under prescribed conditions (Thompson and Wood, 1995)

Predicted environmental concentration (PEC)
See estimated environmental concentration

Predicted no effect concentration (PNEC)
An estimated no-observable effect concentration for an aquatic species of ecosystem based on extrapolated experimental exposure/response data

Pre-emergence
Period before a specified crop or pest has emerged. Generally applied to timing of herbicide applications. Cf post-emergence

Preferential flow
Leaching phenomenon whereby water and a dissolved pesticide percolating down through the soil profile move more rapidly through soil macropores or sand/gravel lens than through the network of smaller pores in the bulk soil

Pre-harvest interval (PHI)
The time interval between the last application of a pesticide to a crop and harvest. See also withholding period

Primary sample
Collection of one or more increments or units initially taken from a population. Note that portions may be combined (composited or aggregated sample) or kept separate (Horwitz, 1990)

Prior informed consent (PIC)
Agreement of the designated national authority in a participating country required before international shipment can proceed of a chemical which is banned or severely restricted in order to protect human health or the environment (FAO, 1986)

Processed food
Product resulting from the application of physical, chemical or biological processes, or combinations of these (e.g. canning), to a primary food commodity, and intended for sale to the consumer, for use as an ingredient in the manufacture of a food product or for further processing

Quantitative structure-activity relationship (QSAR)
Quantitative association between the physio-chemical properties of a pesticide or the properties of its molecular substructures and its biological properties including its non-target toxicity

Random sample
Sub-set of a sampling population that is arrived at by selecting units such that each possible unit has a fixed and determinate probability of selection

Raw agricultural commodity
Part of a crop used as a food or feed commodity directly from the harvested crop without processing

Raw data
All original laboratory records and documentation, or verified copies thereof, including data directly entered in a computer. They are the results from the original activities and observations in a GLP study

Recovery, analytical
Fraction or percentage of a pesticide residue recoverable following extraction and analysis of a matrix containing the pesticide (Gold *et al.*, 1987)

Redox potential
Electrical potential indicating the relative activity of oxidised and

reduced species. The redox potential of an environmental matrix is a measure of the extent to which oxidising species are present to act as terminal electron acceptors in respiration

Re-entry interval   Minimum time between pesticide application and human re-entry to a treated area. Established by a regulatory authority to assure safety of workers from exposure to residues

Reference dose   Expected dose resulting from human exposure to a pesticide at the level at which it is regulated in the environment (US-EPA, 1992). See also acceptable daily intake, tolerable daily intake

Reference material   Material or substance containing pesticide of interest at levels sufficiently homogeneous and well characterised to be used for the calibration of an apparatus or assessment of analytical method performance (Thompson and Wood, 1995). See also certified reference material

Registration   The process whereby the responsible national government authority approves the sale and use of a pesticide following the evaluation of scientific data demonstrating that the pesticide is effective for the purposes intended and not unduly hazardous to human or animal health or the environment (FAO, 1986)

Regulatory method   Validated analytical method which can be applied using commonly available laboratory equipment and instrumentation. A regulatory method has the precision, specificity, limit of determination, etc, needed to test compliance with the regulations

Repeatability   For an analytical method, the closeness of agreement between results of measurements on identical test material subject to the following conditions: same analyst, same instrumentation, same location, same conditions of use, repetition over a short period of time (ISO, 1984)

Reproducibility   For an analytical method, the closeness of agreement between results of measurements on identical test material where individual measurements are carried under changing conditions such as: analyst, instrumentation, location, conditions of use, time (ISO, 1984)

Resistance   Development of tolerance to a pesticide by a target population, generally through natural selection

Respiration   Energy-generating process in an organism where an organic or inorganic compound serves as the electron donor and an inorganic compound (e.g. oxygen) serves as the electron acceptor

Retention sample   Sample which is stored for a specified period in case of a need for re-evaluation of data obtained from the main laboratory samples

Risk   Probability of any defined hazard occurring from exposure to a pesticide under specific conditions. Risk is a function of the likelihood of exposure and the likelihood to harm biological or other systems. See also hazard

Risk assessment   Process of defining the risk associated with a specified use pattern for a pesticide, usually expressed as a numerical probability or as a margin of safety. Quantifying risk ideally requires, identification of hazard, establishment of dose-response relationships in likely target individuals and populations, exposure assessment (using likely exposure patterns as opposed to worst-case estimates) (Duffus, 1993)

Risk management — Decision-making process and procedures used by regulators and others to limit potential risks from use of pesticides. This involves risk assessment, emission control, exposure control and evaluation of the success of the risk mitigation efforts

Rotational crop — Crop grown in sequence of two or more different crops

Run-off

Movement of a pesticide from a treated field by surface water and eroding sediment

Loss of formulation off foliage during spray application, particularly at high volume

Safener — A substance added to a pesticide formulation to eliminate or reduce phytotoxic effects of the pesticide to certain crops. See also adjuvant

Safety factor — See uncertainty factor

Sample — Portion of material selected from a larger quantity of material so that it is representative of the whole. See also aggregate sample, aliquot, composite sample, control sample, increment sample, laboratory sample, primary sample, random sample, retention sample, subsample, test portion and test sample

Sampling plan — Predetermined procedure for the selection, withdrawal, preservation, transportation, and preparation of the portions to be removed from a population as samples (Horwitz, 1990)

SEC — Size exclusion chromatography (cf GPC)

SFC — Supercritical fluid chromatography

SFE — Supercritical fluid extraction

Soil partition coefficient (Kd) — Experimental ratio of a pesticide's concentration in the soil to that in the aqueous (dissolved) phase at equilibrium. It is valid only for the specific concentration and solid/solution ratio of the test. The Kd is a distribution coefficient reflecting the relative affinity of a pesticide for adsorption by soil solids and its potential for leaching movement through soil. See also Koc

Soil incorporation — Application of a pesticide to soil by mixing or injection into the soil body

Soil organic partition coefficient (Koc) — Ratio of a pesticide concentration sorbed in the organic matter component of soil or sediment to that in the aqueous phase at equilibrium. The Koc is calculated by dividing the Kd value by the fraction organic carbon present in the soil or sediment (see soil organic matter)

Soil organic matter — Organic fraction of the soil, including both fresh and aged residues (e.g. humus) of biological origin. Organic carbon refers to that portion of the soil measured as carbon in organic forms, and the organic matter content of soil is assumed to be approximately 1.72-fold that of the organic carbon content

Sorption — Removal of pesticide from solution by soil or sediment via mechanisms of adsorption and absorption

SPE — Solid phase extraction

Specimens	Samples collected from a system for examination, analysis, or storage
Spiked sample (fortified sample)	Control sample with a known amount of pesticide added. Used to test the accuracy (especially the efficiency of recovery) of an analytical method (Thompson and Wood, 1995). See also reference material
Spray drift	Movement beyond the target area of airborne droplets or vapour of pesticide formulation originating from aerial or ground-based spraying operations
Spreader	See wetting agent
Standard solution, primary	Standard prepared by dissolving a weighed amount of an analytical standard pesticide in a known volume of solvent
Standard solution, secondary	Standard prepared by dilution of an aliquot of a primary standard solution with a known volume of solvent, or by subsequent serial dilutions; or a standard solution measured by reference to a primary standard solution
Sticker	Formulant which increases the adhesiveness of a formulation applied to a surface. (FAO, 1995). See also wetting agent
Storage stability test	For a pesticide formulation, a test which measures the chemical and physical stability of the product stored under defined, often worst case, conditions. For pesticide residues, a test which measures stability of residues in stored analytical samples, usually held under frozen conditions at a specified temperature
Subsample	Portion of the sample obtained by selection or division; individual unit of the lot taken as part of the sample; final unit of multistage sampling (Horwitz, 1990)
Surfactant	A formulant for reducing interfacial tension of two boundary surfaces, thereby increasing the emulsifying, spreading, dispersability or wetting properties of liquids or solids (FAO, 1995)
Surveillance	Systematic sampling and residue analysis of commodities, and collation and interpretation of data, in order to ensure compliance with established MRLs. Surveillance may be directed at domestic, imported or exported commodities
Suspension concentrate (SC)	Formulation in which the active ingredient is in the form of a stable dispersion of fine particles in water or organic liquid (GIFAP, 1989)
Synergist	Substance, which, while formally inactive or weakly active, can significantly enhance the activity of the active ingredient in a formulation
Systemic	A *systemic* pesticide is capable of being translocated to sites other than where it was absorbed in sufficient quantities to be biologically effective
Target, biological	Any organism, organ, tissue, cell or cell constituent that is subject to the action of a pesticide or its residue
Technical material	Commercial grade of the pesticide as it comes from the manufacturing plant comprising the active ingredient and associated impurities. It may also contain small quantities of additives necessary for stability

Test guideline	Guideline published by an appropriate authority for the order or conduct of certain tests
Test portion (analytical portion)	Subsample, of proper size for a chemical analysis or other test, removed from the test sample (Horwitz, 1990)
Test sample (analytical sample)	Homogenous sample, prepared from the laboratory sample by mixing, grinding, blending, fine-chopping etc., from which test portions are removed for analysis with minimal sampling error (Horwitz, 1990)
Test substance	The pesticide as a chemical substance or mixture which is under investigation in a GLP Study
Test system	Each system (animal, plant, microbial, other cellular, subcellular; chemical, or physical or a combination thereof) used in a study
Theoretical maximum daily intake (TMDI)	A prediction of the maximum daily intake of a pesticide residue, based on the assumption of levels of residues in food at maximum residue limits and average daily consumption of food per person. The TMDI is expressed in milligrams of residue per person (WHO, 1989)
Threshold	Concentration of a pesticide in an organism or environmental compartment below which an adverse effect is not expected
TID	Thermionic detector (cf NPD, AFID)
TLC	Thin layer chromatography
Tolerable daily intake	Term preferred by the European Commission for acceptable daily intake of environmental contaminants. ADI is reserved for pesticides and food additives where extensive toxicological test data is available
Tolerance, residue	See maximum residue limit
Total diet study	Pesticide residue monitoring to establish the pattern of residue intake by a person consuming a defined diet. Primary sampling is as for a market basket survey but the samples are further processed as for domestic consumption i.e. further trimming and cooking as appropriate to local practice
Total terminal residue	Summation of levels of all the compounds comprising residues of a pesticide in a food. See also pesticide residue
Toxification	See bioactivation
Transformation product	Chemical species resulting from environmental or metabolic processes on a pesticide. See also degradation product, metabolite
Translocation	Movement of a substance within the test system or organism
Transpiration	Vaporisation of water from a leaf into the air
Treated solution	Test solution that has been subjected to reaction or separation procedures prior to measurement of some property
Trigger value	Numerical value for a property of a pesticide, set by regulatory authorities, which determines the sequence and type of tests in a tiered assessment scheme. See also cut-off value
Trophic level	Functionally similar organisms such as algae and plants as primary producers are grouped into trophic levels based on similarities in the patterns of food production and consumption

Ultra low volume (ULV) spray	Signifies that the total volume rate of spray application is very low (5 litres per hectare or less). ULV pesticide formulations are generally specially developed for the purpose and are applied undiluted
Uncertainty factor	Factor in toxicological assessment for extrapolation of data from experimental animals to man (assuming that man may be more sensitive) or from selected individuals to the general population. For example an uncertainty factor is generally applied to the no-observed effect level to derive an acceptable daily intake
UVD	Ultraviolet absorption detection
Validation	In pesticide analysis, the process for establishing that an analytical method or equipment will provide reliable and reproducible results
Vehicle	See carrier
Volatilisation	Evaporation of a pesticide into the atmosphere from a solid or liquid form
Watershed	See catchment
Water dispersible granule (WG)	Formulation containing granules which readily disperse in water to form a suspension (GIFAP, 1989)
Water dispersible powder (WP)	Pesticide in a dry form with surfactant, often mixed with, or coated on, a fine solid carrier, for dispersion in water to form a suspension (GIFAP, 1989)
Water soluble powder	Powder formulation to be applied as a true solution of active ingredient after mixing with water, but which may contain insoluble inert ingredients (GIFAP 1989)
Wettable powder	See water dispersible powder
Wetting agent	Surfactant for use in spray formulations to assist dispersion of a powder in the diluent or spreading of spray droplets on surfaces. May also incorporate functions of a sticker
Withholding period	Minimum permissible time between the last application of a pesticide to a crop (including pasture) and harvesting for human consumption or grazing with livestock. The minimum permissible time between the final application of a pesticide to an animal and the collection of eggs or milk, or slaughter, for human consumption. See also pre-harvest interval
Xenobiotic substance	Natural or synthetic compound which is present in an organism but is not a natural component of that organism. Common usage is for man-made environmental contaminants in organisms (Nagel, *et al.,* 1992)
Xylem	Part of the plant's vascular system adapted to the transport of water and solutes from the roots to aerial parts
Zero tolerance	Limit for a pesticide residue in food or feed, which is assumed to be zero and therefore any detectable residue, is deemed illegal. Zero tolerances are used by some regulatory systems, e.g. USA, where no maximum residue limits have been established for particular pesticide/crop combinations

IUPAC (1996) Pure and Applied Chemistry **68**: 1167-1193.

ABBREVIATIONS OF KEY NATIONAL AND INTERNATIONAL BODIES

AID	:	Agency for International Development
AOAC	:	Association of Official Analytical Chamists
ASTM	:	American Society for Testing and Materials
BIS	:	Bureau of Indian Standards
CAC	:	Codex Alimentarius Commission
CCPR	:	Codex Committee on Pesticide Residues
CIPAC	:	Collaborative International Pesticides Analytical Council
EPA	:	Environmental Protection Agency (USA)
EPPO	:	European and Mediterranean Plant Protection Organization
FAO	:	Food and Agriculture Organization of the United Nations
GEMS	:	Global Environmental Monitoring System
GIFAP	:	Groupement International des Associations National de Fabricants de Produits Agrochimiques (Now GCPE: Global Crop Protection Federation)
IPCS	:	Internationl Programme on Chemical Safety, World Health Organization
IRIS	:	Integrated Risk Information System of the United States
IRPTC	:	International Register of Potentially Toxic Substances
ISI	:	Indian Standards Institution
ISO	:	International Organisation for Standardization
IUPAC	:	International Union of Pure and Applied Chemistry
JMPR	:	Joint FAO/WHO Meeting on Pesticide Residues
OECD	:	Organization for Economic Cooperation and Development
PAN	:	Pesticide Action Network
UNEP	:	United Nations Environment Programme
USEPA	:	United States Environmental Protection Agency
WHO	:	World Health Organization of United Nations

REFERENCES

ASTM (1984), *Standard Practice for Evaluating Environmental Fate Models of Chemicals*. American Society for Testing and Materials, Philadelphia.

Calvert, J.G. (1990), Glossary of atmospheric chemistry terms (IUPAC recommendations). *Pure Appl. Chem.* **62**: 2167-2219.

CIPAC (1980), *Handbook 1A, Analysis of Technical and Formulated Pesticides*. Collaborative International Pesticides Analytical Council. (eds. A. Martin and G. Dobrat). Black Bear Press, Cambridge.

Codex (1989), "Report of the 21rst Session of the Codex Commitee on Pesticide Residues, Codex Alimentarius Commission. The Hague, 10 -17 April, 1989", FAO/WHO, Rome.

Cohen, S. Z., Wauchope, R. D., Klein, A.W., Eadsforth, C. V., Graney, R. (1995), Offsite Transport of Pesticides in water - Mathematical models of pesticide leaching and runoff. *Pure Appl. Chem.*, **67**(12): 2109 - 2148.

Duffus, J. H. (1993), Glossary for chemists of terms used in toxicology. *Pure Appl. Chem.* **65**, 2003-2122.

FAO (1986), *Guide to Codex Recommendations Concerning Pesticide Residues* Part 1: General Notes and Guidelines. Food and Agricultural Organisation, Rome.

FAO (1986), *International Code of Conduct on the Distribution and Use of Pesticides*. Food and Agricultural Organisation, Rome.

FAO (1987), *Manual on the Development and Use of FAO Specifications for Plant Protection Products.* Food and Agricultural Organisation, Rome, Publ. No. 85.

FAO (1990), *Pesticide Residues in Food.* Plant Prod. and Prot. Paper 102, Food and Agricultural Organisation, Rome.

Galen, R. S. and Gambino, S. R. (1975) *Beyond Normality: The Predictive Value And efficiency of Medical Diagnoses,* John Wiley and Sons, New York.

Gold, V., Loening, K. L., McNaught, A. D. and Sehmi, P. (eds.) (1987) *International Union of Pure and Applied Chemistry, Compendium of Chemical Terminology. IUPAC Recomendations,* Blackwell Scientific Publications, Oxford.

GIFAP (1989), Catalogue of Pesticide Formulation Types and International Coding System. *GIFAP Technical Monograph,* **2**: Brussels.

Horwitz, W. (1990), Nomenclature for Sampling in Analytical Chemistry. *Pure Appl. Chem.* **62**: 1193-1208.

IPCS (1989), Glossary of terms for use in IPCS Publications, International Programme for Chemical Safety, World Health Organisation, Geneva.

IUPAC (1992), Glossary of terms used in biotechnology for chemists. *Pure Appl. Chem.* **64**: 144-168.

Keith, L. H. (1991), *Environmental Sampling and Analysis-a Practical Guide.* Lewis Publishers, Boca Raton, Florida.

OECD (1992), The OECD *Principles of Good Laboratory Practice. Environment Monograph,* **45:** Organisation for Economic Co-operation and Development, Paris.

ISO (1984) Metrology: *International Vocabulary of Basic and General Terms in Metrology.* International Organisation for Standardisation, ISO, Geneva.

Mills, I; Critas, T.; Homann, K.; Kallay, N.; Kozo, Kuchitsu (1993), *Quantities, Units and Symbols in Physical Chemistry. IUPAC recommendations* 2nd ed.,

Nagel, B., Dellweg, H., and Gierasch, L. M. (1992), Glossary of terms used in biotechnology for chemists. *Pure Appl. Chem.,* **64**: 144-168.

Rand, G. M. and Petrocelli, S.R. (1985), *Fundamentals of Aquatic Toxicology.* Hemisphere, Washington.

Roberts, T. R. (1984), Non-extractable pesticide residues in soils and plants. *Pure Appl. Chem.,* **56**, 945-956.

Thompson, M. and Wood, R. (1995), Harmonised guidelines for internal quality control in analytical chemistry laboratories. *Pure Appl. Chem.* **67**, 649-666.

US-EPA (1986), *Test Methods for Evaluating Solid Waste.* SW-846, 3rd ed., US Environmental Protection Agency, Washington D.C., 20460.

US-EPA (1992), *Selected Terms and Acronyms.* Environmental Protection Agency, Office of Pesticide Programs, Washington D.C.

Wetzel, R. G. (1983), *Limnology,* 2nd ed. CBS College, New York.

WHO (1989), *Guidelines for Predicting Dietary Intake of Pesticide Residues.* World Health Organisation, Geneva.

Index